Organotransition Metal Chemistry

A Mechanistic Approach

ORGANOMETALLIC CHEMISTRY
A Series of Monographs

EDITORS

P. M. MAITLIS
THE UNIVERSITY
SHEFFIELD, ENGLAND

F. G. A. STONE
UNIVERSITY OF BRISTOL
BRISTOL, ENGLAND

ROBERT WEST
UNIVERSITY OF WISCONSIN
MADISON, WISCONSIN

Organotransition Metal Chemistry

A Mechanistic Approach

Richard F. Heck

Department of Chemistry
University of Delaware
Newark, Delaware

Academic Press   New York and London   1974

A Subsidiary of Harcourt Brace Jovanovich, Publishers

ACADEMIC PRESS, INC.
111 Fifth Avenue, New York, New York 10003

United Kingdom Edition published by
ACADEMIC PRESS, INC. (LONDON) LTD.
24/28 Oval Road, London NW1

Library of Congress Cataloging in Publication Data

Heck, Richard F
    Organotransition metal chemistry.

    (Organometallic chemistry)
    Includes bibliographical references.
    1. Organometallic compounds.   2.   Transition metal
compounds.   I.   Title.
QD411.H37   547'.05'6   73-22378
ISBN 0-12-336150-8

To Annette

Contents

Preface

Organometallic reagents have been used by synthetic organic chemists for more than fifty years. Organomagnesium, lithium, and zinc compounds are very widely used reactants. More recently, thallium, copper, boron, and aluminum organometallics have found synthetic applications. The most recent additions to the list of useful organometallics are the organotransition metal compounds. The chemistry of the transition metals is already so extensive and unique that it is developing into a separate field. With the vast amount of literature being published, organic chemists have had to become more and more selective in their reading, and to a large extent have neglected organotransition metal chemistry. It is the purpose of this volume to provide for the organic chemist a background in and a summary of organotransition metal chemistry. Emphasis has been placed on reaction mechanisms as much as possible. This, of necessity, has involved much speculation since relatively little has yet been done in this area. Some sense is beginning to be made of the large amount of empirical data available, however, and most reactions are now explainable although often not yet predictable.

Organotransition metal reactions are discussed in relation to their reactions with specific functional groups or types of compounds rather than by metals since it is becoming apparent that the basic reactions of most of the transition metals are similar if not identical. This approach stresses the similarities of the metals rather than their differences and allows some basis, at least, for predicting products from reactions which have not been carried out. Clearly, the synthetic applications of transition metal compounds in organic chemistry eventually will be very extensive since even now many unique and synthetically very valuable reactions are known.

This volume is not intended to be comprehensive but only to give a personal view of subjects which appear to be of major importance in this rapidly expanding area of chemistry. It is hoped that it will encourage more organic chemists to use transition metal compounds as synthetic reagents and stimulate more research in the area.

I wish to express my appreciation to those who have read this manuscript and offered many helpful suggestions, particularly Professors J. G. Blann, T. B. Brill, P. M. Maitlis, H. Sternberg, F. G. A. Stone, and I. Wender. Special thanks are also due to Ms. Christine Young for help in preparing the manuscript.

Richard F. Heck

Chapter I

Historical and Chemical Background

A. DEVELOPMENT AND APPLICATIONS OF ORGANOTRANSITION METAL CHEMISTRY

The subject of organotransition metal chemistry as a distinct area of chemistry is less than twenty years old and was of relatively little interest before about 1960. Even though the first recognized organometallic complex, Zeise's salt, $K[PtCl_3(C_2H_4)]$ (1), was prepared in 1827, significant interest in such compounds did not develop until after the structure of ferrocene was determined in 1952 (2) and it was subsequently realized that the transition metals possessed the capability of forming a wide variety of isolable organometallic compounds. Most earlier attempts to prepare these compounds had met with failure because the stabilities of the compounds are very dependent on the particular groups attached to the metal and stable combinations were not found. Considerable industrially oriented work on applications of transition metals in catalytic organic reactions had been carried out prior to the discovery of ferrocene, but very little was known about the organometallic intermediates formed in these reactions.

Although applications in the field of synthetic organic chemistry are and probably will remain the most important for transition metal compounds, the compounds themselves are of considerable theoretical interest as well because of the many new structures and kinds of bonding found in them. Some areas of biochemistry also fall within the field of organotransition metal chemistry since numerous biologically important compounds contain transition metals and the particular biological function may be caused, at least partially, by the metal. Some of the better known biologically active compounds containing transition metals are vitamin B_{12} (cobalt), heme (iron), and cytochrome c (iron). Undoubtedly, many more such compounds remain to be discovered since transition metals occur widely in trace amounts in all types of living organisms. Most of the biological reactions of these compounds, however, probably do not involve carbon–metal-bonded species.

Organotransition metal chemistry is distinguished from the organometallic chemistry of the main group metals by its greater versatility. Although reactive main group organometallics generally add to carbonyl compounds and some activated carbon–carbon double bonds, transition metal compounds frequently react with unactivated, unsaturated organic compounds, often in a catalytic manner. The reactivity of the transition metal compounds is associated with their strong tendency to complex with various organic and inorganic compounds in order to obtain (share) more electrons, usually enough to complete their coordination spheres and attain the electronic configuration of the next higher inert gas. An important, basic difference resulting in different chemistry is the relatively low tendency of transition metals to combine with oxygen or oxygen derivatives compared with the main group metals.

The reactions of primary importance that transition metal complexes bring about at moderate temperatures (usually below 200°C) are combination reactions of various unsaturated molecules (such as olefins, acetylenes, and carbon monoxide) with each other and with other reagents in various specific ways to form products containing new carbon–carbon bonds. Complexing of the transition metal with the unsaturated molecules, in most instances, apparently is the key to getting the unsaturated groups to react. A great many new and unique reactions caused by transition metal compounds are now known, but the field is still in its infancy. The ultimate goal is to learn how to select transition metal catalysts that will specifically combine two or more reactive molecules together in any desired linear or cyclic manner. Progress has been made, but much more is needed in order to realize the full potential of this field of chemistry. The great value of organotransition metal chemistry is that often by its use organic reactions can be brought about very much more easily than would be possible by conventional organic chemistry.

A typical example of the kinds of reactions which now can be carried out is the combination of ethylene and butadiene with an iron catalyst to produce, specifically and in high yield, cis-1,4-hexadiene (3). A more spectac-

$$CH_2{=}CH_2 + CH_2{=}CHCH{=}CH_2 \xrightarrow[\text{AlEt}_3]{\text{Fe[(C}_6\text{H}_5)_2\text{PCH}_2]_2\text{Cl}_2}$$

$$\begin{array}{c} \text{H} \qquad\qquad \text{H} \\ \diagdown \qquad \diagup \\ \text{C}{=}\text{C} \\ \diagup \qquad\qquad \diagdown \\ \text{CH}_2{=}\text{CHCH}_2 \qquad \text{CH}_3 \end{array}$$

ular example is the combination of six relatively simple molecules together, remarkably specifically, to produce a complex organic structure. Allyl chloride, two molecules of styrene, a water molecule, and two carbon monoxide groups from the transition metal reagent, tetracarbonylnickel, react as indicated below to form a disubstituted cyclopentenone derivative in 64% yield (4). Other products are also formed in minor amounts, but the

$$CH_2{=}CHCH_2Cl + 2C_6H_5CH{=}CH_2 + H_2O + Ni(CO)_4 \longrightarrow$$

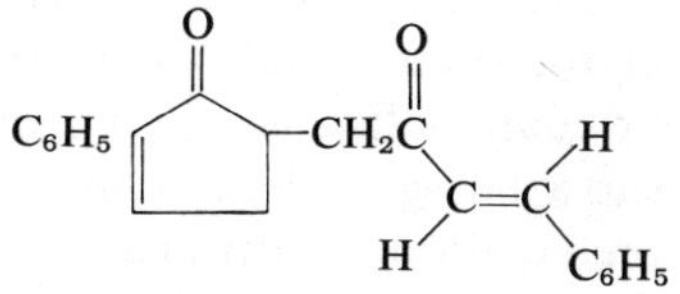

64%

result is still an amazing example of the ability of a transition metal to bring about unusual and complex transformations.

Two other important types of reactions (other than synthesis of more complex molecules) which transition metals may cause are molecular rearrangements and degradation reactions. Degradation may be either of complex molecules into simpler ones or of smaller molecules into "fragments" which then may recombine into more complex structures. Rearrangements vary from simple double-bond shifts to complex skeletal reorganizations. Strained hydrocarbons are often rearranged by transition metal compounds into less strained isomers. A typical example is the rearrangement of cubane to cuneane at room temperature catalyzed by palladium chloride (5).

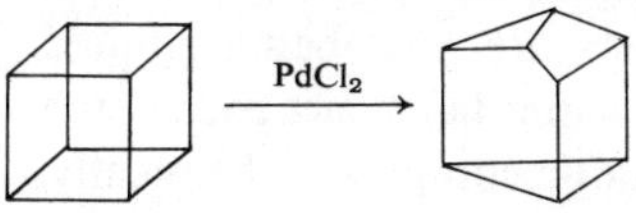

Degradation reactions vary from the relatively simple decarbonylation of aldehydes to hydrocarbons or olefins, to complex multiple bond cleavages and recombination reactions. A particularly interesting example of the last possibility is the conversion of apparently any olefin with titanium tetrachloride at 300°, to trichloro-π-pentamethylcyclopentadienyltitanium in about 25% yield (6).

$$TiCl_4 + C_nH_{2n} \longrightarrow$$

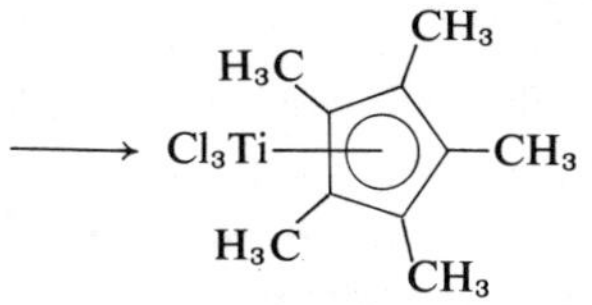

Numerous applications of transition metal catalysis in organic syntheses employ the catalysts in a form which remains as a separate solid phase during the reaction. Such reactions are said to be heterogeneous catalytic reactions. The reactants may be supplied to the catalyst in either the liquid or gas phase. Some reactions may be carried out heterogeneously or with the same or other catalysts homogeneously where all the reagents are in one phase, usually the liquid phase. On the other hand, many reactions are known which can be done only heterogeneously whereas others can be done only homogeneously. These differences result from the fact that different compounds can exist in insoluble phases than exist in solutions. The reactivities of transition metal complexes are often very dependent on the molecules or coordinating groups bonded to the metal and on the number of these surrounding molecules or groups. Heterogeneous catalysts are generally preferred for industrial applications, if there is a choice, because the catalyst remains in a form that can be easily recovered for reuse (or used continuously) and there is no problem of removing the catalyst from the reaction product. Although these practical considerations are economically important and many useful reactions can only be done heterogeneously, heterogeneous reactions are very complicated to study. Much research has been done on heterogeneous catalysis, but relatively little has been learned about reaction mechanisms. The major problem is that heterogeneous catalysts generally are not composed of a single active substance and there is no good way to purify and identify the active component or the reaction intermediates. Fortunately, homogeneous reactions do not suffer from these difficulties and many results obtained by studies of homogeneous catalysis also appear to apply to related heterogeneous reactions. Therefore, we shall be concerned mainly with homogeneous reactions.

New types of catalysts are now being studied which combine the best features of homogeneous and heterogeneous catalysts. These catalysts which have known organometallic complexes chemically bonded to specific groups on a solid support are discussed in more detail in Chapter IV, Section B. In other instances it is sometimes possible to carry out known homogeneous reactions in the gas phase with the catalyst as a second, solid phase, in which case a homogeneous reaction probably occurs in a thin liquid film on the catalyst surface.

B. USEFUL CONCEPTS

The feature which distinguishes transition metals from the rest of the elements and is responsible to a large extent for their unique chemistry is the presence of only partially filled d orbitals. There are five d orbitals, each capable of holding two electrons for a maximum of ten electrons in that subshell. Transition elements and ions exist with from 0 to 10 d electrons and the chemical properties of the metals depend on the number of these present.

The number present in any given metal or ion is readily determined from a periodic chart, although there will often be uncertainty as to when electrons are in d orbitals or the next higher s orbital which is energetically similar. For simplicity, even though in the ground state the higher s orbital may contain electrons, these electrons are generally included with the d electrons for counting purposes. Thus, even though nickel in the zerovalent state, Ni(0), has a ground state configuration of [Ne] $3s^2 3p^6 3d^8 4s^2$, it is conventionally called a d^{10} element. In any real molecule the metal is not in the ground state anyway, so this convention is not unreasonable. By the same rules then, Ni(I) is a d^9 ion and Ni(II) is d^8.

As noted previously, the transition elements have a very strong tendency to obtain more electrons, generally by sharing them with other molecules, in order to reach the electronic configuration of the next higher inert gas. Ni(0), for example, needs eight more electrons to attain the krypton configuration. This Ni(0) can do by sharing electrons, for instance, with four carbon monoxides, since each molecule of carbon monoxide can share a lone pair of electrons. As a general rule, stable transition metal compounds will have the electronic configuration of the next higher inert gas, but there are many exceptions—a few apparently with too many electrons and some with too few. The compounds with apparently too many electrons probably do not use the extra ones. Very rarely are stable complexes formed lacking more than two electrons from the inert gas structure (7). In fact, if the compounds are limited to diamagnetic organometallic complexes of Groups IVB through VIII there appear to be no established exceptions to the "16- or 18-electron rule" (7). Why 16-electron compounds may be formed is discussed below.

The unique reactivity of transition metal compounds very often is the result of the metal coordinating with two or more molecules possessing unshared pairs of electrons which then, in the coordination sphere of the metal, react with each other to form new bonds and new attachments to the metal. The great tendency of transition metals to obtain electrons is shown by the fact that molecularly divided transition metals prepared in high vacuum, so that they do not react with air, will react rapidly at low temperatures with all types of organic compounds, even saturated hydrocarbons to form various hydrides, alkyls, olefin complexes, and carbides (8). Transition metals in the crystalline state are relatively unreactive, on the other hand, because of strong metal–metal bonding (8). Numerous compounds with single metal–metal bonds are known and they are often quite stable. With the multiple bonds which must occur in pure metal crystals the interactions are very strong, producing high melting and boiling points for these elements.

The number of groups which will bond to a transition metal is determined by the number of electrons the metal can obtain from these groups.

Generally, enough groups will be attached (coordinated) to the metal to give it the necessary number of electrons to achieve the next higher inert gas structure. The geometrical arrangement of the groups about the metal, however, is dependent on several factors: the number of d electrons in the metal atom, the electronic properties of the coordinating groups (ligands), and the size of the ligands. Some discussion of these factors is in order since the geometrical arrangement of ligands about a reacting complex determines which coordinated groups are close enough to react.

The s orbitals are spherically symmetrical, whereas the p and d orbitals are directed in space. Each of the p orbitals is directed along one of the three mutually perpendicular axes x, y, and z with the metal at the origin. The d orbitals are more complicated. The five d orbitals are of equivalent energy in the isolated atom. Four of the five have four equal lobes in one plane, while the fifth is mainly directed along one axis. The appropriate shapes, directions, and signs of the wave functions of the lobes are shown in Fig. I-1. In forming bonds between orbitals, only lobes with the same sign of the wave function can interact to give a net overlap.

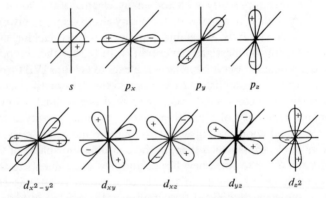

Fig. I-1. Angular probability plots of hydrogenlike orbitals.

Both sigma (σ) and pi (π) bonds occur frequently in organotransition metal compounds. In the molecular orbital formalism a σ bond is defined as one which has no nodal planes (planes where the sign of the wave function passes through zero) containing the internuclear axis, while a π bond has one nodal plane containing the internuclear axis. Possible orbital combinations are shown in Fig. I-2. The π bond between a ligand and a metal is different from the π bond normally encountered between light atoms in that both σ and π interactions are usually involved. Examples are discussed in detail in Section C,2 of this chapter.

σ Combinations

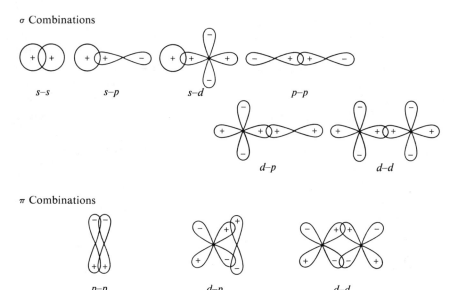

π Combinations

Fig. I-2. Overlap between atomic orbitals which will produce σ and π bonds.

Of particular importance in the bonding of transition metals to CO and olefins may be the empty low-lying antibonding π orbital, π^*, on the ligand. Antibonding orbitals possess an additional nodal plane which is *perpendicular* to the internuclear axis. This type of bonding is depicted later in Figs. I-4 and I-5.

The placing of ligands around a transition metal causes the five previously equivalent d orbitals (energetically degenerate) to change their relative energies since d orbitals containing electrons directed toward ligands (also with electrons) will now be of higher energy than those orbitals which are not directed toward ligands. This effect is known as d-orbital splitting. The relative energies of the d orbitals in various complexes are easily approximated simply by arranging the d orbitals in order of the increasing degree to which they are directed toward the ligands. Some approximate d-orbital energy orders for the most commonly observed ligand geometries are shown in Fig. I-3.

Thus, in the square planar case, for example, the $d_{x^2-y^2}$ orbital is directed toward the four ligands. The d_{xy} orbital interacts more with the ligands than the d_{z^2} orbital does with its central torus in the xy plane and the d_{xz} and d_{yz} orbitals have the least interaction. These diagrams can be used to explain various features of transition metal chemistry. A pertinent example is an explanation of why many d^8 ions tend to have square-planar structures with only four (two electron-donating) ligands forming 16-electron systems rather than the five ligands which would be required for the metals to attain the inert gas, 18-electron structure. The eight metal d electrons will naturally fill the four orbitals of lowest energy. In the square-planar system it is seen that no electrons need to be placed in the highest energy $d_{x^2-y^2}$ orbital. However, if a fifth ligand (with two more electrons) now is added to the complex to form a square pyramid, the d_{z^2} orbital aims directly at the new ligand creating a situation where the electronic repulsions caused by adding the fifth ligand may cost more energy than would be gained by completing the inert gas shell. A similar argument can be used to explain the reluctance of many d^8 ions to form five-coordinate trigonal bipyramids as well.

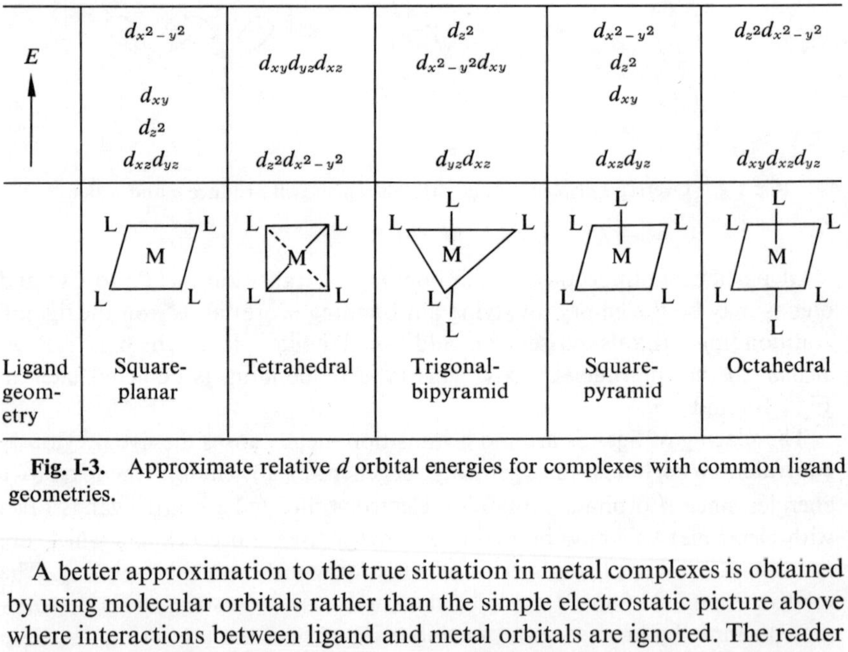

E	Square-planar	Tetrahedral	Trigonal-bipyramid	Square-pyramid	Octahedral
	$d_{x^2-y^2}$	$d_{xy}d_{yz}d_{xz}$	d_{z^2}	$d_{x^2-y^2}$	$d_{z^2}d_{x^2-y^2}$
$\uparrow$	d_{xy}		$d_{x^2-y^2}d_{xy}$	d_{z^2}	
	d_{z^2}			d_{xy}	
	$d_{xz}d_{yz}$	$d_{z^2}d_{x^2-y^2}$	$d_{yz}d_{xz}$	$d_{xz}d_{yz}$	$d_{xy}d_{xz}d_{yz}$

Ligand geometry	Square-planar	Tetrahedral	Trigonal-bipyramid	Square-pyramid	Octahedral

Fig. I-3. Approximate relative d orbital energies for complexes with common ligand geometries.

A better approximation to the true situation in metal complexes is obtained by using molecular orbitals rather than the simple electrostatic picture above where interactions between ligand and metal orbitals are ignored. The reader is referred to several recent publications on ligand field theory for further examples and a more thorough treatment of the subject (9–11).

When eight or less d electrons are present in a complex, there is a choice of where these electrons may go. Unpairing of the electrons within the d orbitals is favored in the free ion, but in a complex, ligands may cause electron pairing to be more favorable. Strongly electronegative ligands, the so-called high-spin, spin-free, or weak-field ligands such as fluoride ion, for example, cause only a small d orbital splitting and unpairing of electrons is unusually favorable. The effect is illustrated with the FeF_6^{3-} ion (a d^5 ion with high-spin ligands) which has five unpaired electrons while $Fe(H_2O)_6^{3+}$ (a d^5 ion with low-spin ligands) has only one unpaired electron. Electron unpairing also may be observed even with normally low-spin ligands if the ligands are polydentate (attached to the metal by more than one group) and complexes are formed in which coordinating groups in the ligand are forced into positions where there are occupied d orbitals. These effects are relatively unimportant in organotransition metal chemistry, however, since organic ligands usually are low-spin, mono- or bidentate ligands. Organotransition metal complexes with odd numbers of electrons likewise are rather rare and in some instances are quite unstable.

The size of ligands relative to the metal can naturally affect the geometry of a complex if the ligands are large enough to prevent neighboring ligands from assuming their most favorable bonding positions. Steric effects of this type are occasionally encountered and examples will be discussed later.

The electronic effects of a variety of σ-bonded transition metal groups have been established by measuring the base strengths of a series of m- and p-(metal substituted-methyl)-pyridines. The variety of substituents used included —Mn(CO)$_5$, —Mo(CO)$_3$Cp,* —W(CO)$_3$Cp, —Co(CN)$_5^{3-}$, and —Fe(CO)$_2$Cp. These metal groups were found to be the most powerful electron-supplying groups known (12, 13). When the metals are π-complexed rather than σ-bonded, however, the electronic effects are reversed. Phenyl-acetic acid, for example, is a stronger acid when it is complexed through the aromatic ring to a tricarbonylchromium group than it is when it is uncomplexed. The electron-withdrawing effect is about the same as if a p-nitro group had been placed in the phenylacetic acid (14).

C. LIGANDS

A wide variety of ligands is encountered in organotransition metal chemistry. In contrast to most inorganic reactions which have been carried out in aqueous media with ionic complexes often containing coordinated water molecules, reactions of interest to organic chemists are generally carried out in nonaqueous solution. The complexes used are usually nonionic, have one or more organic ligands present, and rarely contain coordinated water molecules.

Ligands can be classified most usefully according to the number of electrons they donate to the metal. With some ligands, however, this is an arbitrary decision, depending on the assumed oxidation state of the metal. The major difficulty arises with ligands that are named as neutral ligands by the IUPAC rules (see Section E) but, as such, would have an odd number of electrons, e.g., methyl, ethyl, phenyl, allyl, and cyclopentadienyl. These ligands could be considered as one-electron donors, but this leads to confusion in deter-

* Cp = C_5H_5, cyclopentadienyl.

mining the oxidation state of the metal in complexes containing these ligands. For example, pentacarbonylmethylmanganese would have to be a complex of Mn(0), a paramagnetic d^7 ion, if the methyl group were neutral, but, in fact, it is Mn(I), a diamagnetic d^6 ion. For the purposes of classifying ligands, therefore, neutral groups with odd electrons will be classified as anions having one more electron. With this convention there are no known three-, five-, or seven-electron-donating ligands.

1. One-Electron Donor Ligands

The only significant one-electron-donating ligand encountered in transition metal complexes, excluding neutral ligands with an unpaired electron for the reasons noted above, is the hydride ion when it bridges a transition metal atom and another atom, e.g., in

$$[(CO)_5Cr—H—Cr(CO)_5]^- \quad (Ref.\ 15) \quad and \quad Cp_2Zr\!\!\begin{array}{c} H \\ \diagup \diagdown \\ H \end{array}\!\!B\!\!\begin{array}{c} H \\ \diagup \diagdown \\ H \end{array} \quad (Ref.\ 16)$$

Bridging by other atoms such as carbon is well known with nontransition elements (e.g., aluminum), but is not known in transition metal complexes except in rare, complex, polynuclear compounds where various types of fractional electron-sharing sometimes appear to occur.

2. Two-Electron Donor Ligands

The majority of ligands of interest in transition metal chemistry share two electrons with the metal. Of particular interest in organometallic reactions are carbon monoxide, olefins, acetylenes, phosphines, amines, pyridines, carbonyl compounds, and the anionic ligands, the halides, hydride (nonbridging), and the alkyl, aryl, vinyl, and acyl anions.

Many factors enter into determining metal–ligand bond lengths. The explanation of the variations observed in bond lengths with the same or related ligands in different complexes has been that bond orders between metal and ligand higher than one can occur. Infrared shifts can be conveniently used to measure the presumed amount of multiple bonding, usually referred to as back bonding, of the metal to the ligand. Figures I-4 and I-5 show how carbon monoxide and ethylene may be bonding to transition metals. In carbon monoxide (Fig. I-4) the molecular orbital formed by combining the p orbitals of the carbon and oxygen atoms forms two sets of two filled bonding orbitals between the atoms and two empty antibonding orbitals at the ends of the molecule. (Only one set is shown for simplicity, the other is perpendicular to the plane of the paper). A σ bond is formed between one lobe of an unfilled metal hybrid orbital (e.g., d^2sp^3, dsp^3, dsp^2, sp^3) and a filled sp carbon orbital. A π bond then can be formed between the filled d_{xy} orbital

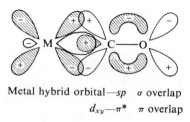

Metal hybrid orbital—sp σ overlap

d_{xy}—π^* π overlap

Fig. I-4. Carbon monoxide bonded to transition metals.

of the metal and the empty, antibonding π-orbital of the carbon monoxide. The relative degree to which this back bonding may occur varies significantly from compound to compound and has been estimated from the carbon–oxygen infrared stretching frequencies of a series of metal carbonyl complexes. The range of absorptions varies from about 2100 cm^{-1} for carbonyls with presumed little back bonding to 1700 cm^{-1} in presumed strongly back-bonded examples.

The bonding in olefin complexes involves the molecular orbital formed from two p orbitals on two olefinic carbon atoms. The π-bonding molecular orbital forms a σ bond with an unoccupied orbital on the metal, while the filled d_{xy} orbital of the metal can form a π-bond with the unoccupied antibonding orbitals of the olefin. Infrared stretching frequencies of the carbon–carbon bond have been used to obtain estimates of the degree of multiple bonding occurring in these complexes. Observed frequencies can be as much as 200 cm^{-1} lower than in the uncomplexed olefin, varying from about 1500 to 1450 cm^{-1} for presumed strongly back-bonded olefins. The assignment of this infrared band to the olefinic group, however, has been questioned and further study appears necessary to establish its origin (17). The position of the carbon–carbon double bond relative to the other ligands in an olefin complex could conceivably be of importance in reactions among coordinated

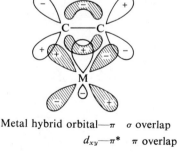

Metal hybrid orbital—π σ overlap

d_{xy}—π^* π overlap

Fig. I-5. Ethylene bonded to transition metals.

ligands. Generally, except in some cases where two or more of the ligands are bonded together (chelating ligands) and restraints are imposed or there are steric problems, rotation of coordinated olefinic groups with respect to the other ligands appears to be a low-energy process (18, 19). The transformation in at least the one example where it has been determined involves rotation with the same side of the olefinic group always facing the metal and not a shift of the metal from one side of the olefin to the other (20).

π-Complexed acetylenic groups also appear to show downward shifts of the triple-bond stretching frequencies from the free acetylene by about 200 cm^{-1} or more.

3. Four-Electron Donor Groups

The four-electron donor system most frequently encountered in transition metal chemistry is the π-allyl anion group. π-Allyl complexes are very important intermediates in a variety of transition metal reactions. The allyl anion may also bond as a two-electron donor in which case it is called a σ-bonded group. The π-bonded allyl anion, donating four electrons to the metal, can adapt various positions relative to the metal depending on the particular complex and the substituents on the allyl carbons. In symmetrical complexes the end carbons of the allyl group will be equidistant from the metal and the middle carbon also will be close enough to be strongly bonded. The three carbons of the allyl group and its five substituents will all be in approximately the same plane with the metal below it. Since the π-allyl group is bent at about a 120° angle, two of the four terminal substituents will be closer to the metal (anti substituents) than the other two (syn substituents). The positions of hydrogen substituents in π-allyl groups are generally determined readily by NMR from a combination of the hydrogen coupling constants and the chemical shifts. The situation can be more complicated in unsymmetrical (e.g., substituted) π-allyl complexes since the allyl group then is no longer symmetrically bonded to the metal.

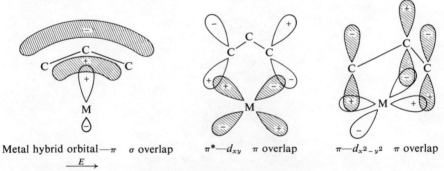

Metal hybrid orbital—π σ overlap π^*—d_{xy} π overlap π—$d_{x^2-y^2}$ π overlap

$\xrightarrow{E}$

Fig. I-6.   Bonding of π-allyl groups.

The bonding situation between the π-allyl group and the metal is a little more complicated than with olefins. The lowest energy molecular orbital of the allyl anion is able to form a strong σ bond with an empty d orbital of the metal, while the allyl antibonding orbital may back bond with a filled metal d orbital. Further bonding may occur between allyl electrons in non-bonding orbitals with another unfilled d metal orbital (Fig. I-6).

Other related four-electron-donating groups which are occasionally encountered in transition metal reactions include π-allyl-type anions, where one or more of the carbon atoms have been replaced by heteroatoms such as nitrogen or oxygen; unsaturated alkyl anions with one or more carbon atoms between the σ-bonding carbon and the π-bonding olefinic group; and the nitrosyl anion (NO^-).

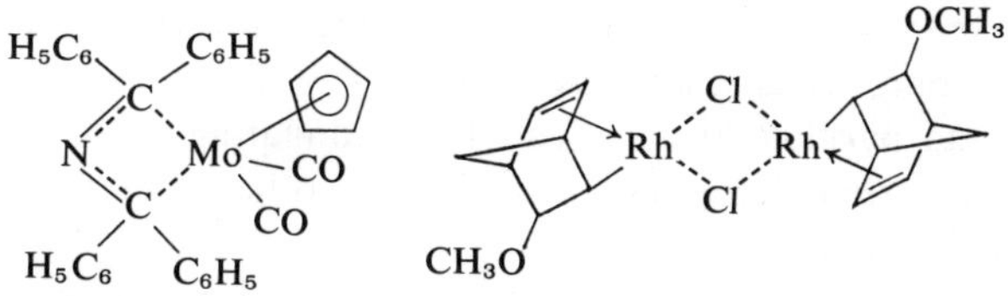

A variety of other types of four-electron-donating ligands are known in which two two-electron-donating ligands are bonded together in various ways. Frequently encountered are π-complexed dienes, both conjugated and nonconjugated. The trimethylenemethane ligand is another type of four-electron donor, but it is only rarely encountered. It could also be classified as a six-electron donor dianion.

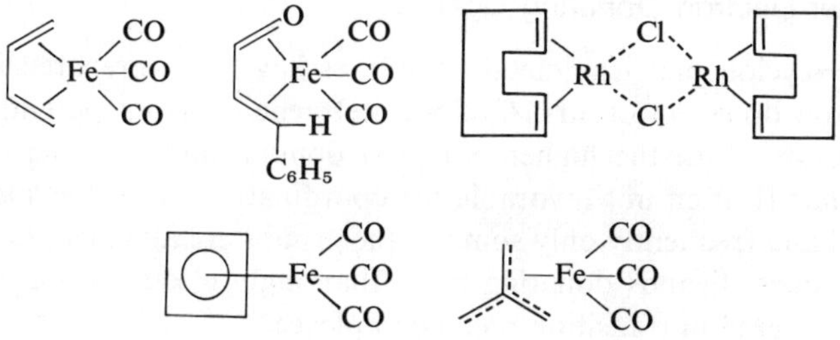

4. Six-Electron Donors

The best known member of this group is the π-cyclopentadienyl anion. The five carbons of the planar ring generally bond equivalently to the metal. Very stable complexes are often formed with this tridentate (tricoordinating) ligand. Ferrocene, [biscyclopentadienyliron(II)], for example, is thermally stable to about 500°, while the biscyclopentadienylcobalt(III) ion is stable to boiling aqueous permanganate (21).

Other six-electron-donating ligands include larger ring and open chain dienyl anion systems, where two double bonds and a carbon anion are in adjacent positions or are more widely separated. By convention, multibonded ligands are often drawn with only a one-line attachment to a metal, even though several bonds may actually be involved, in order to keep the structures easier to visualize.

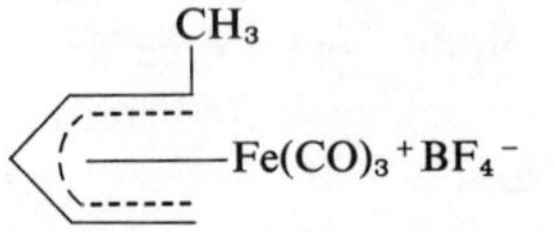

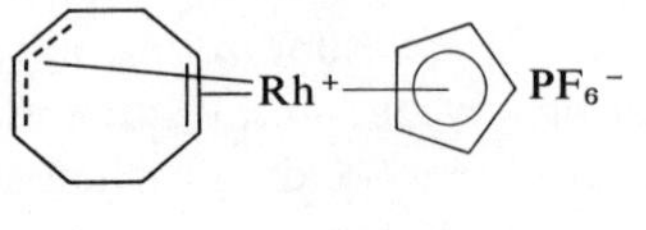

Other well-known six-electron-donating ligands are the aromatic compounds which may share the six π electrons of the aromatic ring with the metal. Heterocyclic compounds with three pairs of π electrons such as pyrrole, thiophene, and pyridine are also capable of donating six electrons to a transition metal. Various linear and cyclic trienes other than aromatic compounds may act similarly.

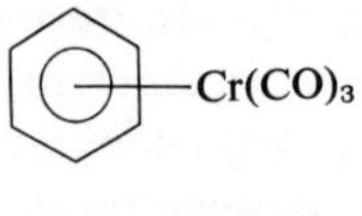

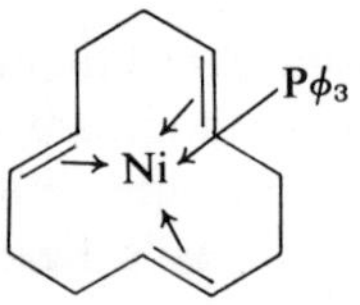

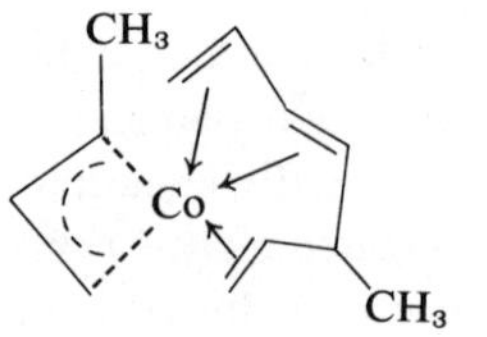

5. Higher Electron-Donating Ligands

Several π-cycloheptatrienylmetal complexes have been prepared in which eight-electron donation occurs. Cyclooctatetraene is also a potential eight-electron donor. With the higher potential donors, however, the geometry of the ligands is often not favorable for coordination of all the sites to the same metal and frequently only some of the groups are attached. Complexes containing single ligands donating more than eight electrons are only very rarely encountered in transition metal complexes.

D. STEREOCHEMICALLY NONRIGID METAL COMPLEXES

Many transition metal complexes with multiunsaturated ligands, where all the π electrons in the ligand are not shared with the metal, possess the ability to undergo internal rearrangements in which the metal moves from one position to another. Complexes that undergo these rearrangements are stereochemically nonrigid molecules. If the rearrangement results in a molecule with an equivalent structure, the molecule is called fluxional. The ease of the rearrangement or whether it occurs at all is determined by the structure of the complex. Most often, stereochemical nonrigidity is detected by NMR studies since the rearrangements are temperature dependent and spectra usually change appreciably over appropriate temperature ranges (22). For example, tetracarbonyl-π-tetramethylalleneiron(0) at $-60°$ shows three different methyl resonances in its NMR spectrum, whereas at 25° there is only one. No exchange occurs with external tetramethylallene (23). Apparently the tetracarbonyliron group is able to move back and forth rapidly between the two olefinic positions in the complex at room temperature. In another

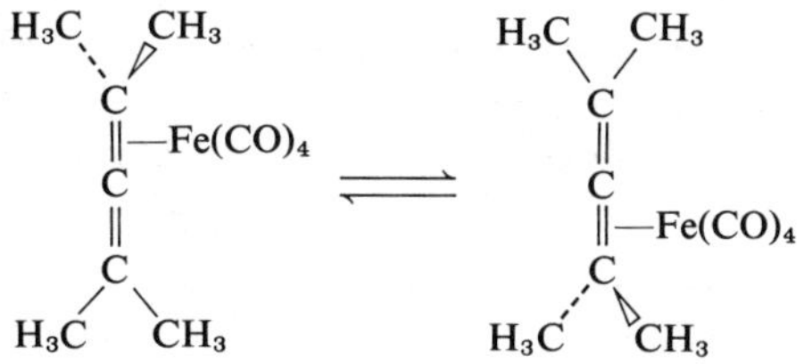

example tricarbonyl-π-cyclooctatetraeneruthenium(0) at 25° shows only a single proton resonance for the cyclooctatetraene protons, even though the 18-electron rule would predict that only two of the four double bonds of the ring should be coordinated to the metal and a more complex spectrum would have been expected. At $-100°$ the NMR spectrum, however, shows the expected complex pattern for the 1,3-diene system attached to the metal. The changes in the spectrum with temperature indicate that the metal moves around the ring by a series of 1,2-shifts (22). Many different types of fluxional molecules are known.

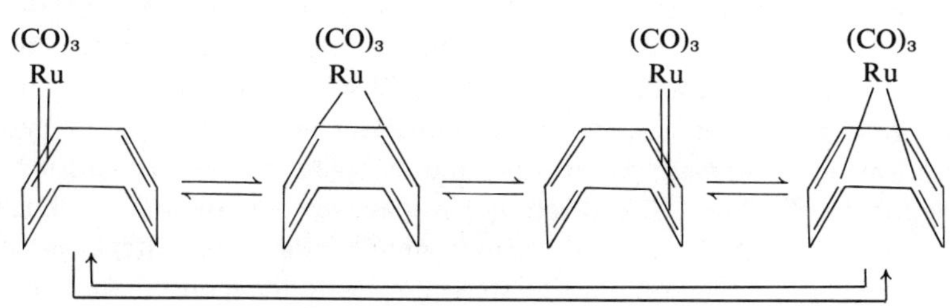

E. NOMENCLATURE OF COORDINATION COMPOUNDS

The field of coordination chemistry has outgrown the existing nomenclature rules and many problems arise in trying to name organotransition metal complexes. Until the rules are revised, however, the original ones must be used as closely as possible (24). In neutral covalent complexes the metal is normally the last part of the name. A prefix of cis, trans, syn, anti, exo, endo, etc., goes first if it applies. Then anionic ligands, all ending in o, are listed in the order H^-, O^{2-}, OH^-, single elements in alphabetical order, polyatomic inorganic anions in alphabetical order, and organic anions in alphabetical order. Next are listed the neutral ligands in the order H_2O, NH_3, other neutral inorganic ligands (in the arbitrary order given in Ref. 24), and organic ligands in alphabetical order. Any cationic ligands are then listed followed by the metal. The neutral ligands are named as the separated molecules or in the case of alkyl or aryl groups and their derivatives as radicals. Ambiguity arises because many ligands can be named as anions or neutral ligands depending on the assumed oxidation state of the metal. The convention in naming complexes has been to treat σ-bonded organic groups as neutral radicals. Other difficulties arise in trying to name complexes containing new types of ligands with various π and σ attachments to the metal. This problem has been solved by a method which is receiving broad usage. This method is to denote the number of atoms of the ligand within bonding distance of the metal with the prefix mono, di, tri, or tetra followed by the word *hapto* (from the Greek word *haptein* meaning to fasten) before the ligand (25). Mono-*hapto*-cyclopentadienyl, written as h^1-C_5H_5, for example, would be the σ-bonded two-electron-donating form of the ligand and penta-*hapto*-, h^5-C_5H_5, the six-electron-donating π form.

F. MECHANISMS OF LIGAND REPLACEMENTS

A basic knowledge of ligand substitution processes is essential in order to understand most organometallic reactions. The first step in a great many organotransition metal reactions is ligand substitution of an "inert ligand" by one that will take part in the reaction of interest. The rates and mechanisms of these substitution reactions therefore may have a major influence on the overall rate of an organometallic reaction. Generally, two basic reaction mechanisms are observed, dissociative and associative. The mechanisms are analogous to S_N1 and S_N2 reactions, respectively, in organic chemistry. Associative mechanisms can be further divided into two categories, one where the entering ligand immediately displaces the coordinated ligand (called an interchange mechanism) and one in which a complex is formed with the new ligand and then later the extra ligand is expelled. Broad generalizations can be made that square-planar complexes usually react by associative mechanisms and that tetrahedral, five- and six- coordinate complexes usually react by dissociative mechanisms, but there are exceptions. Irradiation, however, often will cause first-order dissociations to occur even where they would not occur thermally.

1. Four-Coordinate Complexes

a. Square-Planar Systems

Many ligand replacement reactions have been studied with square-planar complexes. Two-term rate laws are generally found, one term involving solvent, L, as the displacing ligand and the other, the new ligand, Y. The letter L for ligand will be used to designate solvent or some unknown ligand taking part in a reaction. The new ligand, Y (or L), enters the complex from above or below the square plane forming a trigonal bipyramid. The entering and leaving ligands and the one that was originally trans to the leaving group are in the trigonal plane. The leaving group then departs and the complex reverts to square-planar geometry with the new ligand replacing the original one. If solvent replaces the original ligand first, then a second, rapid replacement of solvent occurs, probably by the same mechanism (26).

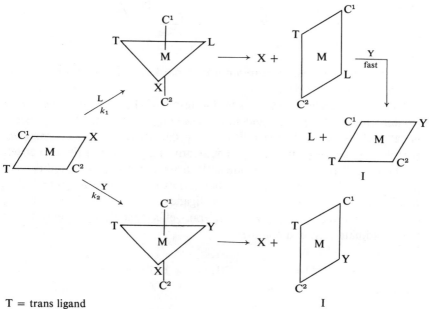

T = trans ligand
C = cis ligand

$$d(I)/dt = k_1[I][L] + k_2[Y][I] = k_1'[I] + k_2[Y][I]$$

Both paths are associative reactions. Clearly, two isomers of L_2X_2M square-planar complexes exist; one has the L's trans to each other and the other cis. The reaction is specific in that cis starting materials yield cis products and trans yield trans products. Reaction rates of cis and trans isomers are often very different. The rate of loss of a ligand is very dependent on the ligand trans to it in square-planar complexes, whereas cis ligands usually have little influence on the rate. This phenomenon is known as the trans effect. For example, the relative rates of replacement of a chloride ligand by pyridine (second-order reactions) from a series of square-planar complexes with trans groups H, CH_3, C_6H_5, and Cl are as follows (27).

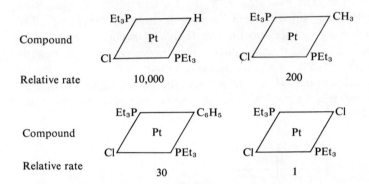

The trans effect appears less important in most complexes with other than square-planar geometries.

The trans effect is believed to be the result of either of two effects or a combination of these, a ground-state weakening of the trans bond between metal and ligand and a stabilization of the activated (five-coordinated) complex by the activating ligand. The relative importance of the two effects to the overall result varies significantly from one ligand to another. The hydride and alkyl effects are probably almost entirely due to ground state weakening of the trans bond, while ligands which are able to strongly π-bond to the metal may exert their trans effect mainly by stabilizing the five-coordinate activated complex.

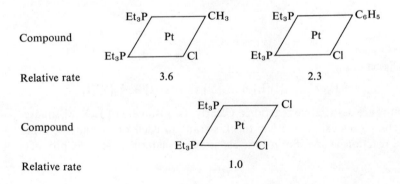

Three of the four cis isomers related to the trans isomers in the above list have also been reacted. Small increases in the same direction are still seen, but the cis effect is only about a fifteenth as large as the trans effect.

It is instructive to note the tremendous reactivity differences between analogous complexes of Ni(III), Pd(II), and Pt(II) in the same chloride displacement reaction by pyridine where only the size of metal is varied.

Compound	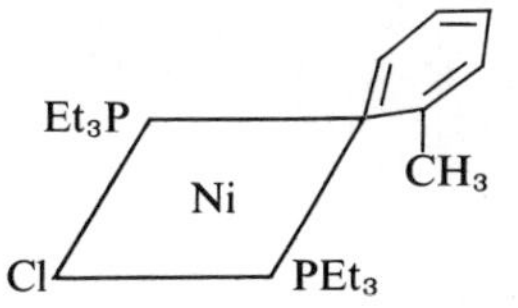	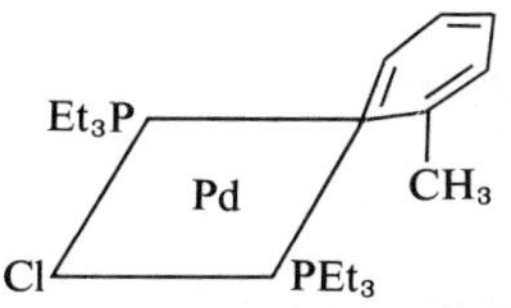
Relative rate	5,000,000	100,000

Compound	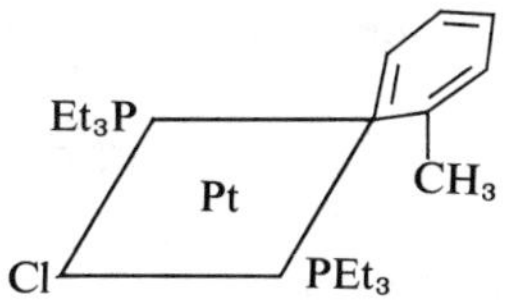
Relative rate	1

Obviously, the rates of reaction strongly depend on the size of the metal. The relative rates correlate with the tendency of the metals to become five-coordinate and overcome the repulsive effects of the d_{z^2} electrons.

Steric effects can be important even in square-planar complexes. The main result of placing larger and larger groups around a square planar metal is to slow down the rate of the ligand substitution process presumably because the entering ligand will find it increasingly difficult to enter a fifth coordination position. The mechanism of reaction changes ultimately to a dissociative reaction when the reacting ligand can no longer directly enter the coordination sphere. A clear demonstration of this effect is seen in the relative rates of chloride replacement by water in the series of amine complexes shown below (28).

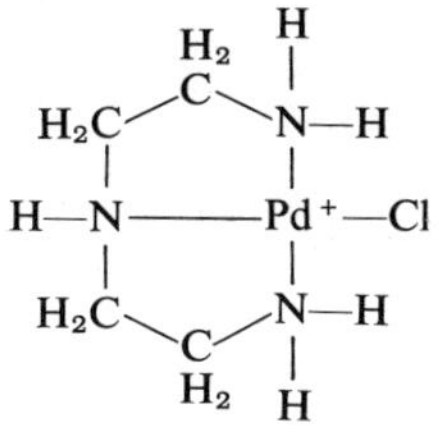

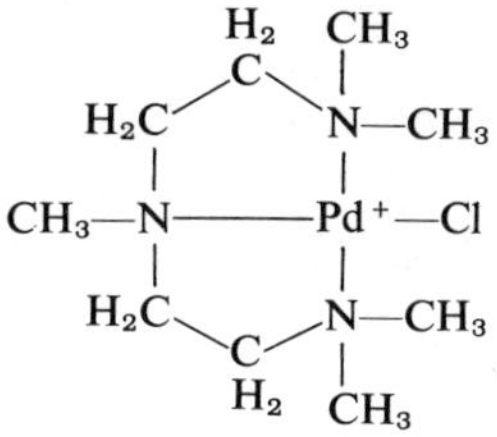

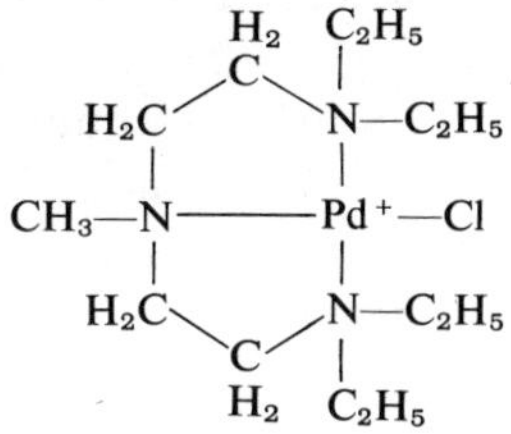

k_{H_2O} (rel) 100,000	400	1

Presumably the slowest compound is reacting exclusively by a dissociative mechanism.

$$
\begin{array}{ccc}
\underset{\substack{\text{H}_2\text{C}\\ \\ \text{CH}_3-\text{N}}}{\overset{\substack{\text{C}_2\text{H}_5\\ \text{H}_2 \mid\\ \text{C}-\text{N}-\text{C}_2\text{H}_5}}{\boxed{\quad\text{Pd}^+\!-\!\text{Cl}\quad}}}
& \xrightarrow{-\text{Cl}^-} &
\underset{\substack{\text{H}_2\text{C}\\ \\ \text{CH}_3-\text{N}}}{\overset{\substack{\text{C}_2\text{H}_5\\ \text{H}_2 \mid\\ \text{C}-\text{N}-\text{C}_2\text{H}_5}}{\boxed{\quad\text{Pd}^{2+}\quad}}}
\end{array}
$$

$$\xrightarrow{\text{H}_2\text{O}} \quad \text{CH}_3-\text{N}\boxed{\quad\text{Pd}^{2+}\!-\!\text{OH}_2\quad}$$

(with the two C—N—C$_2$H$_5$ diethylamino bridging groups on Pd)

b. Tetrahedral Systems

Relatively few examples of ligand replacements in tetrahedral complexes have been studied. From the limited data it appears that dissociative mechanisms are preferred.

Ligand replacement reactions with tetracarbonylnickel show that rate depends only on the nickel carbonyl concentration. Tricarbonylnickel is presumably an intermediate and it reacts rapidly with the new ligand (29).

$$\text{Ni(CO)}_4 \rightleftharpoons \text{CO} + [\text{Ni(CO)}_3]$$

$$\text{Ni(CO)}_3 + \text{L} \rightleftharpoons \text{Ni(CO)}_3\text{L}$$

These reactions are characterized by near-zero or slightly positive entropies of activation. There is always the question with dissociative reactions of whether solvent replaces the dissociated ligand before the new ligand enters. This question is usually answered by noting solvent effects on the rate of the reaction. If the reaction rates become second-order or very slow as the solvating ability of the solvent decreases, then solvation of the intermediate must be important. If solvent replacement is important, then the question arises as to the mechanism of its replacement. This question has rarely been answered. Since nickel carbonyl substitution occurs readily in hydrocarbons, a true tricoordinated intermediate is probably formed. Tricoordinate nickel(0) is coordinately unsaturated since it needs two electrons to have the krypton structure. In some other instances, tricoordinate nickel complexes are stable enough to be isolated.

Ligand substitution reactions of tricarbonylnitrosylcobalt(I) show either first-order or mixed first- and second-order kinetics. The associative reactions therefore are not much less favorable than dissociative ones, at least in this system (30).

$$\text{Co(CO)}_3\text{NO} + \text{L} \longrightarrow \text{Co(CO)}_2(\text{NO})\text{L} + \text{CO}$$

$$-d[\text{Co(CO)}_3\text{NO}]/dt = k_1[\text{Co(CO)}_3\text{NO}] + k_2[\text{Co(CO)}_3\text{NO}][\text{L}]$$

2. Pentacoordinate Complexes

a. Square-Pyramidal Systems

Very few transition metal complexes with this geometry are known and no kinetic studies have been made of their ligand replacement reactions. Presumably, with sterically small ligands, associative mechanisms would be found since enough space would be available for an entering ligand to attack the metal directly.

b. Trigonal-Bipyramidal Systems

An investigation of ligand displacement reactions of one trigonal-bipyramidal iron system has been carried out. Tetracarbonyl(triphenylphosphine)-iron(0) reacts with triphenylphosphine to form the bisphosphine derivatives at a rate independent of the phosphine concentration and therefore the reaction must be of the dissociative type (31).

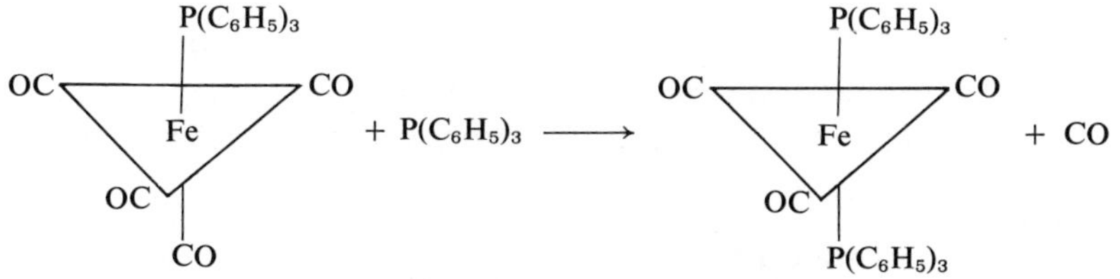

$$d(CO)/dt = k[Fe(CO)_4P(C_6H_5)_3]$$

Similar results were obtained in trigonal-bipyramidal cobalt(I) reactions. Hydridotricarbonyl(trimethylolpropane phosphite*)cobalt(I) reacted with trimethylolpropane phosphite* to form the bisphosphite complex at a rate independent of the phosphite concentration (32).

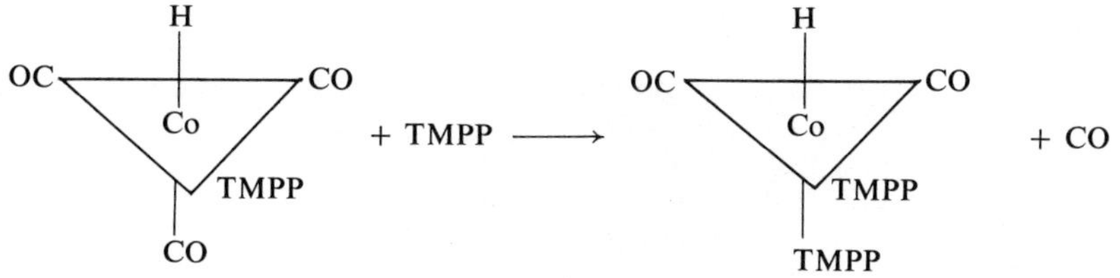

$$d(CO)/dt = k[H(CO)_3(TMPP)Co]$$

The rate constant of this reaction was 10^6 times larger than that of the corresponding acetyl complex, suggesting that there is a very large trans effect in this trigonal-bipyramidal complex (32).

* Correctly named 1-ethyl-3,5,8-trioxa-4-phosphabicyclo[2.2.2]octane.

Comparison of rate constants for the reaction of triphenylphosphine with a series of tetracarbonylacylcobalt(I) derivatives shows how electronic and steric changes may influence reaction rates in trigonal-bipyramidal complexes. The reactions are first-order in the cobalt complex and zero-order in phosphine. Thus, the rates of reaction under the conditions used are a direct measure of the tendencies for the complexes to lose a coordinated carbon monoxide (33). The relative rates of reaction of several complexes with

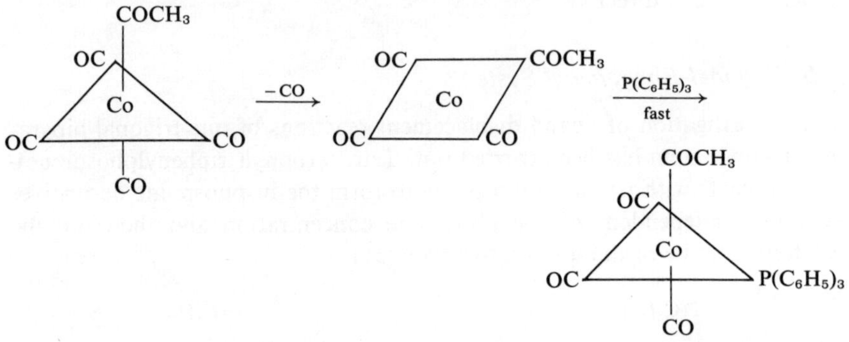

different R groups are given in Table I-1. The first four compounds in the list show the effect of increasing the electron-withdrawing character of the

TABLE I-1
Relative Rates of Dissociation of
Tetracarbonylacylcobalt(I) Derivatives at 0° in
Ether Solution[a]

Compound	Relative rate
$CH_3COCo(CO)_4$	1.0
$ClCH_2CH_2CH_2COCo(CO)_4$	0.8
$CH_3OCH_2COCo(CO)_4$	0.3
$CF_3COCo(CO)_4$	0.1
$CH_3(CH_2)_4COCo(CO)_4$	1.1
$(CH_3)_2CHCOCo(CO)_4$	2.1
$(CH_3)_3CCOCo(CO)_4$	86.0
$C_6H_5COCo(CO)_4$	34.0
$p\text{-}CH_3OC_6H_4COCo(CO)_4$	50.0
$m\text{-}CH_3OC_6H_4COCo(CO)_4$	18.0
$p\text{-}NO_2C_6H_4COCo(CO)_4$	17.0
$p\text{-}CH_3OCOC_6H_4COCo(CO)_4$	20.0
$o\text{-}CH_3C_6H_4COCo(CO)_4$	28.0
$2,4,6\text{-}(CH_3)_3C_6H_2COCo(CO)_4$	0.02
$C_6H_5CH_2COCo(CO)_4$	1.3

[a] From Heck (33).

group. The change from $R = CH_3$ to $CH_2CH_2CH_2Cl$, to CH_2OCH_3, to CF_3 lowered relative rates from 1.0 to 0.8 to 0.3 to 0.1. The decrease is relatively small because of the insulating effect of the acyl carbonyl group, but the trend is clear. As might have been expected, withdrawing electron density from the metal makes dissociation, which is itself a loss of electrons from the metal, a more difficult process. The next three examples in Table I-1 show that as the R group becomes larger the rates of dissociation increase. Relative to $R = CH_3$, the n-butyl group increases the dissociation slightly to 1.1, and the isopropyl group to 2.1, and the t-butyl group to 86. Steric crowding of the ligands is apparently relieved by dissociation of carbon monoxide.

The next series of complexes in Table I-1 are substituted aromatic derivatives. Surprisingly, the phenyl derivative is 34 times more reactive than the methyl compound. One explanation for this effect is that partial or complete phenyl migration from carbonyl carbon to cobalt accompanies the loss of carbon monoxide and the aromatic ring supplies electrons directly to the cobalt. In any case, substitution in the aromatic ring produces the expected electronic effects. The p-methoxyl group increases the dissociation rate, whereas the m-methoxyl, p-nitro, and p-carbomethoxyl groups decrease it. Unexpected decreases occur when o-methyl groups are present. A single o-methyl group decreased the dissociation rate from 34 times the methyl compound to 28 times, while two o-methyls in the 2,4,6-trimethylphenyl derivative decreased the rate to only 0.02 times the methyl derivative. The only reasonable explanation for this deactivation appears to be that the ortho substituents are preventing aryl migration and/or solvation of the metal complex and that one or both of these is important in the dissociation process. o-Methyl groups are known to stabilize other types of transition metal complexes also (34). Separating the phenyl group by one carbon from the acyl group decreases its effect. The benzyl derivative reacts only 1.3 times more rapidly than the methyl compound.

The rate of exchange of radioactive carbon monoxide for normal carbon monoxide in dicarbonyl-π-cyclopentadienylcobalt(I) is a second-order reaction dependent on both the complex and carbon monoxide (35). Presumably, with multidentate ligands such as the π-cyclopentadienyl groups, a partial dissociation of the ligand can occur leaving a vacant coordination position available for a new, external ligand to attack. The adduct can then subsequently expel another ligand. Alternatively, the cyclopentadienyl ligand may be small enough to permit a six-coordinate intermediate to be formed.

3. Hexacoordinate Complexes

Essentially all the most common six-coordinate complexes are octahedrally coordinated. Since these are so common, a great deal of research has been carried out on their reactions. Ligand replacements typically occur by dissociative mechanisms. Reactions of metal carbonyls again provide data of special interest in organotransition metal chemistry.

Ligand replacement reactions of pentacarbonylmethyl- and pentacarbonyl-acetylmanganese(I) have been studied extensively. Discussion of the reactions of the methyl compound will be considered in more detail in Chapter II, Section B,4,a. The addition of ^{13}CO to pentacarbonylmethylmanganese forms the acetylmanganese derivative with ^{13}CO cis to the acetyl group (36). The ^{13}CO exchange with normal CO in the acetyl complex should then also exchange the *cis*-carbonyl group since both this and the preceding reaction must go by way of the same intermediate tetracarbonylacetylmanganese species. If triphenylphosphine is the reagent instead of ^{13}CO, however, a mixture of cis and trans products is obtained (37). Other studies have shown that there is no exchange of acyl carbonyl for coordinated carbonyl under the reaction conditions. The most reasonable mechanism for the reaction involves formation of a five-coordinated intermediate with a cis square-pyramidal structure which relatively slowly isomerizes either to a trigonal bipyramid or a trans square pyramid. The carbon monoxide substitution would occur rapidly with the cis square-pyramidal intermediate, while a slower substitution with the sterically much larger triphenylphosphine would allow time for isomerization of the intermediate to occur (see Scheme I-1).

4. Seven- and Eight-Coordinate Complexes

Seven- and eight-coordination are observed occasionally in inorganic complexes but only rarely in organotransition metal compounds. Several studies have been made with seven-coordinate cyclopentadienylmolybdenum compounds. The halotricarbonyl-π-cyclopentadienylmolybdenum(II) complexes react with phosphines and phosphites with first-order kinetics (38).

$$\pi\text{-}C_5H_5Mo(CO)_3X + PR_3 \longrightarrow CO + \pi\text{-}C_5H_5Mo(CO)_2(PR_3)X$$

$$d(CO)/dt = k[\pi\text{-}C_5H_5Mo(CO)_3X]$$

A comparison of the π-cyclopentadienyl group with the π-indenyl group in the same complex shows the indenyl complex in the first-order reaction to be about 7000 times more reactive (39). This could be the result of steric or electronic effects. The indenyl complex also has an important second-order term in its rate expression. Presumably, the indenyl group may coordinate as a di- or even monodentate ligand, and allow an attacking ligand room to bond to the metal.

Reactions of higher coordinate complexes would be expected to be disso-ciative and have first-order kinetics, at least in the absence of polydentate ligands which could partially dissociate and produce second-order kinetics.

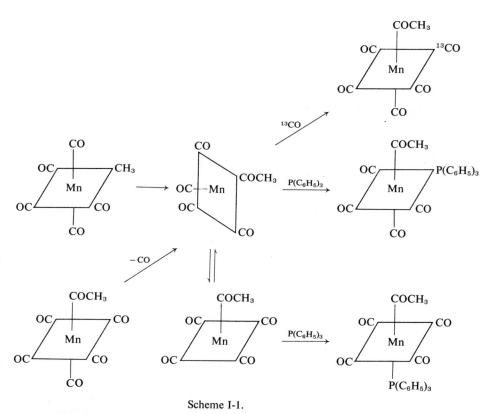

Scheme I-1.

FURTHER READING

R. J. Angelici, Kinetics and mechanisms of reaction of metal carbonyls. *Organometal. Chem. Rev.* **3**, 173 (1968).

F. Basolo and R. G. Pearson, "Mechanisms of Inorganic Reactions," 2nd ed. Wiley, New York, 1967.

D. Benson, "Mechanisms of Inorganic Reactions in Solution." McGraw-Hill, New York, 1968.

F. A. Cotton and G. Wilkinson, "Advanced Inorganic Chemistry," 3rd ed. Wiley (Interscience), New York, 1972.

J. O. Edwards, "Inorganic Reaction Mechanisms." Benjamin, New York, 1964.

R. F. Gould, ed., Homogeneous Catalysis: Industrial Applications and Implications, No. 70. Advan. Chem. Ser., Washington, D.C., 1968.

M. D. Rausch, History and development of organo-transition metal chemistry. *Advan. Chem. Ser.* **62**, 486 (1966).

Chapter II

The Formation of Hydrogen and Carbon Bonds to Transition Metals

Transition metal hydrides and alkyls are involved in nearly all organic transformation caused by transition metal complexes. The hydrides and alkyls may be formed in the reaction mixtures or be prepared separately. Often these compounds are very reactive and unstable in the pure state, however. The high reactivity in most cases is not the result of weak metal–hydrogen or metal–carbon bonds but the result of the complexes having a variety of favorable reaction paths that they can take not involving homolytic cleavage of metal–hydrogen or metal–carbon bonds. Early workers in organotransition metal chemistry ascribed their lack of success in isolating complexes to a low stability of the metal–carbon bond and this view has persisted to some extent until the present. While few pertinent bond energies have been measured, there is no evidence to suggest that dissociation of organotransition metal compounds into radicals, at least at moderate temperatures (40), and without photochemical activation is a common reaction path. Radical chain reactions do occur rarely, but these usually involve a very reactive radical removing a group from the transition metal as the initiating step and not homolytic dissociation. The platinum–carbon bond strength in *trans*-bis(triphenylphosphine)diphenylplatinum(II) has been estimated to be 60 kcal/mole (41), while the cobalt–hydrogen bond in the pentacyanohydridocobalt (III) ion has been estimated to be about 57 kcal/mole (42). Other transition metal–carbon and metal–hydrogen bonds are probably of comparable strengths.

The terms metal hydrides and alkyls (or aryls) are used to describe transition metal complexes with hydrogen and carbon σ-bonded ligands. Unfortunately, the first term, at least, implies that hydrogen behaves as a negative group which is often not the case. Most often the hydrides and alkyl and aryl derivatives react as covalent molecules with little polarization of the metal–hydrogen or metal–carbon bonds.

A. METHODS OF PREPARATION OF HYDRIDO COMPLEXES

1. Exchange Reactions

The metathesis reaction between a transition metal compound, usually a halide although other derivatives also can be used, and a metal hydride, generally a hydride of a main group element, often produces transition metal hydrides. A typical example is the preparation of π-CpRu(CO)$_2$H from the

$$MX + M'H \longrightarrow MH + M'X$$

corresponding iodide and sodium borohydride (43).

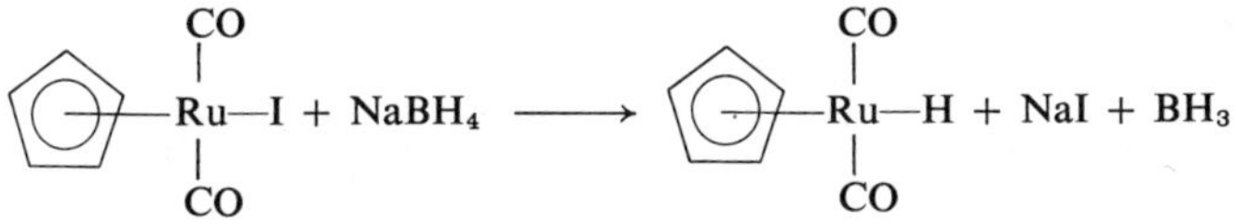

2. Oxidative Addition of H$_2$ or HX

Oxidative addition is a very common reaction of transition metal compounds. In this reaction a metal or metal complex M reacts with a molecule XY to form an adduct XMY, in which the metal is in a formal oxidation state two higher than it was initially. The X and Y groups can originate from a wide variety of XY molecules. For the production of hydrides, of course,

$$M + XY \longrightarrow X—M—Y$$

either X or Y or both must be hydrogen. Dissociation of one ligand must occur before, during, or possibly after the addition if the complex is fully coordinated initially and at the end of the reaction. The hydrogenation of tricarbonylbistriphenylphosphineosmium(0) is a representative example of hydride formation by oxidative addition of hydrogen (44).

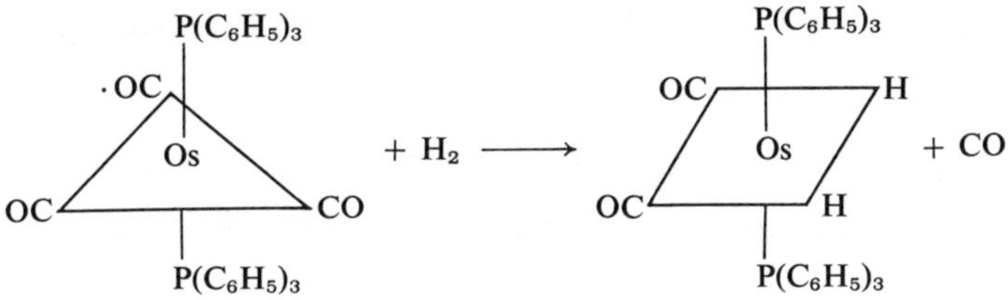

If the reacting complex is stable with fewer than the number of ligands required to give the next higher inert gas structure, then hydrogen addition takes place without prior dissociation or loss of a ligand. The hydrogenation of chlorotristriphenylphosphineiridium(I) is a typical example (45). Cis

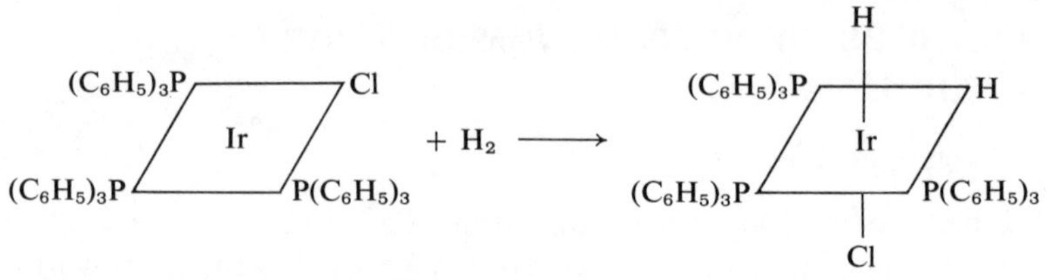

addition of the hydrogen is usually observed.

Reactions with HX molecules are similar except that trans addition is generally found. The following example is typical (46).

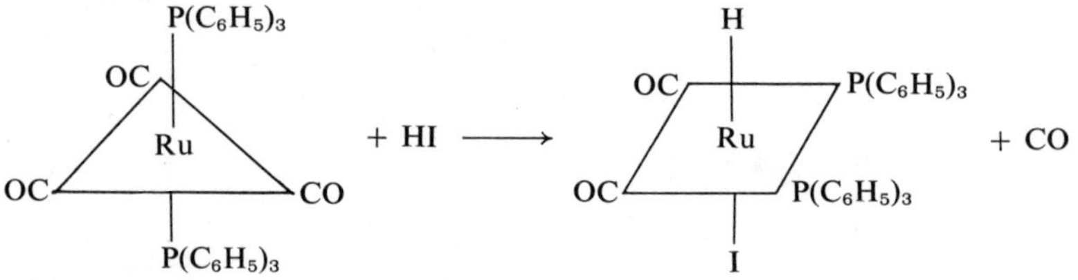

3. Protonation of Neutral or Anionic Transition Metal Complexes

Neutral transition metal complexes with unshared electron pairs are often basic and undergo protonation to form cationic hydrido complexes. This reaction is analogous to oxidative addition except that only the proton is covalently bonded to the metal in the product. Protonation of anionic

$$Ni[P(OEt)_3]_4 + HCl \rightleftharpoons HNi[P(OEt)_3]_4{}^+Cl^- \qquad \text{(Ref. 47)}$$

complexes is also useful for preparing hydrides because anionic complexes are often more readily available by indirect methods than by the obvious direct method of reacting the hydride with a base. Typical is the formation of hydridopentacarbonylmanganese(I) from the sodium salt which, in turn, is produced from decacarbonyldimanganese and sodium amalgam (48).

$$Mn_2(CO)_{10} + 2Na(Hg)_x \xrightarrow{\text{THF}} 2NaMn(CO)_5 \xrightarrow{H_3PO_4} 2HMn(CO)_5$$

4. β-Hydride Elimination Reactions

Transition metal derivatives which have σ-bonded ligands containing β-hydrogens attached to saturated carbon generally undergo metal hydride elimination fairly readily. Examples with alkoxyl (49) and alkyl (50) ligands are well known. The olefin elimination from metal alkyls is frequently involved in catalytic reactions.

$$IrCl_3(PEt_3)_3 + CH_3CH_2O^- \longrightarrow [CH_3CH_2OIrCl_2(PEt_3)_3] \longrightarrow$$

$$HIrCl_2(PEt_3)_3 + CH_3CHO$$

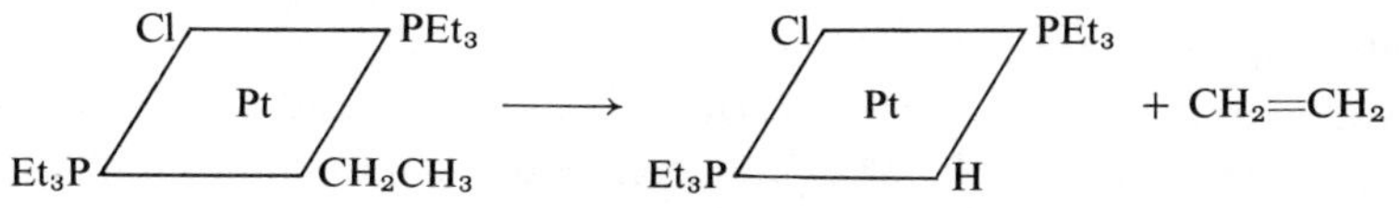

5. Hydride Abstraction

The formation of transition metal hydrides by hydride or hydrogen abstraction is generally limited to high temperature reactions, photochemical reactions, or reactions of complexes which have unpaired electrons. In the latter group an example is the formation of $HCo(CN)_5{}^{3-}$ from $Co(CN)_5{}^{3-}$ and hydrogen (42).

$$Co(CN)_5{}^{3-} + H_2 \longrightarrow HCo(CN)_5{}^{3-} + H\cdot$$
$$H\cdot + Co(CN)_5{}^{3-} \longrightarrow HCo(CN)_5{}^{3-}$$

6. Hydrazine Reduction

Hydrazine is a powerful reducing agent and will reduce many halogen–transition metal derivatives. The reduction often goes to the zerovalent state, but occasionally intermediate hydrides can be isolated.

$$ClRh(CO)[P(C_6H_5)_3]_2 + P(C_6H_5)_3 + N_2H_4 \longrightarrow HRh(CO)[P(C_6H_5)_3]_3 \quad (Ref. 51)$$

B. METHODS OF PREPARATION OF CARBON–TRANSITION METAL BONDS

1. Exchange Reactions

Catalysts prepared by combining metal alkyls, often aluminum alkyls, with transition metal compounds are perhaps the most common type encountered in organotransition metal chemistry. These combinations generally produce alkyl transition metal complexes by exchange (metathesis) reactions. Sometimes the alkyls produced are not stable and further reactions occur to give the true catalytic species. Exchange reactions are generally very useful for producing metal–carbon bonds. The preparation of isolable *trans*-$(PEt_3)_2Pd(Br)CH_3$ is a typical example (52). The use of aluminum alkyls as

$$\underset{Et_3P}{\overset{Br}{\diagup}}\underset{}{\boxed{Pd}}\overset{PEt_3}{\diagup}\underset{Br}{} + CH_3MgBr \longrightarrow \underset{Et_3P}{\overset{Br}{\diagup}}\underset{}{\boxed{Pd}}\overset{PEt_3}{\diagup}\underset{CH_3}{} + MgBr_2$$

exchanging reagents produces aluminum salts as products and these generally are strong Lewis acids. The products may react further and form new complexes which are more reactive than the products obtained by the exchange with other metal alkyls such as those of lithium or magnesium, which do not give strong Lewis acids as reaction products. The effect is exemplified in the reaction of π-Cp_2TiCl_2 with $[Et_2AlCl]_2$, where one chlorine on titanium is replaced by an ethyl group and the other is abstracted, more or less completely, by the dichloroethylaluminum formed in the initial exchange (53). The

$$Cp_2TiCl_2 + [(C_2H_5)_2AlCl]_2 \longrightarrow [Cp_2TiC_2H_5]^+[C_2H_5AlCl_3]^-$$

removal, or at least partial removal, of the chloride group reduces the electron density on the titanium thus increasing its tendency to coordinate with unsaturated molecules and undergo further reactions.

The mechanism of the exchange reaction probably can be either associative or dissociative depending on the reagent used, but little is really known about it. With Lewis acid-type alkylating agents, halogen or oxygen group removal from coordinately saturated complexes probably accompanies the alkylation in a four-centered, front-sided displacement mechanism.

2. Oxidative Addition Reactions

Oxidative addition of alkyl or aryl halides is reminiscent of Grignard reagent formation. A transition metal complex or the metal itself reacts with the halide to form an adduct complex in which the metal oxidation state is increased by two. Again, a coordinately unsaturated complex is usually required for this reaction. In the following example, either CO dissociation or partial dissociation of the cyclopentadienyl ring probably precedes the addition reaction.

$$\text{⬠}-Co(CO)_2 + CF_3I \longrightarrow \text{⬠}-\underset{CO}{\overset{CF_3}{Co}}-I + CO \qquad \text{(Ref. 54)}$$

The formation of stable complexes by this reaction with more than the number of ligands predicted by the inert gas rule is very rare. In some cases ionic intermediates are formed which may or may not lose a ligand subsequently to form a completely covalent product. The reactivity of transition metal complexes in oxidative addition reactions varies widely depending on the complex and its ligands. Compounds which can undergo oxidative

$$\text{Cp–Rh} \overset{CO}{\underset{PMe_2Ph}{}} + CH_3I \longrightarrow \left[\text{Cp–Rh–CH}_3 \overset{CO}{\underset{PMe_2Ph}{}} \right]^+ I^- \qquad (\text{Ref. 55})$$

addition reactions include alkyl and aryl halides, acid halides, sulfonyl halides, vinyl halides, nitriles (R and CN added), esters (R′ and OCOR added) and acyl metal complexes (RCO and the metal group added forming an acyl metal compound with a metal–metal bond). Complexes are known which will react with all these reagents; others will only react with some. Interestingly, some will react with alkyl and not aryl halides and some with aryl and not alkyl halides. Presumably heterogeneous hydrocarbon cracking and reforming reactions involve oxidative additions of hydrocarbons, initially forming alkyl metal hydrides which then undergo metal hydride or alkyl metal eliminations with formation of olefins. Many reactions of this and other types then probably keep occurring ultimately producing the thermodynamically most favored products.

$$RCH_2CH_2CH_3 + M \rightleftharpoons \overset{MH}{RCH_2CHCH_3} \nearrow \overset{RCH=CHCH_3}{\underset{H}{\overset{|}{M}}} \searrow \overset{CH_2=CHCH_3}{\underset{R}{\overset{|}{M}}H}$$

It appears that oxidative additions may occur by ionic, radical, or covalent mechanisms depending on the reagents involved, although very few mechanistic studies have been carried out.

3. Alkylation of Anions

Second-order displacement reactions will occur between nucleophilic anionic metal complexes and organic halides or other compounds with displaceable groups. The reaction, of course, is only useful with organic compounds that are reactive in S_N2 displacements; tertiary, aromatic, and vinylic halides generally cannot be used. A typical example is the formation of pentacarbonylmethylmanganese(I) from sodium manganese carbonylate and methyl iodide (56). The nucleophilicity of metal anions varies greatly

$$NaMn(CO)_5 + CH_3I \longrightarrow CH_3Mn(CO)_5 + NaI$$

from complex to complex as the metal and/or ligands are changed. Some idea of how the reactivity may vary is seen in Table II-1, which gives relative rates of reaction of a series of anionic complexes with methyl iodide (57). Other compounds than halides which are useful reactants are acid halides, acid

TABLE II-1
Relative Rates of Reaction of Anions with Methyl Iodide[a]

Anion	Relative reaction rate (liter/mole sec)
$\pi\text{-}C_5H_5Fe(CO)_2^-$	7×10^7
$\pi\text{-}C_5H_5NiCO^-$	5.5×10^6
$(CO)_5Re^-$	2.5×10^4
$\pi\text{-}C_5H_5W(CO)_3^-$	5×10^2
$(CO)_5Mn^-$	77
$\pi\text{-}C_5H_5Mo(CO)_3^-$	67
$\pi\text{-}C_5H_5Cr(CO)_3^-$	4
$(CO)_4Co^-$	1
$(CO)_5(CN)Cr^-$	$\ll 0.01$

[a] From Dessy *et al.* (57).

anhydrides, epoxides in hydroxylic solvents, toluenesulfonates, and oxonium salts. Oxonium salts are extremely reactive reagents and even alkylate weakly nucleophilic metal anions.

4. Metal Complex Addition Reactions

A wide variety of metal complex addition reactions are commonly encountered in transition metal chemistry. The reactions can be represented in general terms as the addition of a metal compound M—Z to an unsaturated molecule Y: which has at least two electrons available for reaction.

$$M\text{—}Z + Y: \longrightarrow M\text{—}Y\text{—}Z$$

The reaction is conveniently discussed in four parts depending on whether the metal complex adds to one, two, three, or four atoms of the Y: group. These four parts can be further divided according to reaction mechanism, although this requires considerable speculation at the present time since the mechanisms in general have not been very thoroughly investigated.

a. 1:1 Addition Reactions

The addition of a metal compound to a divalent carbon compound forms a new carbon–metal bond. The reaction is important mainly for one reactant, carbon monoxide, but it has been observed with other divalent species also such as isocyanides (58) and carbenes and with other types of compounds in which one of its atoms can easily increase its valence by two. This is an oxidative addition of the metal compound to the reactant. Scheme II-1 gives typical examples which produce new metal–carbon ligands (48, 59–62). The

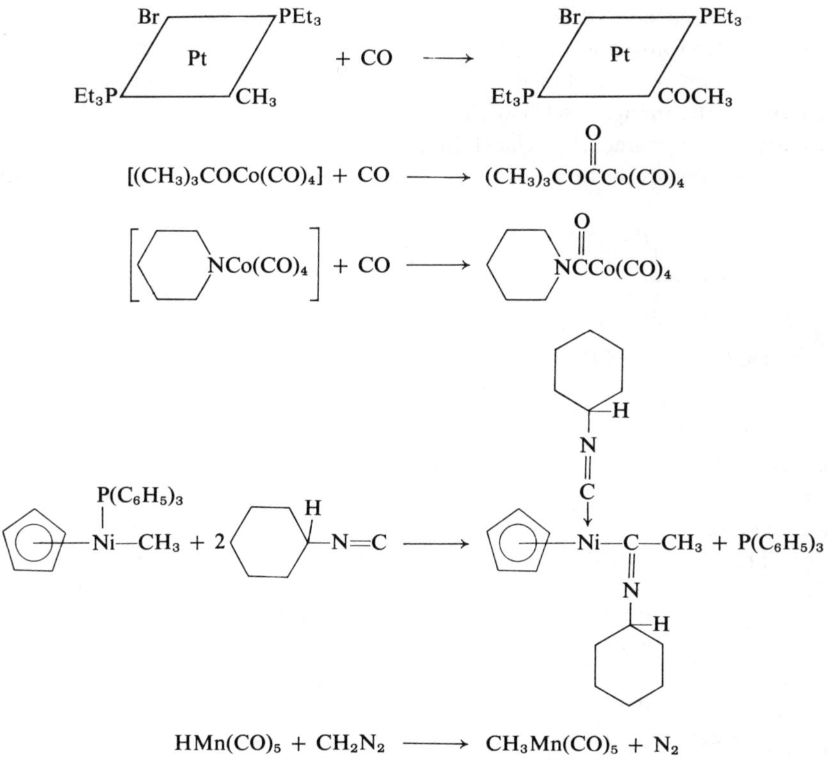

$$[(CH_3)_3COCo(CO)_4] + CO \longrightarrow (CH_3)_3CO\overset{O}{\overset{\|}{C}}Co(CO)_4$$

$$HMn(CO)_5 + CH_2N_2 \longrightarrow CH_3Mn(CO)_5 + N_2$$

Scheme II-1.

last example very probably does not involve free methylene, but the net result is the same as if it did.

The mechanism of addition to carbon monoxide, often referred to inaccurately as the carbon monoxide insertion reaction, has been studied in considerable detail. The addition has been shown to be a 1:2 shift reaction of a group from the metal to the carbon of the carbon monoxide. The reaction is often first-order and independent of external carbon monoxide concentration showing that migration of the organic group may occur spontaneously without a simultaneous nucleophilic attack. Some reactions show large solvent effects, however, and even become second-order in hydrocarbons indicating that solvent may assist the shift, probably by coordinating with the complex. The reversible reaction of pentacarbonylmethylmanganese(I) with carbon monoxide to form pentacarbonylacetylmanganese(I) has been thoroughly studied using ^{13}C-labeled carbon monoxide. The characteristic carbon–oxygen stretching frequency of the carbonyl group is appreciably shifted when carbon-12 is replaced by carbon-13 and by a careful analysis of the infrared absorptions, the position of the ^{13}C-labeled carbonyl in the complexes can be determined. The (probably) first-order reaction of pentacarbonylmethylmanganese(I) with ^{13}C-labeled carbon monoxide led to the formation of pentacarbonylacetylmanganese(I) with the labeled carbon monoxide exclusively in a position cis to the acetyl group (36). The reverse

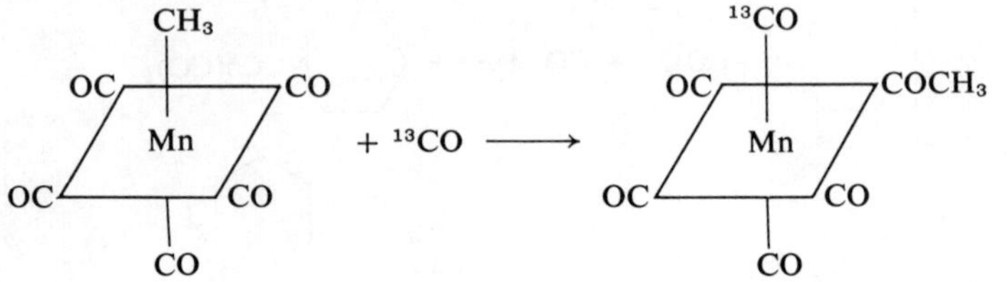

reaction of pentacarbonylacetylmanganese(I) with the label in the acetyl carbonyl group to form pentacarbonylmethylmanganese(I) and carbon monoxide gave a product with the labeled carbonyl cis to the methyl group (36). The related reaction of [cis-^{13}CO]pentacarbonylmethylmanganese(I)

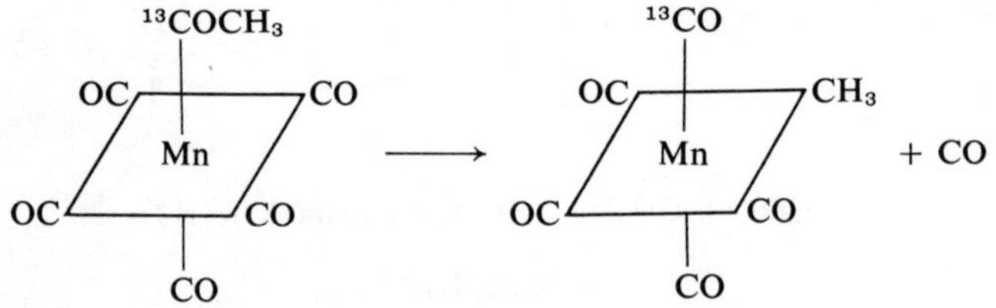

with CO to give three isomeric acyl complexes demonstrates conclusively that the methyl group is shifting rather than the carbonyl group moving.

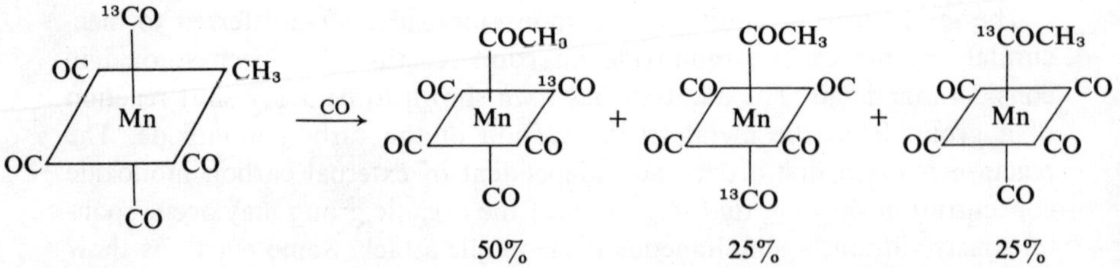

Three ^{13}CO isomers, the cis-, trans-, and acyl-^{13}CO-labeled compounds, are obtained in the statistical proportions expected (2:1:1) from a 1:2 shift of the methyl group. The trans isomer only can be obtained if the methyl group is migrating (36). It follows from the mechanism proposed that ligands other than carbon monoxide should also react and form acyl metal complexes. These reactions are well known. Tetracarbonylmethylcobalt(I), for example, reacts with triphenylphosphine to form the acyl derivative (63).

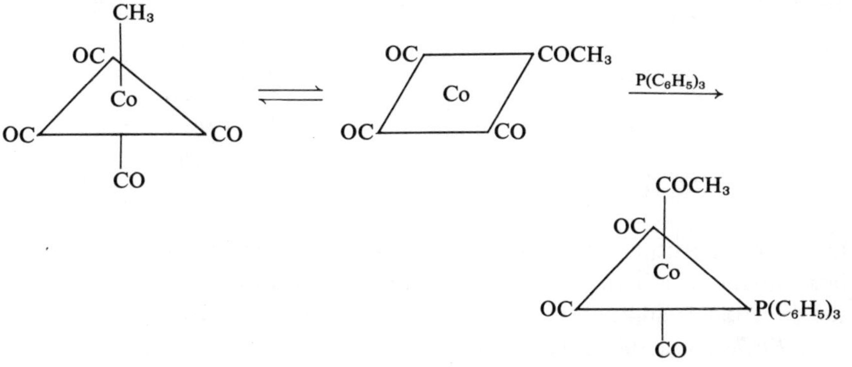

b. 1:2 Addition Reactions

The 1:2 addition reactions of metal complexes to unsaturated organic compounds, i.e., olefins, carbonyl compounds, imines, and acetylenes, may occur by at least three mechanisms—ionic, radical, or covalent.

i. Ionic 1:2 Additions. The best known examples of this type of reaction occur with nontransition metal compounds. The addition of mercuric acetate to olefins in alcoholic solution is an example of a 1:2 ionic addition (64).

$$Hg(OAc)_2 \rightleftharpoons HgOAc^+ + OAc^-$$

$$HgOAc^+ + \;\rangle C{=}C\langle\; \longrightarrow \left[AcOHg{-}\overset{|}{\underset{|}{C}}{-}\overset{|}{\underset{|}{C}}{+} \right] \xrightarrow{CH_3OH} AcOHg{-}\overset{|}{\underset{|}{C}}{-}\overset{|}{\underset{|}{C}}{-}OCH_3$$

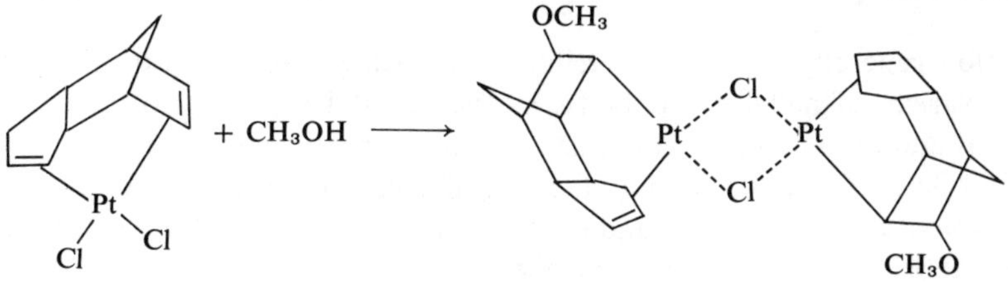

Examples probably occur in transition metal chemistry, but they have not been specifically identified. A number of related additions are known which result from anions or neutral species attacking certain coordinated ligands. These are 1:2 ionic additions, but the formal covalent metal compound added probably was never actually formed. The reaction of dichlorodicyclopentadieneplatinum(II) with methanol is an example of this type of addition. The reaction converts a π-bonded ligand into a σ-bonded one (65). A more obscure example is the reaction of the dicarbonyl-π-propylene-π-cyclopenta-dienyliron(II) cation with borohydride anion (66). In this case only part of the anion (one hydrogen) is added. The addition of the anionic reagent to

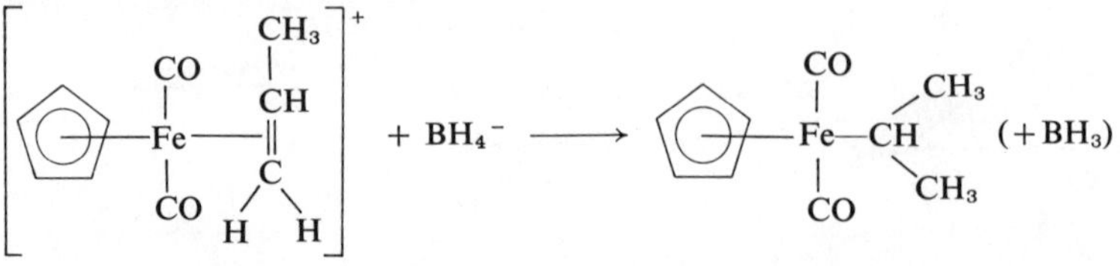

the more substituted and more electron-deficient position of the double bond might have been expected, but coordination with a metal may reverse the direction of addition. The reasons for this are not yet clear.

ii. Radical 1:2 Additions. Radical 1:2 addition reactions are also much more common with main group compounds, e.g., silicon and tin hydride additions, than they are with transition metal derivatives. An example of a transition metal reaction is the addition of the hydridopentacyanocobalt(III) ion to 2-methacrylonitrile (67). The addition of the first group is expected

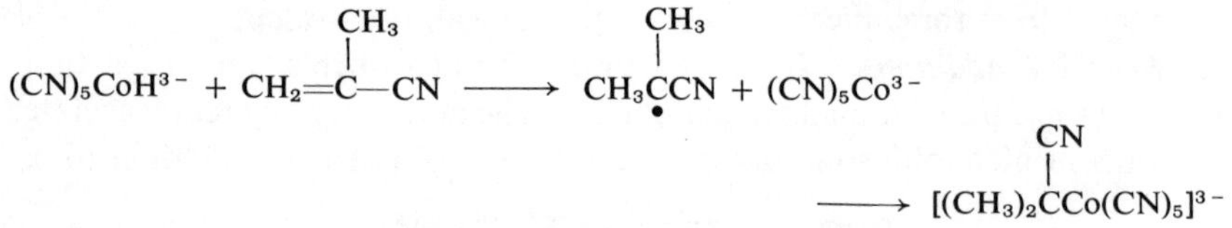

to occur at the position which yields the most stable, intermediate radical.

iii. Covalent 1:2 Additions. Most transition metal additions seem to fall in this group. The most common example is metal hydride addition to olefins. This reaction is often part of a more complicated sequence or part of a catalytic cycle and the intermediate alkyl may not be easily isolable or even detectable. A typical example giving an isolable adduct is the addition of hydridochlorobis(triethylphosphine)platinum(II) to 1-octene (50). The

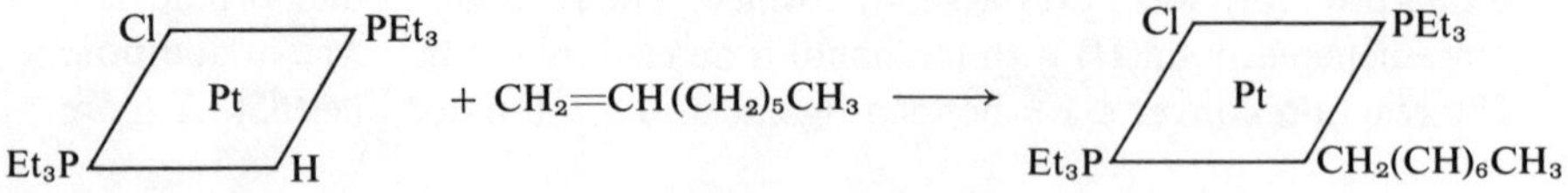

reaction generally is easily reversible. 1:2 Covalent additions occur with a wide variety of metal compounds: hydrides, halides (may also be ionic or radical additions), alkyls, acyls, amides, alkoxides (usually ionic), cyanides, carboxylates (may be ionic) and metal–metal-bonded compounds. Cis addition of the metal compound is generally observed. Directions of addition often appear to be sterically controlled, but all the factors involved are not well understood. Covalent additions generally occur with a preliminary coordination of the unsaturated compound, followed by a shift of the group adding from the metal to the coordinated, unsaturated compound. In coordinately saturated complexes, dissociation is usually the initial step of the reaction.

c. 1:3 Additions

Few 1:3 additions are known, but there is no reason why many reactions of this type with 1:3 dipolar compounds should not be possible. Some possibilities for producing metal–carbon bonds exist, but the examples studied do not form them. Reactions with azides are the best known. Generally, the reactions appear to be 1:3 oxidative additions to the metal complex, followed by loss of nitrogen possibly with formation of a nitrene complex which then may react to form a more stable product. For example (68, 69),

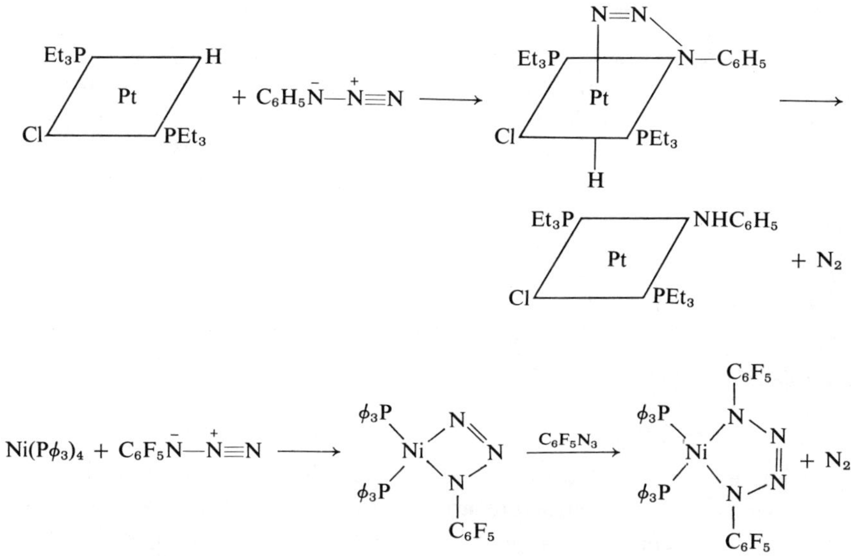

$$Ni(P\phi_3)_4 + C_6F_5\bar{N}-\overset{+}{N}\equiv N \longrightarrow$$

d. 1:4 Addition Reactions

1:4 Additions are encountered in reactions of transition metal complexes with conjugated dienes and probably with α,β-unsaturated carbonyl compounds. For example, the reaction of hydridopentacarbonylmanganese(I) with butadiene gives the 1:4 addition product (70). The product is a mixture of cis and trans isomers and the mechanism of the addition is not known.

$$HMn(CO)_5 + CH_2=CHCH=CH_2 \longrightarrow CH_3CH=CHCH_2Mn(CO)_5$$

An ionic or possibly radical addition is most likely.

The addition of tetracarbonylmethylcobalt(I) to acrolein is an example of a 1:2 methyl shift followed by either a 1:4 addition of the acylcobalt complex to an unsaturated carbonyl compound with π-allyl formation or of a 1:2 addition to the aldehyde carbonyl with subsequent π-allyl formation (71).

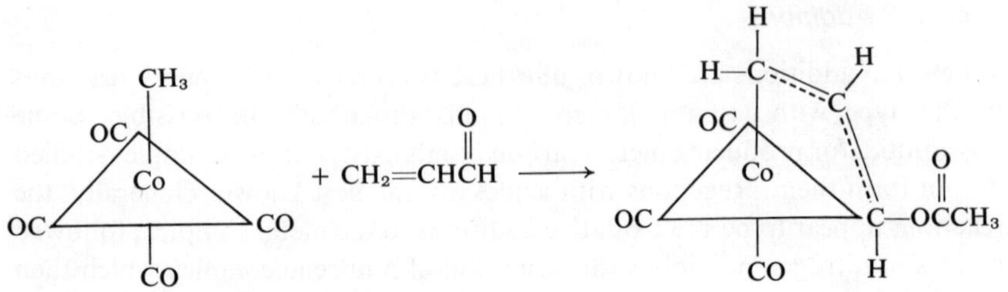

5. Metalation Reactions

The direct substitution of a hydrogen on carbon by a transition metal compound can occur with surprising ease in some instances. *N,N*-Dimethylbenzylamine, for example, reacts at room temperature with lithium tetrachloropalladate to form a chelated, ortho-metalated complex in high yield (72). The presence of the coordinating amine group greatly facilitates the

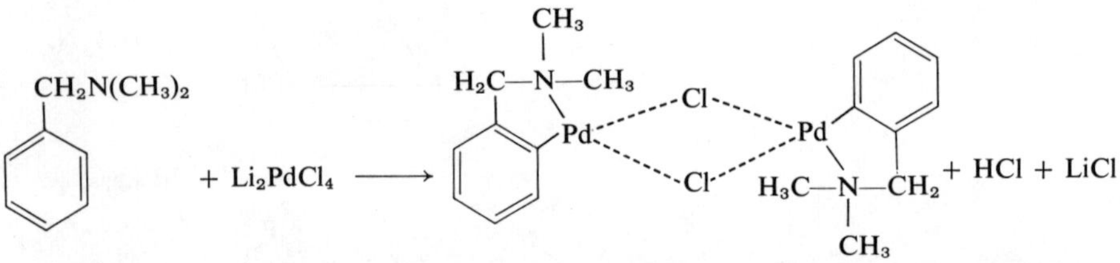

reaction and it also stabilizes the product. Nonchelated arylpalladium compounds of similar structure are not stable. Substitution at saturated carbon also occurs with appropriate compounds. For example, benzylic metalation takes place when 8-methylquinoline is reacted (73). In the absence

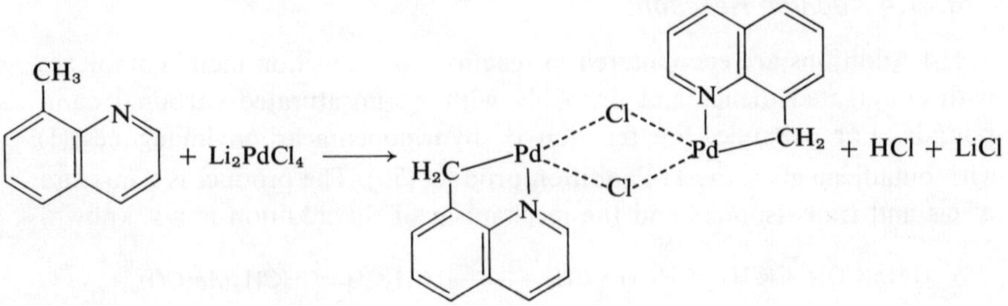

of chelating groups metalation apparently also occurs, but at higher temperatures, and the metal alkyls or acyls are not stable under the reaction conditions. Heating toluene with palladium chloride and sodium acetate in acetic
acid, for example, produces a mixture of isomeric bitolyls (74), probably
produced by decomposition of intermediate isomeric tolylpalladium chlorides.
Sodium acetate accelerates the reaction, probably by either or both of two
effects: (1) by producing palladium acetate which is a more reactive metalating
agent or (2) by lowering the acidity and lowering the rate of the reverse
reaction of the tolylpalladium chloride with the product hydrogen chloride.

   Similar metalations occur with other reactive transition metal complexes.
For example, pentacarbonylmethylmanganese(I) and benzylideneaniline
form an ortho-metalated chelated complex and, presumably, carbon monoxide
and methane (75, 76). Ortho-metalated cobalt complexes appear to be

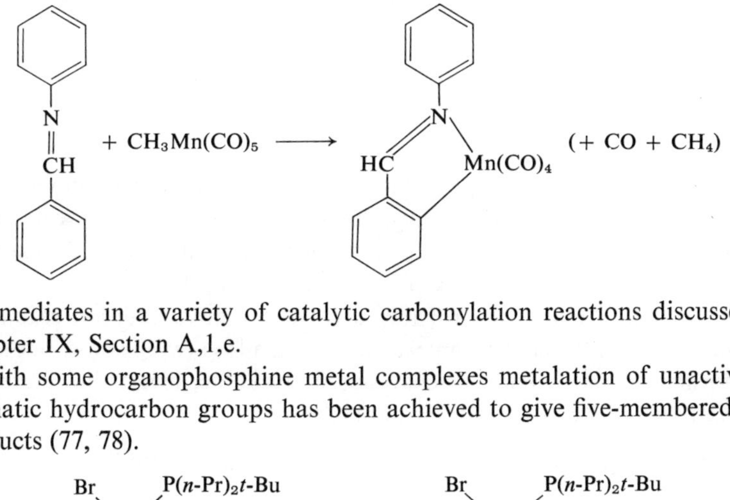

intermediates in a variety of catalytic carbonylation reactions discussed in
Chapter IX, Section A,1,e.

   With some organophosphine metal complexes metalation of unactivated
aliphatic hydrocarbon groups has been achieved to give five-membered ring
products (77, 78).

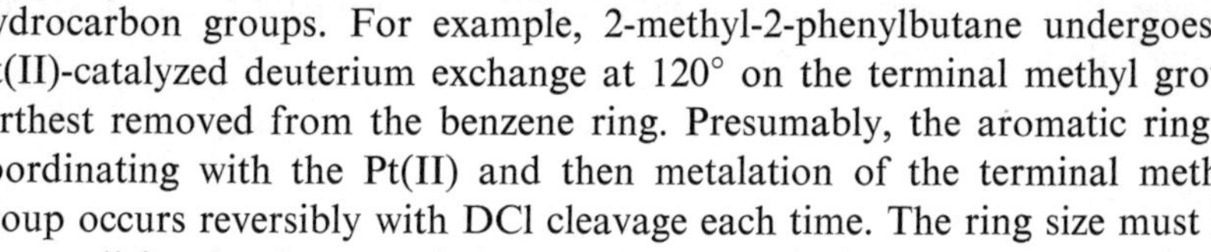

   Even weak coordination can appreciably facilitate metalation of saturated
hydrocarbon groups. For example, 2-methyl-2-phenylbutane undergoes a
Pt(II)-catalyzed deuterium exchange at 120° on the terminal methyl group
farthest removed from the benzene ring. Presumably, the aromatic ring is
coordinating with the Pt(II) and then metalation of the terminal methyl
group occurs reversibly with DCl cleavage each time. The ring size must be
too small for the closer methyls to undergo metalation (79).

   A similar selective deuterium exchange occurs with long-chain olefins and
a homogeneous platinum catalyst (80). High-temperature, nonselective,

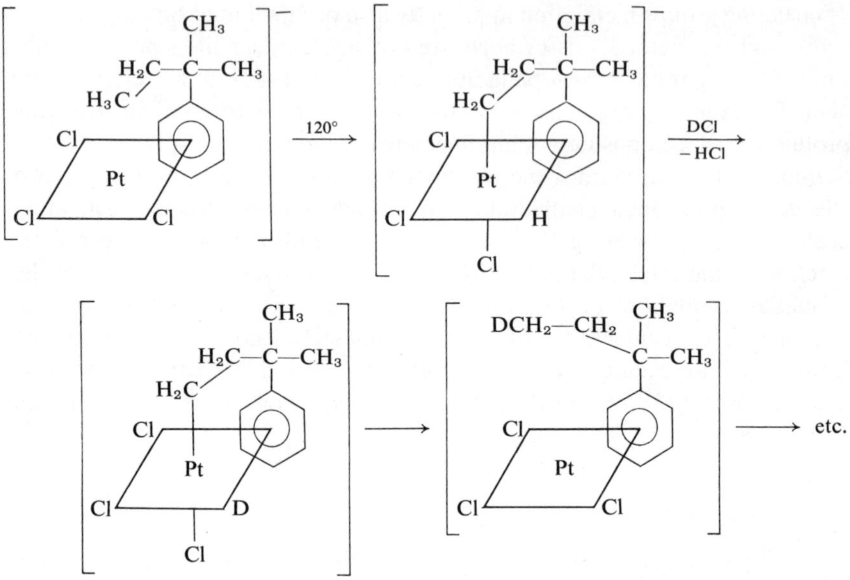

aliphatic metalation is very probably involved in the catalytic cracking of petroleum. The possibility of finding more selective, low-temperature methods for achieving the isomerization or functionalization of saturated hydrocarbons has created considerable interest in this area of catalysts.

The metalation reactions probably involve oxidative addition reactions, first forming higher valent complexes which then lose two ligands, but since no information on reaction mechanisms is available, metalation has been discussed separately as a substitution reaction.

6. Reactions of Metal Complexes with Diazo Compounds

Most hydrido transition metal compounds will react with diazo compounds to form nitrogen and the alkyl metal derivative. For example (81),

$$CpMo(CO)_3H + CH_2N_2 \longrightarrow CpMo(CO)_3CH_3 + N_2$$

This reaction may also be considered a 1:1 addition of the hydride to a carbene. Similar reactions occur with halotransition metal complexes producing halomethyl derivatives.

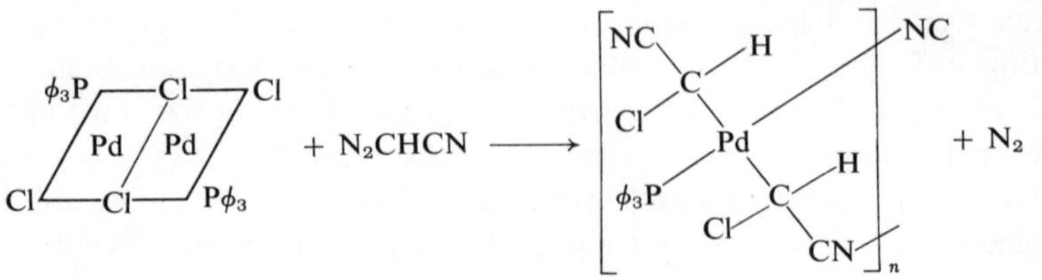

7. Decomposition of Acyl, Arylsulfonyl, Carboxy, and Aryldiazonium Metal Complexes

Few examples of these reactions are now known, but they will probably turn out to be quite general. The decarbonylation of pentacarbonylacetylmanganese(I) to the methyl compound has already been mentioned above. Decarbonylation is particularly useful for preparing aryl metal compounds, since they cannot usually be obtained directly by arylation of the metal anion, whereas the related acylation occurs easily (83). Lowering the carbon mon-

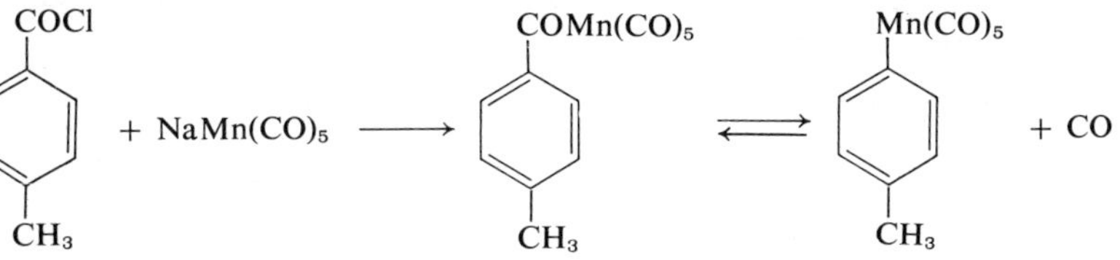

oxide concentration and raising the temperature generally favors the decarbonylation. Decarbonylation has also been achieved with the aid of chlorotris(triphenylphosphine)rhodium(I) which forms a very stable carbonyl derivative in the reaction. It is not clear whether the reagent removes a carbonyl group directly from the molecule being decarbonylated or it just reacts efficiently with carbon monoxide that has already dissociated from it. In any case the reaction appears to go more easily than the thermal decarbonylation. The use of an optically active acyl group in the example below showed the reaction to go with retention of configuration about the alkyl carbon (84).

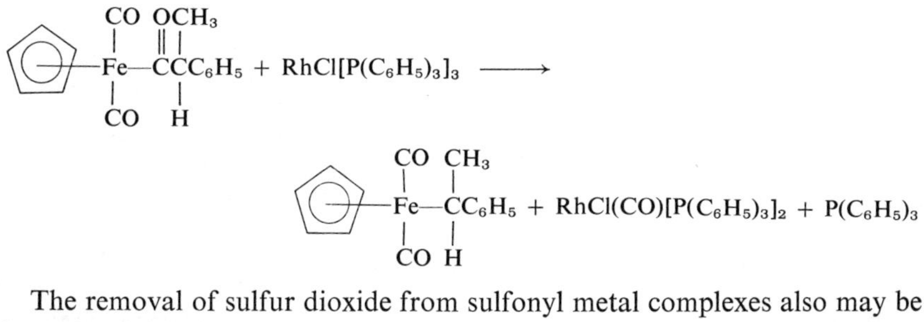

The removal of sulfur dioxide from sulfonyl metal complexes also may be brought about thermally. Since the sulfonyl derivatives can be made from sulfonyl chlorides this reaction provides another possible route to carbon–transition metal compounds (85). The mechanism of the SO_2 elimination

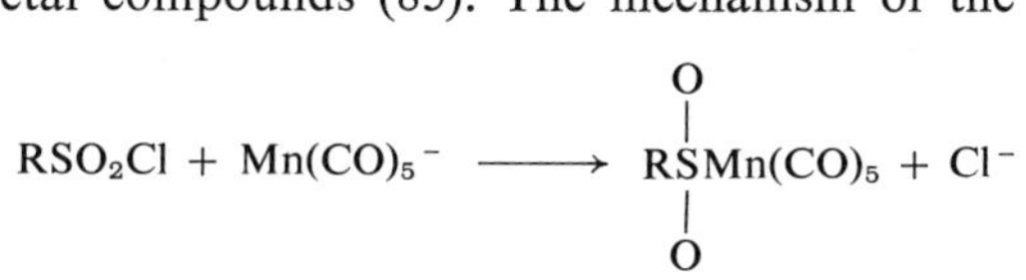

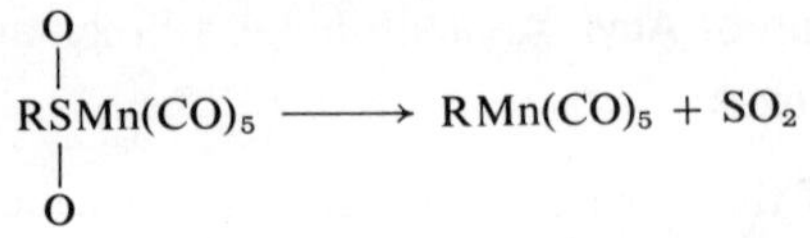

$$\underset{\underset{O}{\overset{\displaystyle O}{\big|}}}{RSMn(CO)_5} \longrightarrow RMn(CO)_5 + SO_2$$

may be similar to the decarbonylation reaction with final replacement of coordinated SO_2 by CO, but this has not been established (however, see Chapter X, Section D).

The elimination of carbon dioxide from transition metal carboxylic acid derivatives is rare. It does occur, however, when the elimination is facilitated by having the carboxylate group beta to an activating group such as cyano or carboalkoxy (86).

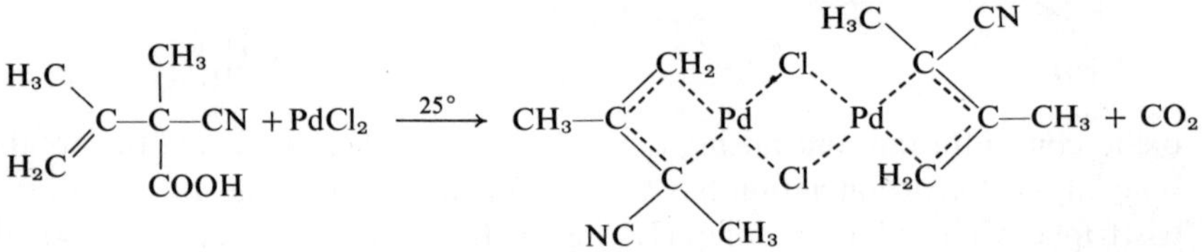

Decomposition of aryldiazonium metal derivatives to an aryl metal complex only has been carried out once, but it should be a general reaction (87).

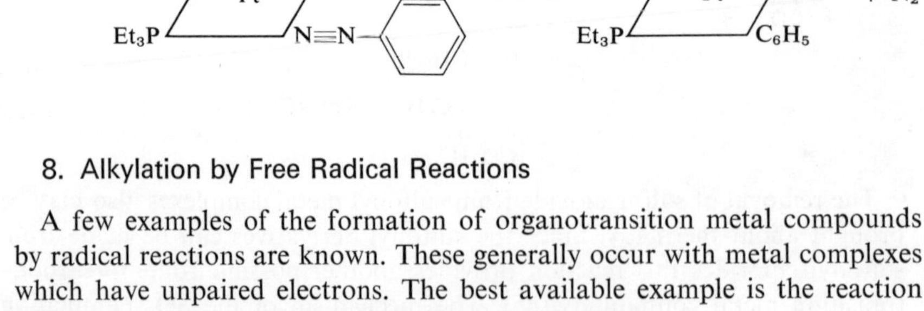

8. Alkylation by Free Radical Reactions

A few examples of the formation of organotransition metal compounds by radical reactions are known. These generally occur with metal complexes which have unpaired electrons. The best available example is the reaction of pentacyanocobalt(II) ion with methyl iodide (88).

$$(CN)_5Co^{3-} + CH_3I \longrightarrow (CN)_5CoI^{3-} + CH_3\cdot$$

$$CH_3\cdot + (CN)_5Co^{3-} \longrightarrow (CN)_5CoCH_3^{3-}$$

9. Coupling of π-Bonded Ligands

Two π-bonded ligands on a transition metal may react forming a new σ bond between the ligands and two bonds to the metal. This reaction is a kind of oxidative addition and, therefore, the metal must be one capable of increasing its oxidation state easily by two. Interesting examples of this reaction occur with π-bonded dienes which usually form at least one π-allyl group in the coupling (89–91). Coupling of acetylenic ligands generally forms

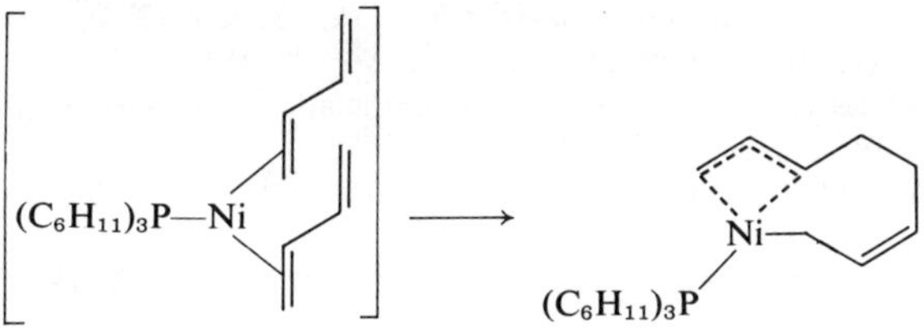

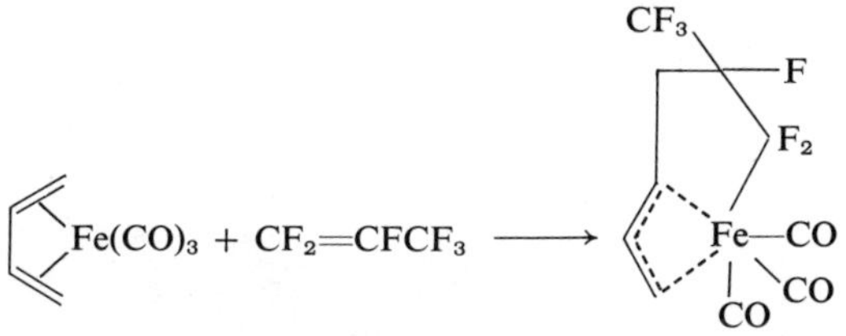

$$\left[(C_6H_{11})_3P\text{—Ni} \right] \longrightarrow$$

$$\underset{\displaystyle (C_6H_{11})_3P}{\text{Ni—}}$$

$$\overset{CF_3}{\underset{}{}} \quad \begin{array}{c} \text{—F} \\ \text{F}_2 \end{array}$$

$$\text{Fe(CO)}_3 + CF_2{=}CFCF_3 \longrightarrow \underset{CO}{\overset{}{\text{Fe—CO}}}\ \text{CO}$$

metalacyclopentadiene complexes. The following example is typical (92).

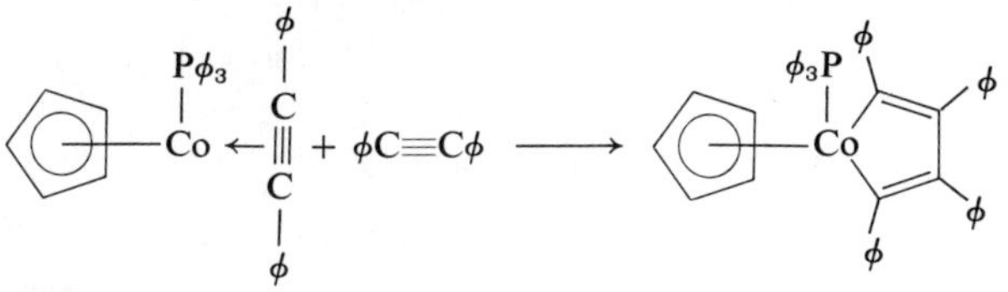

$$\underset{\underset{\phi}{\overset{|}{\text{C}}}}{\overset{P\phi_3}{\underset{|}{\text{Co}}}} \overset{\phi}{\underset{}{\overset{|}{\text{C}}}} + \phi C{\equiv}C\phi \longrightarrow \underset{}{\overset{\phi_3P}{\text{Co}}}\ \phi$$

Many ligand coupling reactions are known with the highly reactive perfluorinated olefins, acetylenes, imines, and ketones. The following are some examples (93).

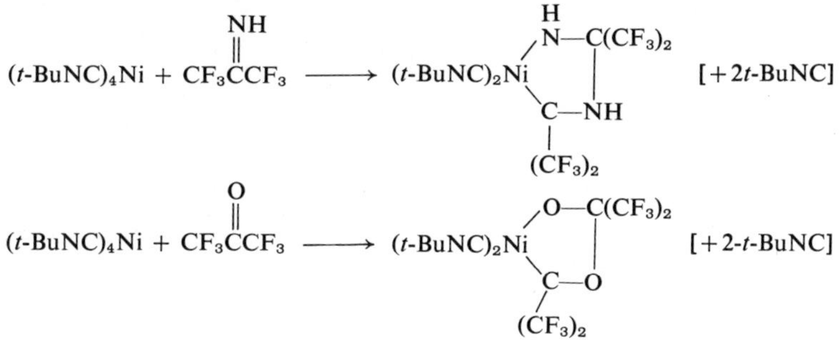

$$\overset{NH}{\underset{}{\|}}$$
$$(t\text{-BuNC})_4\text{Ni} + CF_3\overset{}{C}CF_3 \longrightarrow (t\text{-BuNC})_2\text{Ni}\begin{array}{c} N\text{—}C(CF_3)_2 \\ | \\ C\text{—NH} \\ | \\ (CF_3)_2 \end{array} \quad [+2t\text{-BuNC}]$$

$$\overset{O}{\underset{}{\|}}$$
$$(t\text{-BuNC})_4\text{Ni} + CF_3\overset{}{C}CF_3 \longrightarrow (t\text{-BuNC})_2\text{Ni}\begin{array}{c} O\text{—}C(CF_3)_2 \\ | \\ C\text{—O} \\ | \\ (CF_3)_2 \end{array} \quad [+2\text{-}t\text{-BuNC}]$$

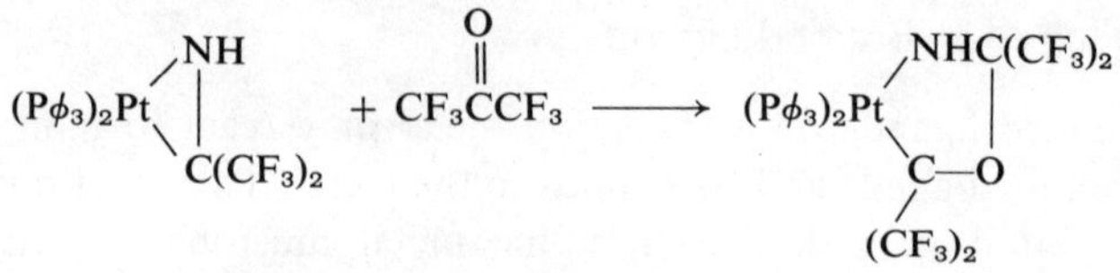

$$(P\phi_3)_2Pt \begin{array}{c} {}^{NH} \\ | \\ C(CF_3)_2 \end{array} + CF_3\overset{\overset{\displaystyle O}{\|}}{C}CF_3 \longrightarrow (P\phi_3)_2Pt \begin{array}{c} {}^{NHC(CF_3)_2} \\ | \\ C-O \\ {}_{(CF_3)_2} \end{array}$$

FURTHER READING

J. Halpern, Oxidative addition. *Accounts Chem. Res.* **3**, 386 (1970).

R. F. Heck, Insertion reactions. *Advan. Chem. Ser.* **49**, 181 (1965).

R. B. King, Metal carbonyl anions in organometallic syntheses. *Accounts Chem. Res.* **3**, 417 (1970).

M. F. Lappert and B. Prokai, Insertion reactions. *Advan. Organometal. Chem.* **5**, 225 (1967).

E. L. Muetterties, "Transition Metal Hydrides." Dekker, New York, 1971.

Chapter III

Reactions of Transition Metal σ- and π-Bonded Derivatives

A. REACTIONS IN WHICH THE METAL IS RETAINED

1. σ Derivatives

a. Hydrides

Transition metal hydrides generally initiate organometallic reactions by adding to one of the unsaturated reactants present. Ionic, radical, and covalent additions appear to be possible, but few mechanistic studies have been completed. These additions produce new σ-bonded complexes, usually metal alkyls or, in rare instances, complexes with other atoms than carbon directly attached to the metal. Two subsequent reactions are now possible: either further addition reactions may occur or the σ-bonded group can be lost from the metal by one of the reactions described in Section B,3 of this chapter. For example (94, 94a),

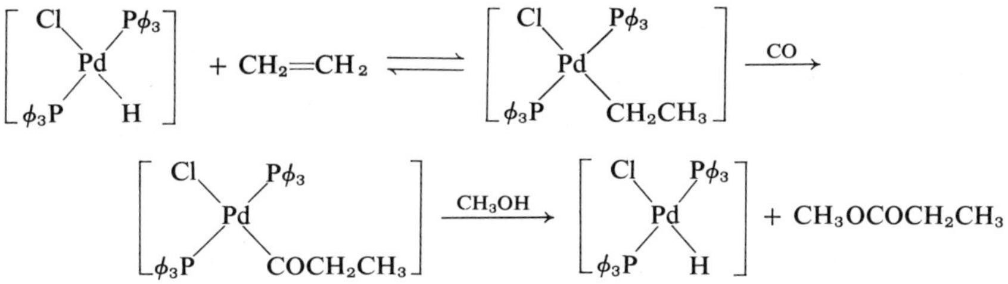

$$\begin{bmatrix} Cl & P\phi_3 \\ & Pd \\ \phi_3P & H \end{bmatrix} + CH_2{=}CH_2 \rightleftarrows \begin{bmatrix} Cl & P\phi_3 \\ & Pd \\ \phi_3P & CH_2CH_3 \end{bmatrix} \xrightarrow{CO}$$

$$\begin{bmatrix} Cl & P\phi_3 \\ & Pd \\ \phi_3P & COCH_2CH_3 \end{bmatrix} \xrightarrow{CH_3OH} \begin{bmatrix} Cl & P\phi_3 \\ & Pd \\ \phi_3P & H \end{bmatrix} + CH_3OCOCH_2CH_3$$

$$[HCo(CO)_3] + CH_3CHO \longrightarrow [CH_3CH_2OCo(CO)_3] \xrightarrow{H_2}$$
$$CH_3CH_2OH + [HCo(CO)_3]$$

45

b. Carbon-Bonded Derivatives

Two reaction courses are commonly observed with carbon σ-bonded transition metal complexes: (1) Ligand substitution by a reactant occurs and then the carbon group moves from metal to the ligand (covalent addition) or (2) an external reagent attacks the complex, either at the σ-bonded carbon group (S_N2) or at the metal (oxidative addition). For example (34, 94b, 94c),

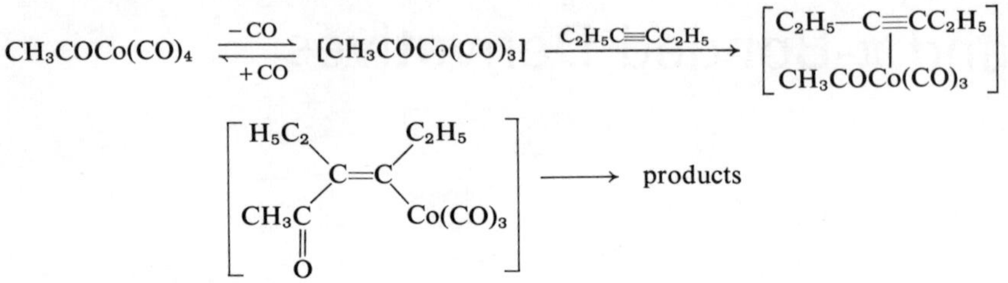

$$CH_3COCo(CO)_4 \underset{+CO}{\overset{-CO}{\rightleftharpoons}} [CH_3COCo(CO)_3] \xrightarrow{C_2H_5C{\equiv}CC_2H_5} \left[\begin{array}{c} C_2H_5{-}C{\equiv}CC_2H_5 \\ | \\ CH_3COCo(CO)_3 \end{array} \right]$$

$$\longrightarrow \text{products}$$

$$C_6H_5CH_2COCo(CO)_4 + C_6H_5NH_2 \longrightarrow C_6H_5CH_2CONHC_6H_5 + HCo(CO)_4$$

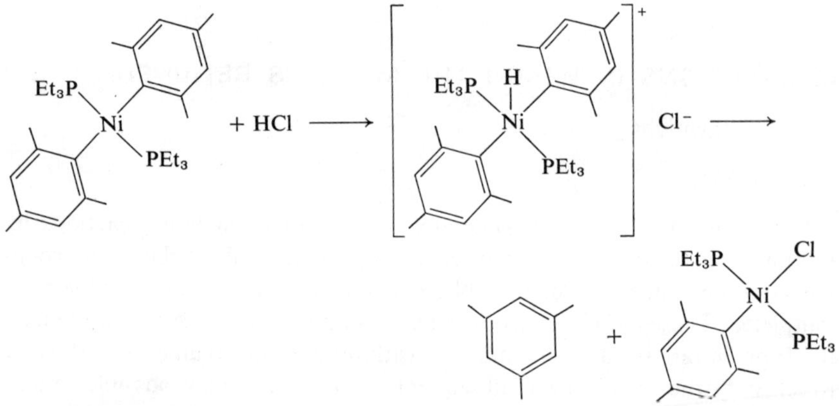

c. Reactions of σ-Bonded Compounds Other than Metal Hydrides and Metal–Carbon Derivatives

These derivatives often undergo addition reactions with unsaturated carbon compounds forming metal–carbon σ-bonded products. The subsequent reactions are then usually the same as with other carbon–metal derivatives. The following example is typical (94d).

$$Pd(OAc)_2 + CH_2{=}CH_2 \xrightarrow{L} [L_2Pd(OAc)CH_2CH_2OAc] \longrightarrow$$

$$[L_2Pd(OAc)H] + CH_2{=}CHOAc$$

Relatively few useful organic reactions, other than oxidation by electron transfer, involving transition metal complexes but not involving metal–carbon bonds in at least one stage of the reaction, are known.

2. π Derivatives

The reactions by which π-bonded ligands are converted into σ complexes have been discussed in Chapter I. There are some other synthetically useful substitution reactions of π-bonded ligands which sometimes can be carried out. The major problem is usually that the metal complex is not stable enough to withstand the reaction conditions necessary to effect reactions at the π-bonded ligand. In cases where the complexes are stable, the usual kinds of organic reactions may be carried out, but the reactivity of the ligand is usually greatly modified by coordination with the metal. Two examples illustrate this point. π-Butadienetricarbonyliron(0) can be acetylated with acetyl chloride and aluminum chloride to form the 1-acetyl derivative (95), while the free ligand would be polymerized under the reaction conditions. In

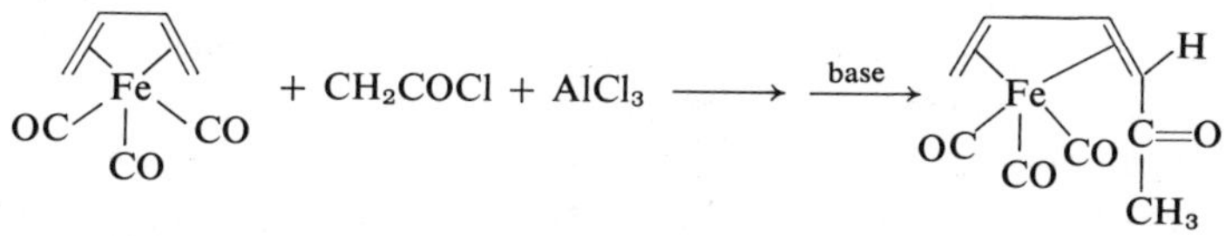

another example, the complex formed between octacarbonyldicobalt(0) and 5-hydroxy-2-methyl-1-penten-3-yne is stable enough to allow an acid-catalyzed hydration of the uncomplexed olefinic group to be carried out (96).

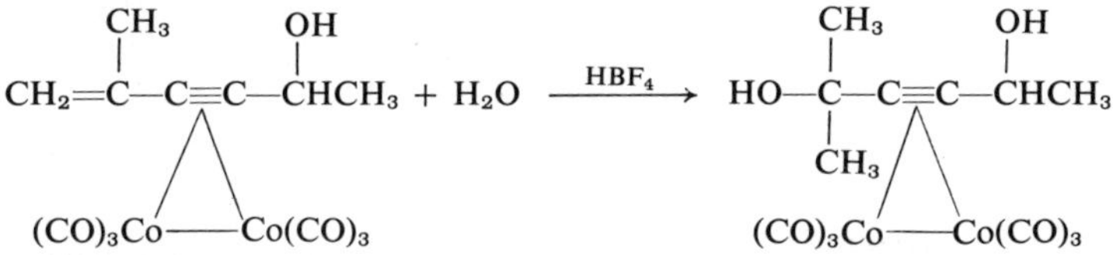

Undoubtedly, much more complicated reactions would have occurred with this ligand under the reaction conditions if it were not complexed with cobalt. Complexation of acetylenic groups with cobalt, as above, provides protection for the acetylenic group through a variety of other reactions as well, which could not be cleanly carried out if the acetylenic group were free. Since the acetylene can be regenerated from these complexes after completion of the reactions by mild oxidation, these reactions are of considerable synthetic interest (96, 97).

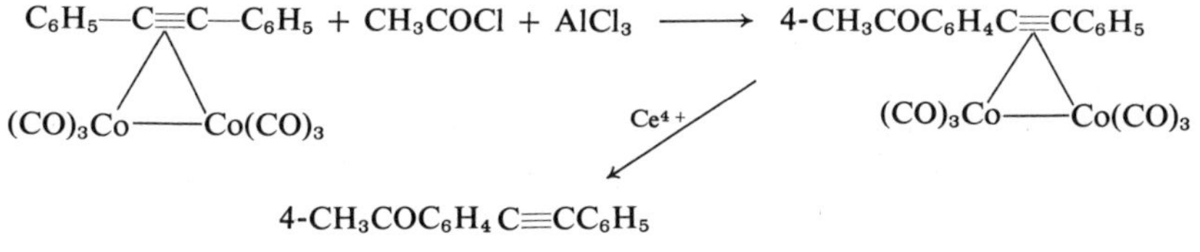

B. METHODS FOR REMOVING ORGANIC GROUPS FROM TRANSITION METAL COMPOUNDS

The final step in an organometallic synthesis of an organic compound is the removal of the desired organic compounds from the metal it is attached to at the conclusion of the reaction. In catalytic reactions, of course, this must occur spontaneously.

1. Replacement of π-Bonded Groups

Often in catalytic reactions the last step is simply a replacement of the π-bonded organic product by one of the starting reactants. The product may leave by dissociation, as is most common, or be displaced by a reactant. In cases where the product π-complex is stable under the reaction conditions the organic group may be removed by adding more strongly coordinating ligands such as phosphines or amines. For example, *cis*-bicyclo[6.1.0]-2,4,6-nonatrienetricarbonylmolybdenum(0) rearranges into a bicyclo[4.2.1]-2,4,7-nonatriene complex at 125°. The new triene can then be removed from the molybdenum by reaction with the chelating triamine, bis(2-aminoethyl)amine (98). The rearrangement occurs to give the most favorable combination of a

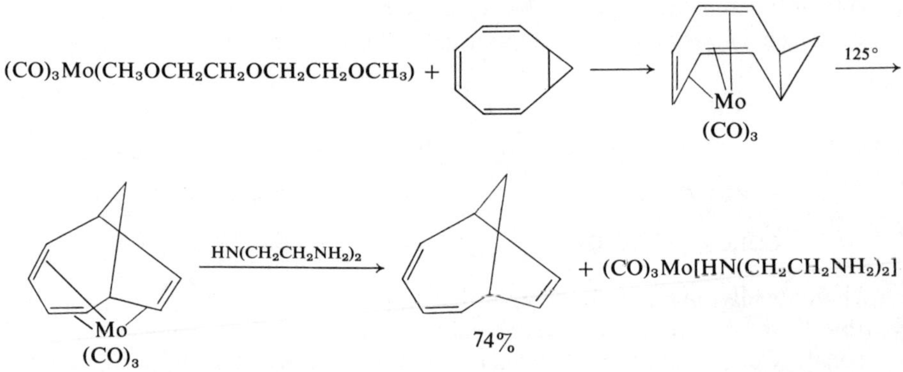

less strained hydrocarbon and a good ligand for the metal. In the absence of the metal the hydrocarbon rearranges differently on heating.

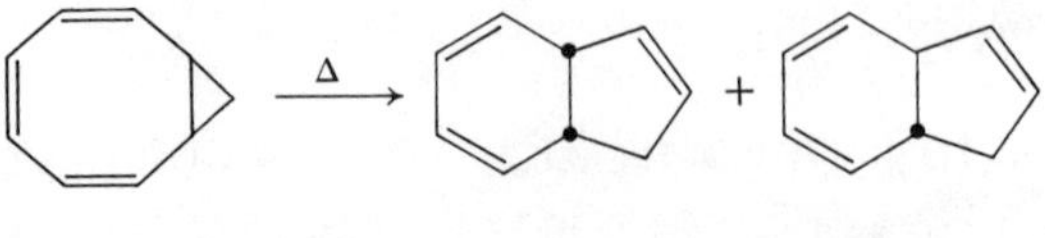

It cannot always be assumed that because an organic group is displaced from a complex by reaction with a strongly coordinating ligand that the organic group was necessarily only π-bonded in the complex. More complicated reactions also may be caused by the new ligands.

2. Reductive Elimination Reactions

Reductive elimination is the reverse of oxidative addition. The reaction involves a loss of two σ-bonded ligands, together as a molecule, while the metal decreases its coordination number by two and its oxidation state by two. The loss of the ligands, in some examples, seems to be accelerated by the presence of strongly coordinating ligands in the solution. These ligands may be effective because they cause a second-order displacement to occur by way of a higher coordinated intermediate. This explanation seems reasonable for square-planar complexes but less so for complexes having more than four ligands where first-order reactions are the rule. Another explanation may be that the new ligand replaces one or more of the nonreacting ligands initially present and the resulting complex more easily undergoes the reductive elimination reaction. An example of the reaction is the formation of 1,5-hexadiene from diallylnickel(II) and carbon monoxide (99).

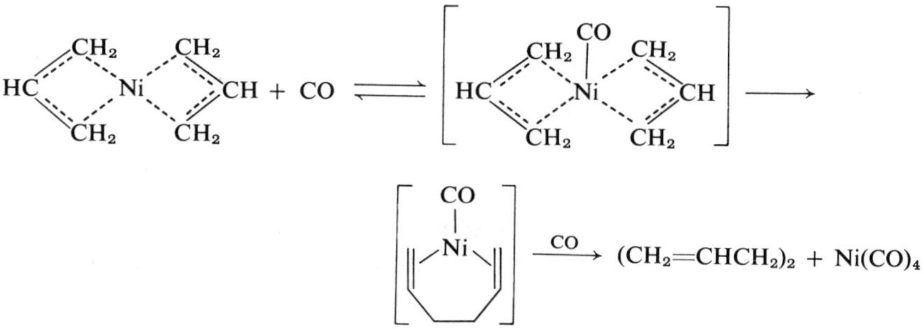

3. β-Elimination Reactions

Elimination of a metal group and a β-hydrogen occurs frequently in organotransition metal reactions. Essentially all alkyl transition metal complexes with hydrogens beta to the metal will undergo β-hydride eliminations on heating. Presumably, these eliminations (and other β eliminations) usually occur with coordinately unsaturated intermediates which produce olefin–metal hydride (or other metal derivative) π-complexes initially, which then may dissociate or undergo other reactions. The mechanism of thermal decomposition of cis-bistriphenylphosphinedi-n-butylplatinum(II) has been investigated and it provides a good example of the β-hydride elimination reaction (100). In this example the final step is a reductive elimination of butane.

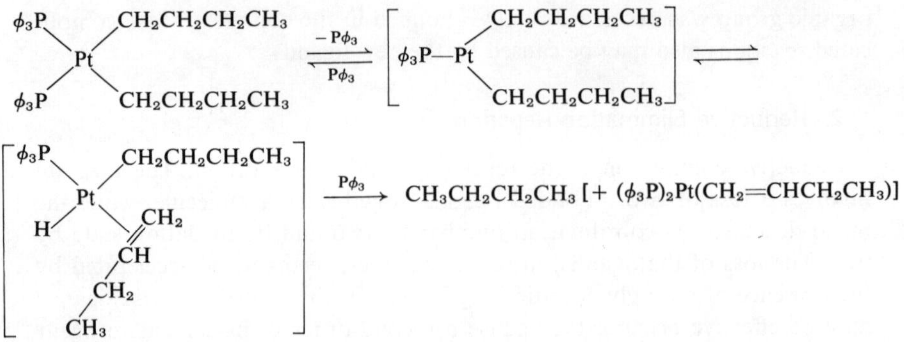

$$\phi_3P \diagdown Pt \diagup CH_2CH_2CH_2CH_3 \underset{P\phi_3}{\overset{-P\phi_3}{\rightleftharpoons}} \left[\phi_3P-Pt \diagup CH_2CH_2CH_2CH_3 \atop CH_2CH_2CH_2CH_3 \right] \rightleftharpoons$$

$$\left[\begin{array}{c} \phi_3P \diagdown \\ Pt \diagup CH_2CH_2CH_2CH_3 \\ H \diagup \qquad CH_2 \\ \qquad CH \\ \qquad CH_2 \\ \qquad CH_3 \end{array} \right] \xrightarrow{P\phi_3} CH_3CH_2CH_2CH_3 \ [+ \ (\phi_3P)_2Pt(CH_2{=}CHCH_2CH_3)]$$

The elements of a metal hydride often can be eliminated from a σ complex with hydride-abstracting agents. The triphenylmethylcarbonium ion (as the tetrafluoroborate salt) is the usual abstracting agent which has been used. This reaction generally leads to the formation of a cationic, π-olefin complex as a product. Typical is the reaction of π-cyclopentadienyldicarbonyl-1-propyliron(II) with triphenylmethyl tetrafluoroborate to form the π-propylene cation (66).

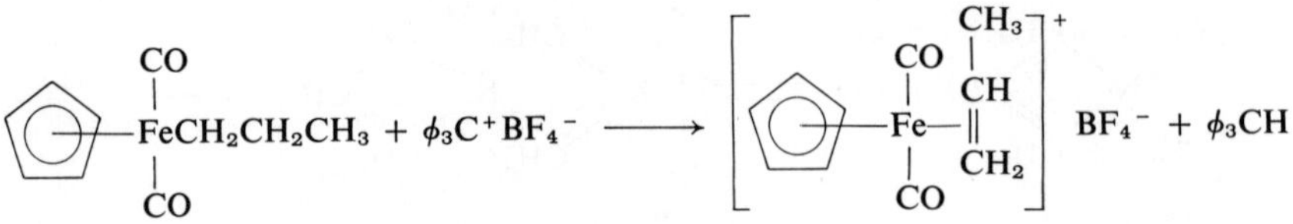

$$\text{(Cp)}-FeCH_2CH_2CH_3 + \phi_3C^+BF_4^- \longrightarrow \left[\text{(Cp)}-Fe{-}\!\!\begin{array}{c} CO \\ | \\ | \\ CO \end{array}\!\!\begin{array}{c} CH_3 \\ | \\ CH \\ || \\ CH_2 \end{array} \right]^+ BF_4^- + \phi_3CH$$

The elimination may also be base-catalyzed in examples where the hydrogen is relatively acidic, e.g., the sodium methoxide-promoted elimination of tetracarbonylcobaltate anion from tricarbonyl-π-(1-acetylmethylallyl)cobalt-(I) (101).

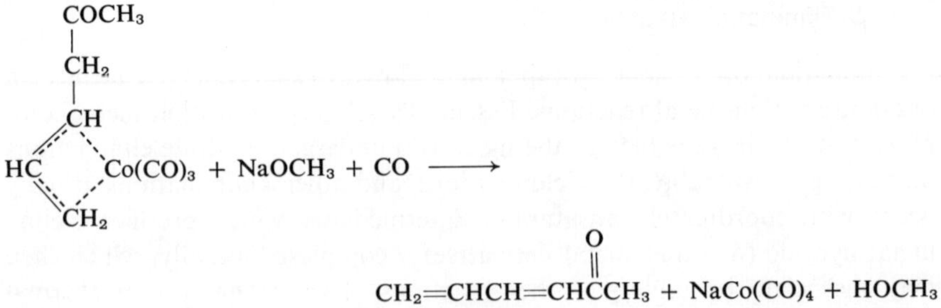

$$\begin{array}{c} COCH_3 \\ | \\ CH_2 \\ | \\ CH \\ HC \diagdown\!\!\diagup Co(CO)_3 \ + \ NaOCH_3 \ + \ CO \ \longrightarrow \\ CH_2 \end{array}$$

$$CH_2{=}CHCH{=}CHC\overset{\displaystyle O}{\overset{||}{C}}CH_3 + NaCo(CO)_4 + HOCH_3$$

Elimination of β-halogen with the metal is also apparently a general reaction, but relatively few examples are known. The catalytic vinyl chloride–acetate interconversion with palladium acetate probably involves addition of palladium acetate followed by elimination of chloropalladium acetate (102).

$$CH_2\!=\!CHCl + Pd(OAc)_2 \longrightarrow [AcOPdCH_2CH(Cl)OAc] \longrightarrow$$
$$[AcOPdCl] + CH_2\!=\!CHOAc$$

Probably, β eliminations of various other groups will be found when the proper compounds are investigated.

4. Exchange Reactions

σ-Bonded organic groups in organotransition metal complexes may sometimes be conveniently removed by exchange reactions with other metallic or nonmetallic compounds. In the first case, the product will be a new organometallic which may itself undergo further reaction. The examples below illustrate two possibilities.

The first example is the octacarbonyldicobalt-catalyzed addition of a silicon hydride to an olefin (103). The initial step is the formation of hydrido-tetracarbonylcobalt(I), the true catalyst, and a silylcobalt derivative. This reaction probably goes by an oxidative addition–reductive elimination sequence. The hydridocarbonylcobalt adds (reversibly) to the olefin in a 1:2 covalent manner to form an alkylcobalt derivative. The alkyl group is then transferred back to silicon by a hydride exchange. Hydridotetracarbonyl-cobalt(I) is known to give isomeric mixtures of alkyl derivatives by addition in both directions to terminal olefins. The fact that the silicon exchange reaction gives only terminal substitution must mean that only the linear cobalt alkyl undergoes the exchange reaction at a significant rate and that it is selectively removed from the mixture.

$$n\text{-Bu}_3SiH + Co_2(CO)_8 \longrightarrow n\text{-Bu}_3SiCo(CO)_4 + HCo(CO)_4$$

Catalytic cycle:

$$HCo(CO)_4 + RCH\!=\!CH_2 \rightleftharpoons RCH_2CH_2Co(CO)_4 + R\overset{\displaystyle CH_3}{\underset{\displaystyle |}{C}}HCo(CO)_4$$

$$RCH_2CH_2Co(CO)_4 + n\text{-Bu}_3SiH \longrightarrow RCH_2CH_2Si(Bu\text{-}n)_3 + HCo(CO)_4$$

In the second example, an exchange reaction apparently occurs between an organopalladium compound and lead tetraacetate. The exchange produces an organolead triacetate which decomposes by reductive elimination into an organic acetate and lead diacetate (104). The reaction actually carried out also involves an organometallic synthesis forming the organopalladium compound. In this synthesis phenylmercuric acetate is reacted with palladium-(II) acetate in the presence of ethylene. The first two reactants undergo an exchange forming very reactive, solvated phenylpalladium acetate, which then rapidly adds to the ethylene present forming solvated 2-phenylethyl-palladium acetate. The last compound may then be expected to undergo exchange with the lead tetraacetate present and ultimately form 2-phenylethyl acetate. Since phenyl acetate is not formed, the lead tetraacetate exchange must occur much more readily with aliphatic palladium compounds than with aromatic ones or the aryllead triacetate must be stable under the reaction conditions.

$$C_6H_5HgOAc + Pd(OAc)_2 \underset{-L}{\overset{L}{\rightleftharpoons}} [C_6H_5PdL_2OAc] + Hg(OAc)_2$$

$$[C_6H_5PdL_2OAc] + CH_2{=}CH_2 \longrightarrow [C_6H_5CH_2CH_2PdL_2OAc]$$

$$[C_6H_5CH_2CH_2PdL_2OAc] + Pb(OAc)_4 \underset{}{\overset{-L}{\rightleftharpoons}} [C_6H_5CH_2CH_2Pb(OAc)_3] + Pd(OAc)_2$$

$$[C_6H_5CH_2CH_2Pb(OAc)_3] \longrightarrow C_6H_5CH_2CH_2OAc + Pb(OAc)_2$$

L = solvent

5. Oxidative Cleavage

Often transition metal–carbon compounds are less stable when the oxidation state of the metal is raised. For example, the oxidation of iron(0) in tricarbonyl-π-cyclobutadieneiron(0) to the ferric state (in the presence of other ligands) causes the cyclobutadiene ligand to dissociate (105). The very reactive free cyclobutadiene produced cannot be isolated in a pure state, but it is a useful reagent for further synthetic reactions. With σ-bonded

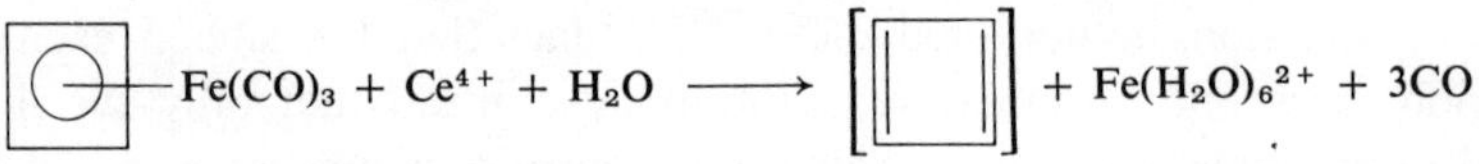

$$\text{Fe(CO)}_3 + Ce^{4+} + H_2O \longrightarrow \quad + Fe(H_2O)_6^{2+} + 3CO$$

organometallics, the oxidizing agent may end up bonded to the organic group. Carbon–transition metal bond cleavage with iodine, for example, generally yields iodides as products. The above reaction does not require an

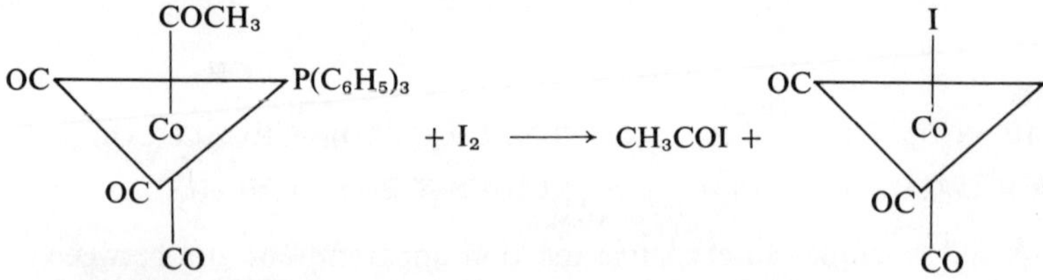

$$+ I_2 \longrightarrow CH_3COI + \qquad\qquad \text{(Ref. 106)}$$

initial dissociation of a ligand, as might have been expected, however, since the rate of the reaction is much faster than its dissociation rate. The true mechanism has not been established.

6. Thermal Decomposition

Heating organotransition metal compounds generally leads to loss of organic ligands. In many instances, however, this is not a useful reaction since multiple, complex reactions or general decomposition may occur. Sometimes the organic group is degraded into the components that it can be prepared from. For example, pyrolysis of tetracarbonylpropionylcobalt(I) produces mainly ethylene, hydrogen, and carbon monoxide (107). More

$$CH_3CH_2COCo(CO)_4 \longrightarrow CH_2{=}CH_2 + 5CO + H_2 + Co$$

useful reactions occur in other examples such as in the thermal decomposition of $Co_2(CO)_4[HC{\equiv}CC(CH_3)_3]_3$ which forms 1,2,4-tri-*t*-butylbenzene (108, 109).

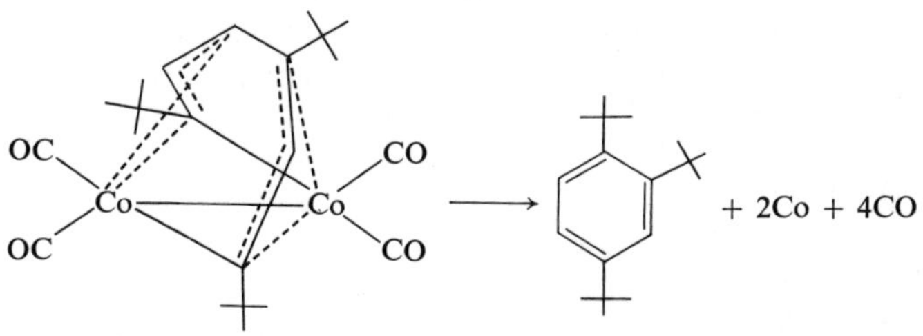

Pyrolysis of complexes containing only π-bonded groups may give the free ligand in very low to good yields depending on the particular complex. Pyrolysis of dichloro-π-tetramethylcyclobutadienenickel(II) at 250°, for example, yields the dimer of the unstable tetramethylcyclobutadiene (110).

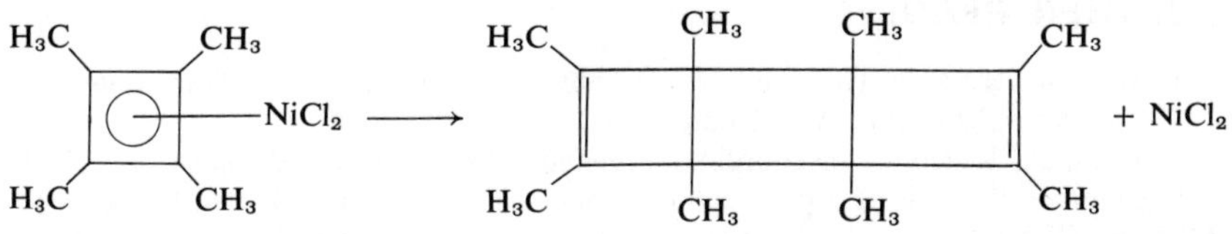

Ligand displacement reactions (with good ligands), however, generally give better yields of products than pyrolysis.

7. Displacement Reactions on Carbon

The direct displacement of the metallic portion of a complex by attack of an external anion, another nucleophilic agent, or an electrophilic reagent on the carbon of the metal–carbon σ bond often can be achieved. The reaction occurs with particular ease with acyl metal complexes and nucleophilic reagents. The reaction of pentacarbonylacetylmanganese(I) with sodium methoxide is a typical example (111).

$$CH_3COMn(CO)_5 + NaOCH_3 \longrightarrow CH_3COOCH_3 + NaMn(CO)_5$$

Examples of direct displacements at σ-bonded saturated carbon are much less common, but undoubtedly many such reactions could be done if they were of interest. The racemization of ethyl 2-(pentacarbonylmanganese) propionate with pentacarbonylmanganese anion probably involves a reversible nucleophilic displacement at the optically active carbon (112).

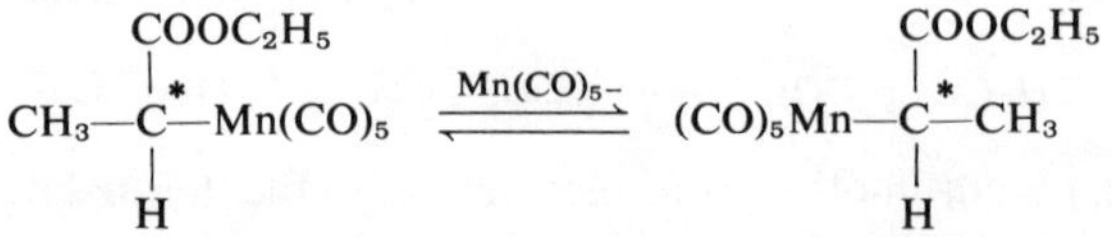

$$
\begin{array}{ccc}
\underset{\overset{|}{\underset{H}{}}}{\overset{COOC_2H_5}{\overset{|}{CH_3-\overset{*}{C}-Mn(CO)_5}}} & \underset{\longleftarrow}{\overset{Mn(CO)_5-}{\longrightarrow}} & \underset{\overset{|}{\underset{H}{}}}{\overset{COOC_2H_5}{(CO)_5Mn-\overset{*}{C}-CH_3}}
\end{array}
$$

An example of an electrophilic cleavage is the reaction of π-cyclopentadienyldicarbonyl-(*threo*-1,2-dideutero-3,3-dimethylbutyl)iron(II) with mercuric chloride. The reaction was shown to go with retention of stereochemistry (113). Such reactions probably would be better classified as exchange reactions, however.

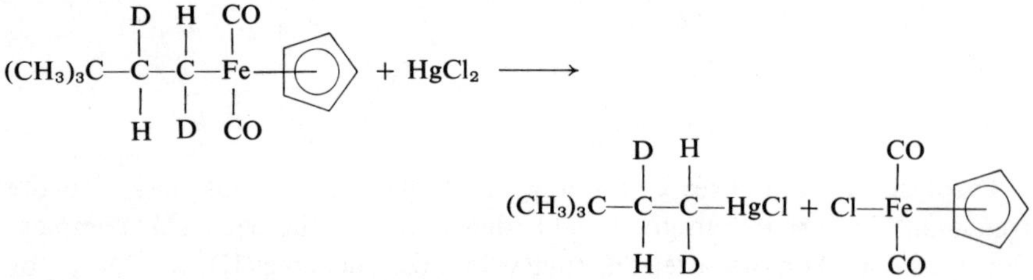

$$
(CH_3)_3C-\overset{\overset{D}{|}}{\underset{\underset{H}{|}}{C}}-\overset{\overset{H}{|}}{\underset{\underset{D}{|}}{C}}-\overset{\overset{CO}{|}}{\underset{\underset{CO}{|}}{Fe}}-\text{Cp} + HgCl_2 \longrightarrow
$$

$$
(CH_3)_3C-\overset{\overset{D}{|}}{\underset{\underset{H}{|}}{C}}-\overset{\overset{H}{|}}{\underset{\underset{D}{|}}{C}}-HgCl + Cl-\overset{\overset{CO}{|}}{\underset{\underset{CO}{|}}{Fe}}-\text{Cp}
$$

FURTHER READING

J. P. Candlin, K. A. Taylor, and D. T. Thompson, "Reactions of Transition Metal Complexes," Elsevier, Amsterdam, 1968.

J. P. Collman, Reactions of coordinated ligands. *Transition Metal Chem.* **2**, 1 (1966).

J. P. Collman, Patterns of organometallic reactions. *Accounts Chem. Res.* **1**, 136 (1968).

M. L. H. Green, "Organometallic Compounds," Vol. II. Methuen, London, 1968.

R. F. Heck, Addition reactions. *Advan. Organometal. Chem.* **4**, 243 (1966).

R. F. Heck, Addition reactions of metal complexes. *Accounts Chem. Res.* **2**, 10 (1969).

G. Henrici-Olive and S. Olive, Influence of ligands on activity and specificity. *Angew. Chem., Int. Ed. Engl.* **10**, 105 (1971).

G. W. Parshall and J. J. Mrowca, σ-Alkyl and Acyl-transition metal derivatives. *Advan. Organometal. Chem.* **7**, 157 (1968).

Chapter IV

Homogeneous Hydrogenation Reactions

A. OLEFIN REDUCTION

Hydrogenation is a prime example of a reaction that can be carried out both homogeneously and heterogeneously. The two types of reactions are so similar that there can be little doubt but that the basic features of the mechanisms of both reactions are the same.

The homogeneous hydrogenation of olefins is one of the best understood and one of the simplest of transition metal-catalyzed reactions. Two basic steps are involved: (1) metal hydride addition to the olefin and (2) hydrogenolysis of the metal alkyl bond with reformation of the metal hydride catalyst.

$$M—H + \ \overset{\diagdown}{\diagup}C{=}C\overset{\diagup}{\diagdown} \ \longrightarrow \ -\underset{\underset{M}{|}}{\overset{|}{C}}-\underset{\underset{H}{|}}{\overset{|}{C}}- \ \xrightarrow{H_2} \ M—H + -\underset{\underset{H}{|}}{\overset{|}{C}}-\underset{\underset{H}{|}}{\overset{|}{C}}-$$

In detail, however, the reaction involves several additional steps. The hydride catalyst will probably exist, at least, mainly as a coordinately saturated complex. If this is the case, then dissociation generally will be required initially in order to have a position available for coordination of the reacting olefin. The olefin π complex formed could then undergo a shift of hydrogen from the metal to one carbon of the coordinated double bond with concurrent formation of a metal–carbon σ-bond to the other carbon of the double bond. This step would produce a coordinately unsaturated metal alkyl. This complex will then probably react reversibly with another potential ligand in the solution before reacting with hydrogen by oxidative addition. The elimination of the

product hydrocarbon by a second hydrogen shift from metal to carbon (a reductive elimination) now leaves the coordinately unsaturated hydride available for coordination with either another olefin molecule to continue the catalytic cycle or some other coordinating species in the solution. The true

$$L_nM\text{---}H \rightleftharpoons L + L_{n-1}M\text{---}H$$

situation may be even more complicated than these equations indicate since both the olefin–metal hydride π complex and the dihydridoalkyl metal complex intermediates generally may exist in more than one isomeric form. Only isomers which have the reacting hydrogen and the carbon groups cis to each other can directly undergo the necessary reactions. It must be emphasized, however, that complete identification of all the possible intermediates in any of these reactions (and most other transition metal reactions) has not been made. This is an extremely difficult task because at least some of the intermediates are very reactive and unstable, and exist only in low concentrations. Although this detailed mechanism is believed to be basically correct for at least several of the known homogeneous hydrogenation catalysts (and presumably for heterogeneous ones as well), there are variations possible and some of these may be observed as more cases are investigated. It is not certain, for example, that π complexing of the olefin to the hydride is always necessary before addition can occur, nor is it clear whether oxidative addition of hydrogen is always necessary before hydrogen cleavage of the metal alkyl bond can occur. Four-centered transition states could be imagined as alternatives.

One of the most thoroughly studied homogeneous hydrogenation catalysts is hydridocarbonyl(tristriphenylphosphine)rhodium(I) which has been investigated by Wilkinson and co-workers (114). This catalyst readily causes hydrogen reduction of terminal olefins to hydrocarbons at 25°C and atmospheric pressure, but internal olefins are very much less reactive.

The NMR spectrum of the catalyst at $-30°$ shows the hydride proton at 19.69τ as a double quartet with $1:3:3:1$ area ratios. The three triphenylphosphine groups are apparently all equivalent causing splitting of the hydride resonance into a quartet ($J_{P\text{---}H} = 14$ Hz) and the ^{103}Rh causing a doubling of the quartet ($J_{Rh\text{---}H} = 1$ Hz). At room temperature the spectrum consists of only a broad line because of rapid phosphine dissociation and exchange. The catalyst is a d^8 complex and dissociation of one triphenylphosphine group to form a square-planar, four-coordinate intermediate is not surprising for the reasons given in Chapter I, Section B. The apparent similarity of the infrared spectrum of the dissociated catalyst (30° spectrum) and the starting complex (−30° spectrum) suggests that the hydrogen and CO groups are trans in both compounds. The dissociated complex is then presumed to reversibly form a π complex with the olefin. An alkylrhodium species is now believed to be formed by a reversible 1:2 hydrogen shift from the metal to the olefin. Although the alkyl has not been detected directly in this case, the observed rapid exchange of the hydride hydrogen (deuterium) for olefinic vinyl hydrogens suggests that it is an intermediate. Oxidative addition of hydrogen and another shift of hydrogen from rhodium to carbon would produce the hydrocarbon product with reformation of the catalyst.

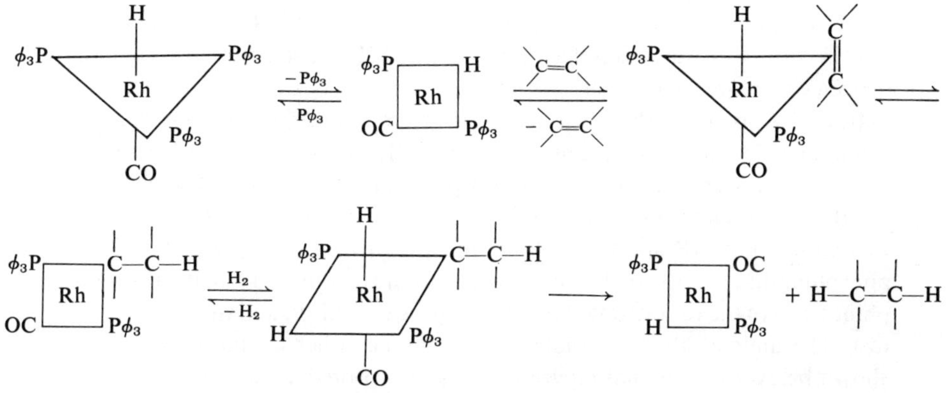

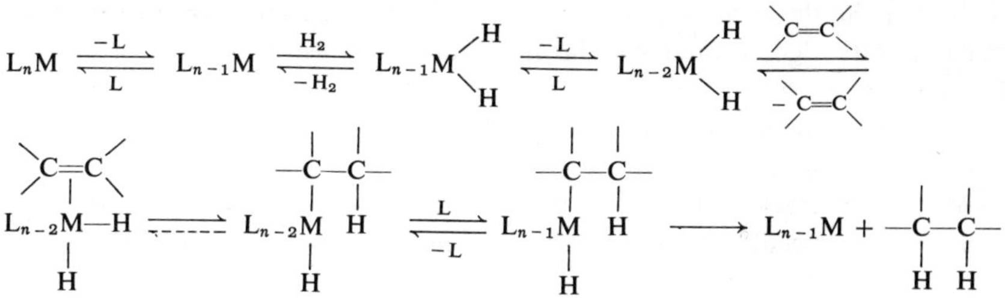

A variation of this mechanism has been suggested with other catalysts. In these examples the oxidative addition of hydrogen occurs before the olefin coordination. The dihydride formed by the oxidative addition first coordinates with and then adds to the olefin forming a hydridoalkylrhodium species. This complex then undergoes reductive elimination of the hydrocarbon with reformation of the same coordinately unsaturated metal complex that was formed in the initial dissociation of the catalyst, thus completing the catalytic cycle. An example of this mechanism is believed to be the homogeneous hydrogenation catalyzed by chloro(tristriphenylphosphine)rhodium(I) also studied by Wilkinson and his co-workers (115). This catalyst is more reactive and generally useful than the one just discussed.

The square-planar, four-coordinate d^8 catalyst in this example is stable as the coordinately unsaturated specie, but probably dissociation of one of the triphenylphosphine groups can occur to a small degree with the dissociated compound rapidly forming a chloride bridged dimer even in dilute solution (116, 117). The dissociated form likely reacts with hydrogen next and then with olefin. The dimer also reacts with hydrogen forming an unsymmetrical bridged dihydride. The dissociated catalyst also seems to react with ethylene forming an isolable trans square planar π complex. Carbon monoxide reacts similarly. The unsymmetrical dimer dihydride, or the diphosphine-dihydride mononuclear species may react with triphenylphosphine and form isolable dihydrido-chlorotristriphenylphosphinerhodium(I). The NMR spectrum of the dihydride at 25° showed two metal hydride absorptions at 28τ and 21.5τ of equal area, but the latter one was a doublet with $J_{P-H} \sim 20$ Hz. A structure with one hydrogen cis to three phosphines and the other cis to two and trans to the third is indicated. Apparently there was no significant coupling of cis-phosphorus or of rhodium-103 with the hydride protons. Generally trans-phosphorus has a much larger coupling constant than a cis-phosphorus in the same metal complex. At 25° the NMR shows line broadening because of phosphine ligand exchange. From the ^{31}P NMR spectrum at 30° and $-25°$, it also can be deduced that phosphine dissociation from the dihydride occurs readily and that it is the phosphine trans to the very strongly trans activating hydride ligand that is reversibly lost. The undetectable dissociated form may be either a square pyramid as shown below or a trigonal bipyramid. It seems probable that this dissociated form coordinates with the olefin next and that this is followed by a hydride transfer from rhodium to the coordinated olefin. If there were a shift of the hydrogen and no movement of the olefinic group a new square-pyramidal complex would be formed in which the remaining hydride group would be trans to the alkyl group. This complex could easily rearrange by way of a trigonal pyramid and react with phosphine again to place the hydride and alkyl groups cis to each other in a position to react. A reductive elimination of hydrocarbon with reformation of the initial catalyst would then complete the cycle (see Scheme IV-1). This mechanism suggests that cis addition of the hydrogen should occur and this has been observed. Reduction of dimethyl maleate with deuterium gave the meso-dideutero ester, whereas dimethyl fumarate gave the ($\pm$)-dideuterosuccinate ester. This catalyst also reduces acetylenes and dienes in contrast to the first one which only forms complexes with these reagents.

As in the previous example, the mechanism in Scheme IV-1 contains considerable speculation, but enough is known to be reasonably certain at

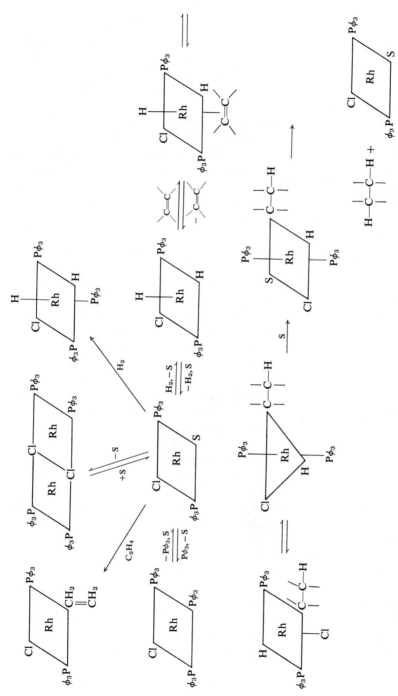

Scheme IV-1.

least that it is not far from the actual mechanism. The alternative mechanism which cannot be completely ruled out, however, is the one in which olefin coordination occurs first and then the oxidative addition of hydrogen takes place (118). Although it appears unlikely since ethylene is known to inhibit its own hydrogenation. Although the detail given may not be completely justified by the facts, it is instructive to attempt to construct as complete a mechanism as possible in these examples since there is generally considerably less known about other transition metal-catalyzed reactions.

The important effect that small amounts of impurities can have on some transition metal catalysts is illustrated by the effect of oxygen on the chloro-(tristriphenylphosphine)rhodium(I) catalyst. The oxidized material is still a hydrogenation catalyst, but at the same time it also causes olefin isomerization which is not significant when the unoxidized catalyst is employed (119).

A variety of homogeneous hydrogenation catalysts are now known. All these very probably operate basically by one or the other of the above mechanisms. A partial list includes $HRuCl(P\phi_3)_3$ (selective for terminal olefins) (120), $HIr(CO)(P\phi_3)_3$ (121), $HCo(CO)_2[P(n-C_4H_9)_3]_2$ (122), $HRu-(OCOCH_3)(P\phi_3)_4$ (123), $Ru_2(OCOCH_3)_4$ (reduces internal olefins easily) (124), $IrCl(P\phi_3)_3$ (10 times more active than the rhodium analog, but olefin isomerization also occurs) (125), and $\{H_2Rh(P\phi_3)_2[(CH_3)_3CO]_2\}PF_6$ (126). The last compound is of special interest because it reduces ketones fairly easily (127), whereas the others listed do not, and because of its method of preparation and the fact that it has two acetone molecules as ligands. The compound was prepared by reacting π-norbornadienebis(triphenylphosphine)rhodium(I) hexafluorophosphate with hydrogen in acetone solution. The diene ligand is reduced to the hydrocarbon and only solvent is available to fill the empty coordination positions created in the reduction. This reaction appears to be a generally useful way for preparing very reactive solvated homogeneous hydrogenation catalysts. In general, the catalysts become more reactive as more, weakly bonding ligands are placed in the coordination sphere. There is a limit to the number of poor ligands that can be used, however, since the complexes often become unstable and decompose to free metal when too many are present.

Many variations in the phosphine ligands in the above catalysts have been studied. The triarylphosphines are best and changing the substituents in the aryl groups may produce significant effects upon hydrogenation rates. 1-Hexene, for example, hydrogenates about 60 times faster with the tris-*p*-anisylphosphine catalyst than with the tris-*p*-chlorophenylphosphine complex (128). The rates are generally lower when the aryl groups are replaced by alkyls. The effect, in at least some instances, is probably mainly steric. The larger ligands dissociate more readily and produce a more reactive, solvated catalyst. A ligand variation of considerable synthetic potential is the use of optically active phosphine ligands to effect asymmetric reductions.

Two sources of optical activity in the ligand have been employed: the use of an optically active carbon group as one of the three substituents on the phosphorus and the use of opticelly active phosphines with three different substituents.

In the first group optically active neomenthyldiphenylphosphine was prepared by the reaction of lithium diphenylphosphide with optically active menthyl chloride and converted into the analog of one of Wilkinson's catalysts—chlorotris(neomenthyldiphenylphosphine)rhodium(I). The catalyst was less active than the triphenylphosphine analog, but at 60° and with 300 psi of hydrogen, reduction of olefins occurred and in appropriately substituted cases optical activity was induced in the product. 3-Phenyl-2-butenoic acid under these conditions (in 24 hr) gave an 80% yield of 3-phenylbutyric acid which was 61% optically pure (129).

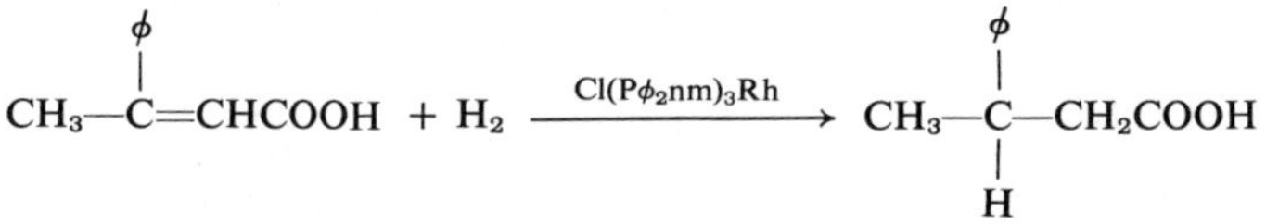

$$CH_3-\overset{\displaystyle \phi}{\underset{\displaystyle |}{C}}=CHCOOH + H_2 \xrightarrow{Cl(P\phi_2nm)_3Rh} CH_3-\overset{\displaystyle \phi}{\underset{\displaystyle |}{\underset{\displaystyle H}{C}}}-CH_2COOH$$

nm = neomenthyl

In complexes with the second type of phosphine ligand, a whole series of optically active phosphines has been used. Instead of preforming the chloro-trisphosphinerhodium catalyst, mixtures of rhodium trichloride and the phosphine were used. As would be expected from the mechanism proposed, only two phosphine groups per rhodium are required. Faster reduction is also observed with the 2:1 ratio.

The degree of optical purity obtained in the reduction product depends markedly on the groups attached to phosphorus and on the structure of the olefin being reduced. Reduction of 2-N-benzoylamino-3-(3-methoxy-4-hydroxyphenyl)acrylic acid with a catalyst containing two optically active m-anisylmethylphenylphosphines per rhodium atom gave only 1% optical purity in the reduced product, while the use of o-anisylcyclohexylmethyl-phosphine gave 90% optical purity under the same conditions (130). Use of

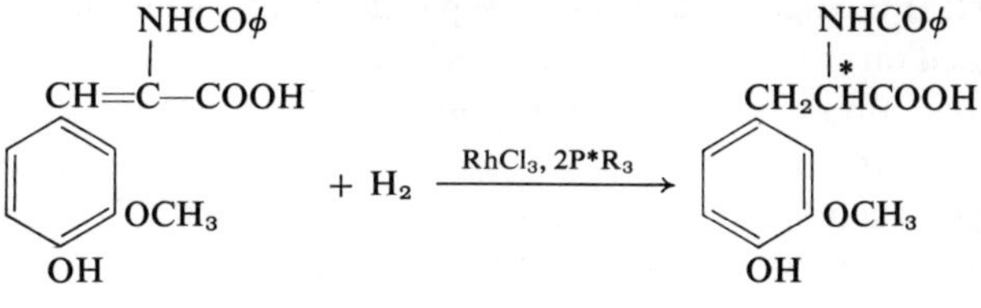

the same catalyst to reduce 2-N-acetylaminoacrylic acid gave a product with only 60% optical purity (130).

Similar catalysts with asymmetric chelating diphosphines have also been studied (131).

Catalysts for asymmetric reduction, of course, are not limited to phosphine complexes. All that is necessary is for the ligand to remain attached while the hydrogenation is occurring. A catalyst with coordinated, optically active formamide derivatives has proved to be capable of the asymmetric reduction of methyl 3-phenyl-2-butenoate. The catalyst is produced by reducing trichlorotripyridylrhodium(III) with sodium borohydride in the presence of the optically active formamide derivative. With (R)- or (S)-1-phenylethyl-formamide as the amide derivative, 50% optical purity was obtained in the hydrogen reduction of methyl 3-phenyl-2-butenoate (132). A review chapter

$$py_3RhCl_3 + NaBH_4 + 2\phi\overset{\overset{\displaystyle NHCHO}{\displaystyle |*}}{CHCH_3} \longrightarrow py_2(\phi\overset{\overset{\displaystyle NHCHO}{\displaystyle |*}}{CHCH_3})RhCl_2(BH_4)$$

$$\phi\text{-}C\overset{\overset{\displaystyle CH_3}{\displaystyle |}}{=}CHCOOCH_3 + H_2 \xrightarrow{py_2(\phi\overset{\overset{\displaystyle NHCHO}{\displaystyle |*}}{CHCH_3})RhCl_2(BH_4)} \phi\overset{\overset{\displaystyle CH_3}{\displaystyle |*}}{CHCH_2COOCH_3}$$

has been written on asymmetric reductions (133).

B. HOMOGENEOUS–HETEROGENEOUS CATALYSTS

A new type of catalyst system which combines features of both homogeneous and heterogeneous catalysts is now being developed and one area of application is in hydrogenation. The usual heterogeneous catalyst is made by precipitating a metal or metal compound on a support material and the catalyst is generally not chemically bound to the support. The precipitated catalyst will be of a different thickness at different places on the support and the ligands around the metal will vary. Consequently, the activity of the catalyst is also likely to vary from place to place on the surface. The new catalyst system employs an organic polymer with chemically binding ligands as a support. If the ligands are widely separated, one metal catalyst molecule will be bonded on the average to only one ligand. The use of strongly bonding ligands in the support, such as phosphine groups, attaches the metal very strongly to the support, but the other ligands on the metal are free to exchange and undergo reactions. All the catalyst sites are essentially chemically the same, except for minor differences because of irregularities in the polymer surface.

Polystyrene copolymerized with a little divinylbenzene in order to insolu-bilize it can be prepared in a form containing pores or holes of relatively constant size within the same sample. Since the pores represent the major part of the surface of the support, the size of the pores will determine which molecules can fit inside and be reduced. The attachment of the chlorotris-(triphenylphosphine)rhodium(I) catalyst by phosphine exchange to a polystyrene with diphenylphosphinomethyl groups and the use of the proper pore size allows selective hydrogenation on the basis of the size of the olefin molecules. Small olefins can enter the pores easily and reach the catalyst, whereas larger olefins cannot (134). These catalysts have the practical advantage that the catalyst remains insoluble and there is not a problem of removing it from the reaction mixture or of recovering it. Potentially, with these systems, there is the possibility of preparing very reactive catalysts which would be unstable in solution with respect to disproportionation and precipitation of the metal since these reactions cannot occur if the metal atoms are held apart by firm attachment to the polymer support. The advantages of this type of catalysis should be important in other reactions than hydrogenation and more applications will most likely be found.

C. REDUCTION OF CONJUGATED DIENES

Some of the above-mentioned soluble catalysts will also reduce conjugated dienes and others form stable complexes with them. The mechanism of reduction could be identical with that of the olefin reduction with some catalysts, but it is different with others since some do not reduce simple olefins. The reduction in the latter cases stops after the addition of two hydrogen atoms.

Benzenetricarbonylchromium(0) is a very selective catalyst for the reduction of conjugated dienes to olefins by a 1:4 addition of hydrogen (135–137). The reduction of methyl sorbate, for example, takes place with this catalyst at 150° with 700 psi of hydrogen in cyclohexane solution to produce exclusively cis-methyl 3-hexenoate in up to 98% yield. It is presumed that the π-complexed benzene and a carbonyl group in the catalyst are replaced by a coordinated diene and two hydrogen atoms giving a dihydride which then transfers the hydrogen atoms to the ends of the diene system. The hydrogens may be transferred one at a time giving a π-allyl intermediate or, perhaps less likely, both together. The latter possibility seems unlikely because the chromium atom would be losing three ligands simultaneously, whereas with the stepwise route solvent could enter and only one ligand would be lost in the first step and two in the second. The 1:4 addition then occurs because the second hydride group is cis to the 4-position of the original diene (see Scheme IV-2).

Cobalt cyanide complexes form hydrides with hydrogen which will reduce activated olefins and conjugated dienes. The reduction of butadiene can give either 1-butene as the major product or trans-2-butene depending on the ratio of cyanide ion to cobalt in the solution (138). When there is more than

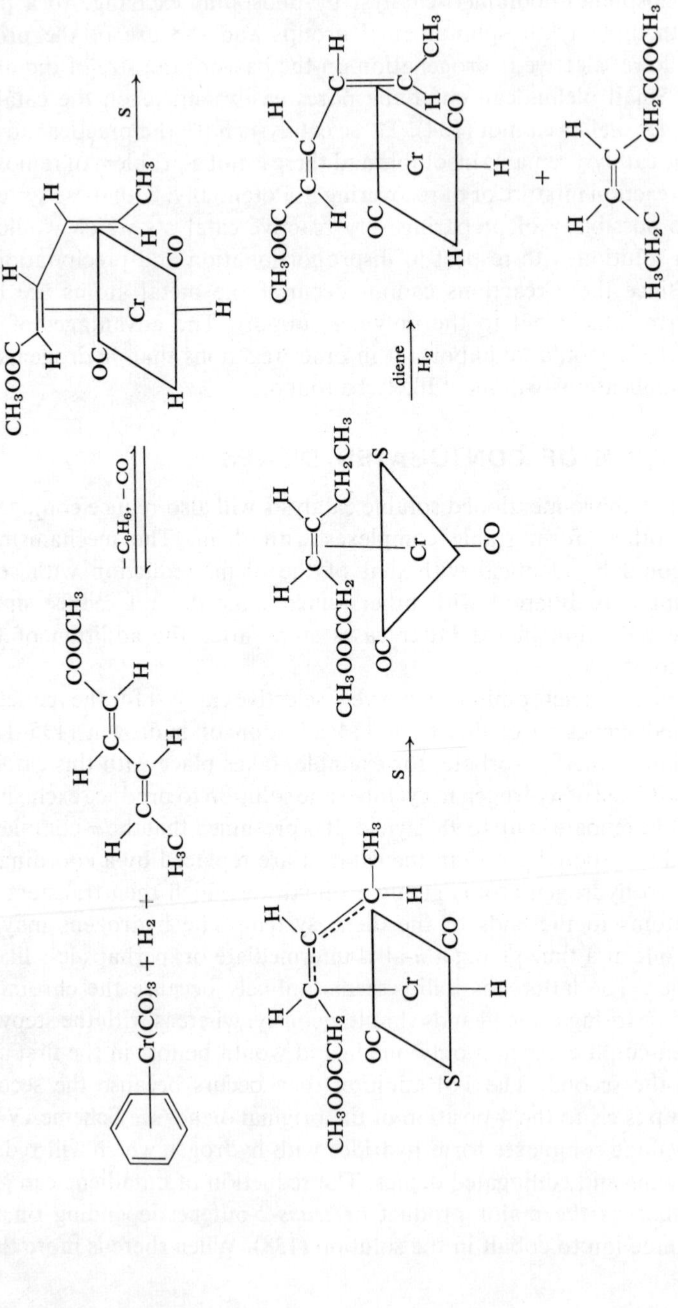
Cr(CO)3 + H2 +
H
H3C
C=C
H
C=C
H
COOCH3
S
CH3OOCCH2
H
C
C
H
C
CH3
H
OC
Cr
CO
H
S
−C6H6, −CO
CH3OOCCH2
H
C=C
H
CH2CH3
OC
Cr
CO
S
CH3OOC
H
H
CH3
OC
Cr
H
H
CO
S
diene
H2
CH3OOC
H
C
C
H
H
C−H
C
CH3
H
OC
Cr
CO
H
+
H
H3CH2C
C=C
H
CH2COOCH3

Scheme IV-2.

a 5:1 ratio of cyanide to cobalt, 1-butene is formed apparently by way of the pentacyano-2-butenylcobalt(III) ion since the NMR spectrum of the reaction mixture of the hydride with butadiene shows the presence of the 2-butenyl group. Reaction of the 2-butenylcobalt compound with a hydrido-pentacyanocobalt(III) ion with hydrogen substitution at the double bond could give 1-butene. A reaction between two triple negatively charged ions would ordinarily be very unfavorable, but, in this example, the charges would be partially distributed into the ligands and the attack at the double bond would allow the hydrogen transfer to occur without bringing the charges very closely together.

$$\tfrac{1}{2}H_2 + C_4H_6 + Co^{2+} + 5CN^- \longrightarrow [(CN)_5CoCH_2CH{=}CHCH_3]^{3-}$$

$$[(CN)_5CoCH_2CH{=}CHCH_3]^{3-} + HCo(CN)_5{}^{-3} \longrightarrow 2Co(CN)_5{}^{3-} + CH_2{=}CHCH_2CH_3$$

With a cyanide-to-cobalt ratio of less than five the reduction produces *trans*-2-butene. The NMR spectrum of this reaction mixture indicates the presence of a *syn*-π-(1-methylallyl) group attached to cobalt. The likely ion would have four cyanide ligands also with a charge of minus two. Reaction of this complex with hydrogen or hydridopentacyanocobalt(III) ion could easily give *trans*-2-butene. The exact mechanism of the final reduction, however, is not known.

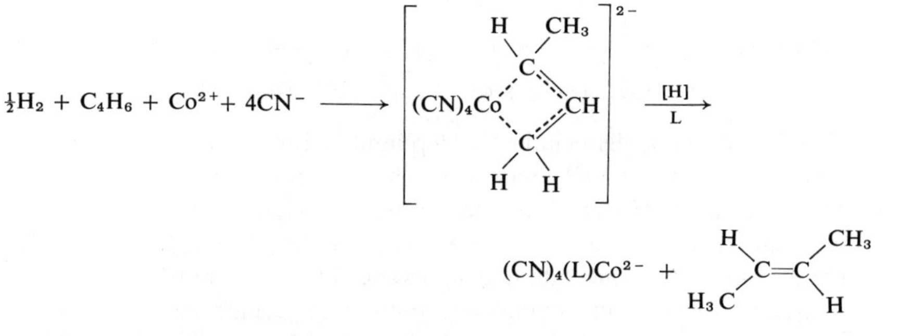

$$\tfrac{1}{2}H_2 + C_4H_6 + Co^{2+} + 4CN^- \longrightarrow$$

$$(CN)_4(L)Co^{2-} +$$

Other homogeneous catalysts which selectively reduce dienes to olefins are *trans*-$HPt(SnCl_3)(P\phi_3)_2$ (139–141), π-$C_5H_5M(CO)_3H$ (M = Cr, Mo, W) (142), and $Fe(CO)_5$ (143).

D. REDUCTION OF KETONES AND ALDEHYDES

Organic carbonyl compounds are, in general, much more difficult to reduce than are olefins, dienes, or acetylenes with homogeneous (or heterogeneous) transition metal catalysts. This fact allows quantitative reduction of unsaturated ketones and aldehydes to saturated carbonyl compounds with many soluble catalysts. There are, however, a few homogeneous transition metal catalysts available which will reduce carbonyl compounds reasonably well.

Hydridotetracarbonylcobalt(I) will reduce aldehydes to alcohols at about 180° under about 1000 psi of a 1:1 mixture of carbon monoxide and hydrogen. The carbon monoxide is necessary because the catalyst decomposes in its absence. Although carbon monoxide is necessary to keep the catalyst in solution, it is also an inhibitor presumably because coordinately unsaturated intermediates are involved and carbon monoxide competes with hydrogen and/or the carbonyl compound for these intermediates. The rate of the reduction is dependent on the aldehyde, cobalt, and hydrogen pressure and inversely proportional to the square of the carbon monoxide pressure.

$$d(RCH_2OH)/dt = k[RCHO][Co][H_2]/[CO]^2$$

On the basis of this dependence the following mechanism has been proposed (94a).

$$HCo(CO)_4 \rightleftharpoons HCo(CO)_3 + CO$$

$$RCHO + HCo(CO)_3 \rightleftharpoons \underset{\underset{H}{\overset{\displaystyle\downarrow}{\diagdown\!Co(CO)_3}}}{R{-}\overset{\displaystyle\overset{H}{|}}{C}{=}O} \rightleftharpoons RCH_2O\,Co(CO)_3$$

$$RCH_2OCo(CO)_3 + H_2 \longrightarrow RCH_2OCoH_2(CO)_3 \longrightarrow RCH_2OH + HCo(CO)_3$$

$$RCH_2OCo(CO)_3 + CO \rightleftharpoons RCH_2OCo(CO)_4$$

The initial step is dissociation, a step inhibited by carbon monoxide. This is another example of a d^8 complex which shows the tendency to lose its fifth ligand. The dissociated, hydridotricarbonylcobalt(I) can next coordinate with the carbonyl compound. A "side-on" coordination (a π-bonding ligand) is proposed in the intermediate that undergoes hydrogen transfer from cobalt to carbon. An "end-on" complex (a σ-bonding ligand) with the aldehyde probably is a more stable form of the intermediate. The tricarbonylalkoxy-cobalt(I) compound formed by the hydrogen shift then can undergo oxidative addition with hydrogen and the dihydride can shift one hydrogen from cobalt to oxygen forming alcohol and regenerating the catalyst hydridotricarbonyl-cobalt(I). The tricarbonylalkoxylcobalt(I) intermediate may react with carbon monoxide as well as with hydrogen and form the tetracarbonyl. This side reaction is probably the second place that carbon monoxide inhibits the reduction. This reduction is discussed further in Chapter IX, Section A,2,a.

Two other homogeneous carbonyl reduction catalysts which are more convenient to use are trihydrido(tristriphenylphosphine)iridium(III) and the solvated dihydridobisphosphinerhodium(III) cation mentioned previously.

The iridium complex with hydrogen reduces aldehydes to alcohols easily at 50° at atmospheric pressure (144). The mechanism of this reduction is not

$$\text{RCHO} + \text{H}_2 \xrightarrow[\text{HOAc, 1 atm, 50°C}]{(\text{P}\phi_3)_3\text{IrH}_3} \text{RCH}_2\text{OH}$$

clear since it appears that acetic acid is a necessary component of the reaction mixture. This catalyst will also reduce activated olefins but more slowly than it does aldehydes. More information has been obtained on the rhodium catalyst. A mechanistic study of the hydrogen reduction of acetone with dihydridobisacetonebis(dimethylphenylphosphine)rhodium(I)   hexafluorophosphate in aqueous acetone solution has led to the proposal of the reaction mechanism given in Scheme IV-3. The reduction is very much accelerated by the presence of small amounts of water in the reaction mixture; 1% in the acetone solution was optimum. The water is presumed to replace acetone reversibly in the coordination sphere of the catalyst. The probable monoacetoneaquodihydride intermediate probably first undergoes a hydrogen shift; the hydrogen moving from rhodium to the carbon of the carbonyl group. "Side-on" coordination of the carbonyl group may be required for the shift to occur. The alkoxyrhodium intermediate now formed can undergo a second hydrogen transfer to form a five-coordinate complex containing coordinated isopropanol, perhaps in more than one step. It is believed that water is a catalyst for the transfer of the hydrogen from the position trans to the alkoxy group. The coordinated apical water molecule could give a hydrogen to the cis-alkoxy group and the hydride group could then, or simultaneously, move to the cis-hydroxy group formed in the first shift giving a five-coordinate intermediate. A final reaction with hydrogen and loss of isopropanol would form the starting complex again. The function of water in this reaction may be the same as the acetic acid in the preceding example. Reduction of aldehydes in contrast to ketones with the rhodium catalyst is a poor reaction because the catalyst decarbonylates aldehydes readily and forms a catalytically inactive carbonyl derivative. Decarbonylation reactions will be discussed in more detail in Chapter IX.

Asymmetric reduction of carbonyl groups is also possible when optically active ligands are present in the above catalysts. Reduction of simple ketones with the Schrock and Osborn-type catalyst (127) with R-benzylmethylphenylphosphine ligands gave only slightly optically active alcohols, however (145). The effect has also been demonstrated with a less reactive catalyst, a cobaloxime derivative, in the reduction of (easily reduced) benzil. The catalyst, formed from two moles of dimethylglyoxime and one mole of a cobalt(II) salt, has the four nitrogen atoms coordinated in a square-planar arrangement around the metal and a neutral fifth ligand occupies an apical position. The fifth ligand may be a variety of substances. For the asymmetric reduction,

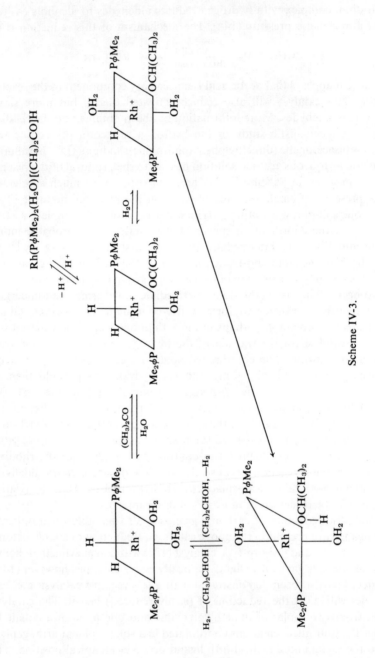

Scheme IV-3.

the optically active alkaloid quinine was employed as the fifth ligand. This complex is capable of reacting with hydrogen presumably forming a six-coordinate monohydride which at room temperature and atmospheric pressure reduced benzil to benzoin with 61.5% of the theoretical optical purity (146). The mechanism of this reduction is not known. If the reduction involves metal hydride addition to the carbonyl, it would seem that prior coordination of the carbonyl group in this case is unlikely since probably there is not room for the hydride and the carbonyl ligand on the same side of the nitrogen–cobalt plane. With this mechanism it is also difficult to explain the relatively high optical activity induced in the product by the quinine when it is very probably coordinated at the position trans to the hydride ligand, where it seemingly should have little influence on the addition reaction. This catalyst with pyridine as the fifth ligand will reduce methyl

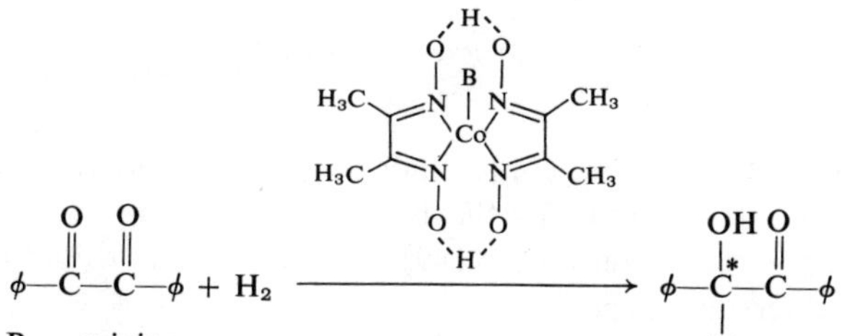

p-nitrocinnamate selectively in 78% yield to methyl 4-aminocinnamate and azobenzene in 94% yield to hydrazobenzene (146).

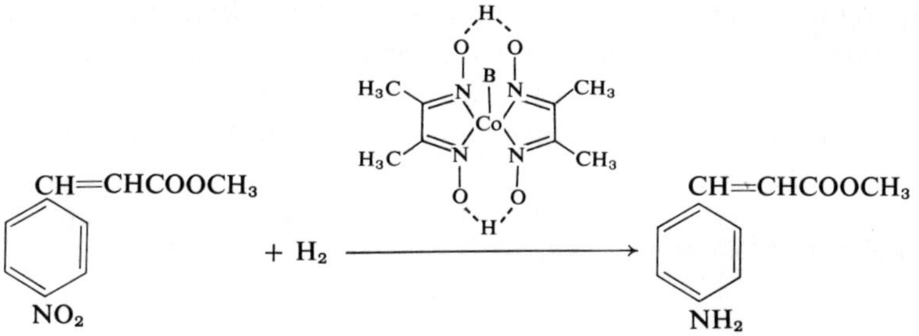

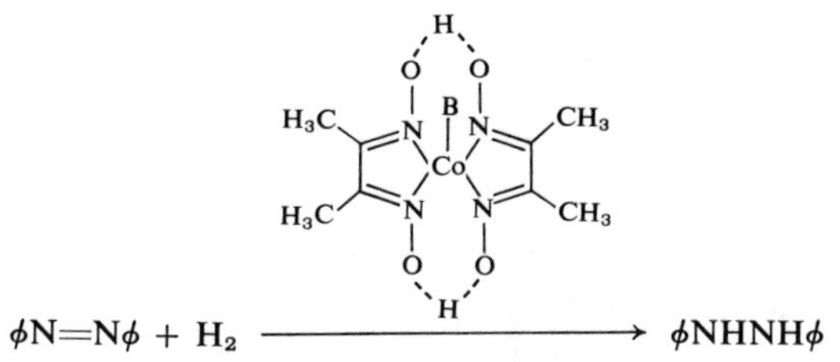

A very good review of the practical aspects of homogeneous hydrogenation has been written (147).

E. REDUCTION BY HYDROGEN TRANSFER

The source of the hydrogen in a reduction may be other substances than hydrogen gas. Generally, these reactions do not go by way of hydrogen gas as an intermediate except perhaps for reductions involving both carbon monoxide and water. In this case the catalyst may simply be causing the reaction of carbon monoxide with water to form carbon dioxide and hydrogen.

$$CO + H_2O \rightleftharpoons H_2 + CO_2$$

The high-yield, rhodium carbonyl-catalyzed reduction of nitro compounds to amines with carbon monoxide and water may be of this type (148).

$$RNO_2 + 3CO + H_2O \xrightarrow[150°, 120\ atm]{Rh_6(CO)_{16}\ +\ pyridine} RNH_2 + 3CO_2$$

A similar reduction of nitro compounds occurs with dodecacarbonyltriiron with methanol as the source of the hydrogen. The fate of the methanol, however, has not been determined (149). Likely products would be dimethyl carbonate and ferrous carbonate.

$$ArNO_2 + Fe_3(CO)_{12} + CH_3OH \longrightarrow ArNH_2 + [(CH_3O)_2CO + FeCO_3]$$

Reduction of various unsaturated organic groups by transition metal compounds and acids, in some instances at least, appears to involve metal hydrides formed by oxidative addition. The metal hydrides then add to the unsaturated group (probably via a π complex) and finally acid cleaves the metal alkyl formed. The last step may proceed by an oxidative addition of the acid, with subsequent hydrogen shift from metal to alkyl. This mechanism

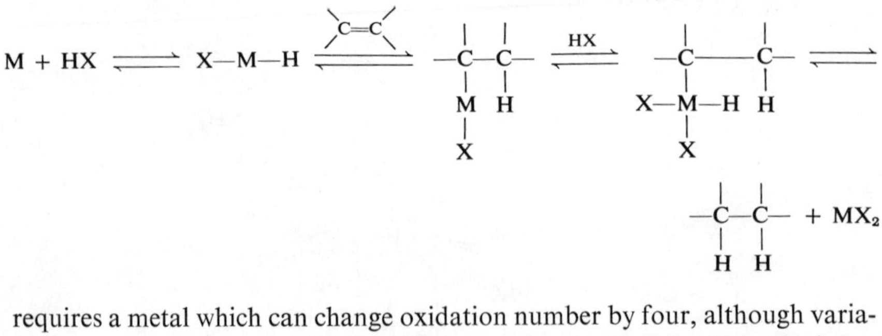

requires a metal which can change oxidation number by four, although variations are possible where, for example, an exchange of X for H in the intermediate olefin adduct occurs and only a two-electron change is required.

$$\underset{\underset{X}{\overset{|}{\underset{|}{M}}}{\overset{|}{\underset{|}{C}}}\!\!-\!\!\overset{|}{\underset{|}{C}}\!\!-\;+\;X\!\!-\!\!M\!\!-\!\!H\;\xrightarrow{-MX_2}\;\underset{\underset{H}{\overset{|}{\underset{|}{M}}}{\overset{|}{\underset{|}{C}}}\!\!-\!\!\overset{|}{\underset{H}{\overset{|}{C}}}\!\!-\;\longrightarrow\;M\;+\;\underset{H}{\overset{|}{\underset{|}{C}}}\!\!-\!\!\underset{H}{\overset{|}{\underset{|}{C}}}\!\!-$$

Cis addition of the hydrogen would be expected from either of these paths, but it is not always observed. An example of probably the former possibility is the reduction of diphenylacetylene with trifluoroacetic acid and bistriphenylphosphine-π-diphenylacetyleneplatinum(0), a stable coordinately unsaturated complex. Initially an isolable *cis*-vinylplatinum intermediate complex is formed which then reacts further with trifluoroacetic acid with isomerization to give *trans*-1,2-diphenylethylene (150). The mechanism of this transformation is discussed further in Chapter VII, Section B,1.

The use of isopropanol as a hydrogen transfer reagent has been demonstrated in the iridium(III)-catalyzed reductions of α,β-unsaturated ketones, acetylenes, and saturated ketones.

Benzalacetophenone can be converted into (phenylethyl)phenylketone in 95% yield by refluxing it in isopropanol with a catalyst prepared from iridium trichloride, dimethyl sulfoxide (a fairly good ligand, bonding through oxygen), and hydrogen chloride (151). The iridium complex is most likely converted into the monohydride by reaction with isopropanol by way of an isopropoxy complex. The hydride then may add to the double bond of the unsaturated ketone. The adduct which can be isolated, apparently has the internal ketone group coordinated to the iridium. On further heating with isopropanol containing hydrogen chloride, the saturated ketone is formed. Since reduction of this adduct requires hydrogen chloride, an oxidative addition of the acid may be involved, although this would require an Ir(V) species, at least, as a short-lived intermediate. The adduct could then undergo a hydrogen shift from iridium to carbon. The iridium trichloride formed could be re-reduced

$$HCl + (CH_3)_2SO + IrCl_3 \longrightarrow \{HIrCl_4[(CH_3)_2SO]\}\cdot 2(CH_3)_2SO$$

$$\{HIrCl_4\cdot[(CH_3)_2SO]_2\}\cdot 2(CH_3)_2SO + CH_3CHOHCH_3 \longrightarrow$$
$$HIrCl_2[(CH_3)_2SO]_3 + 2HCl + (CH_3)_2SO + CH_3COCH_3$$

$$HIrCl_2[(CH_3)_2SO]_3 + \phi\overset{\overset{\textstyle O}{\|}}{C}CH{=}CH\phi \xrightarrow{-(CH_3)_2SO}$$

$$\underset{(CH_3)_2SO\quad\quad OS(CH_3)_2}{\underset{O\!-\!-\!-\!Ir\!-\!Cl}{\phi\!-\!C\overset{\displaystyle CH_2}{\underset{\displaystyle}{\diagup\;\diagdown}}CH\phi}}\;\;\;\xrightarrow[(CH_3)_2SO]{HCl}\;\;\phi\overset{\overset{\textstyle O}{\|}}{C}CH_2CH_2\phi + IrCl_3\cdot 3(CH_3)_2SO]_3$$

$$IrCl_3\cdot 3(CH_3)_2SO + (CH_3)_2CHOH \longrightarrow HIrCl_2\cdot[(CH_3)_2SO]_3 + CH_3COCH_3 + HCl$$

by isopropanol to the metal hydride and the catalytic cycle would be completed.

A completely analogous catalytic reduction of diphenylacetylene in the presence of hydrogen chloride, isopropanol, and hydridodichlorotris(dimethyl sulfoxide)iridium(III) has been observed (152). The intermediate vinyliridium complex can be isolated. The adduct and final product are the expected cis isomers.

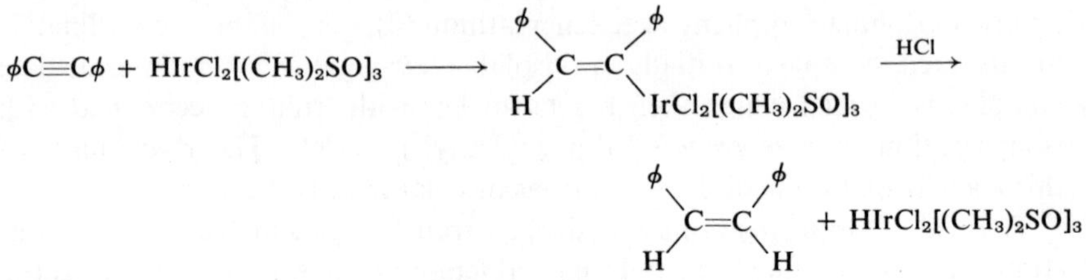

$$\phi C\equiv C\phi + HIrCl_2[(CH_3)_2SO]_3 \longrightarrow \underset{H}{\overset{\phi}{\diagdown}}C=C\underset{IrCl_2[(CH_3)_2SO]_3}{\overset{\phi}{\diagup}} \xrightarrow{HCl}$$

$$\underset{H}{\overset{\phi}{\diagdown}}C=C\underset{H}{\overset{\phi}{\diagup}} + HIrCl_2[(CH_3)_2SO]_3$$

Ketones, also, may be reduced with an excess of refluxing isopropanol and phosphoric acid (instead of hydrogen chloride) with iridium trichloride as catalyst. Reduction of cyclohexanones produces axial hydroxy derivatives quite selectively, probably for steric reasons. The hydride addition probably occurs mainly from the less substituted side of the cyclohexanone ring (153).

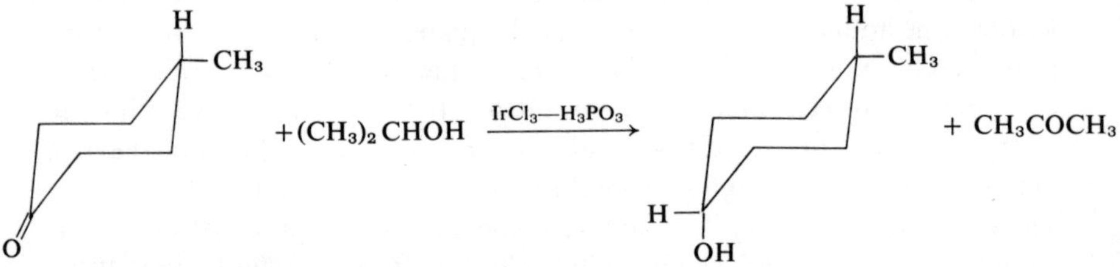

$$\text{(structure with H, CH}_3\text{, O)} + (CH_3)_2 CHOH \xrightarrow{IrCl_3-H_3PO_3} \text{(structure with H, CH}_3\text{, OH)} + CH_3COCH_3$$

F. DEHYDROGENATION

The removal of hydrogen from organic compounds in a few instances can be achieved with homogeneous catalysts. A potentially useful example is the two-step reaction of an olefin with palladium chloride to produce a conjugated diene. The first step is the reaction of the olefin with palladium chloride in the presence of sodium acetate and acetic acid at 100° to form a π-allylic palladium chloride dimer, and the second step is vacuum pyrolysis of the π-allylic complex with removal of the diene as it is formed. Considerable double-bond isomerization is observed, however, in cases where isomers are possible (154). 1-Octene in this reaction, for instance, gives an 88% yield of a 2:1 mixture of the 2,4- and the 1,3-dienes.

$$CH_2=CH(CH_2)_5CH_3 + PdCl_2 + NaOAc \xrightarrow[-HOAc]{-NaCl}$$

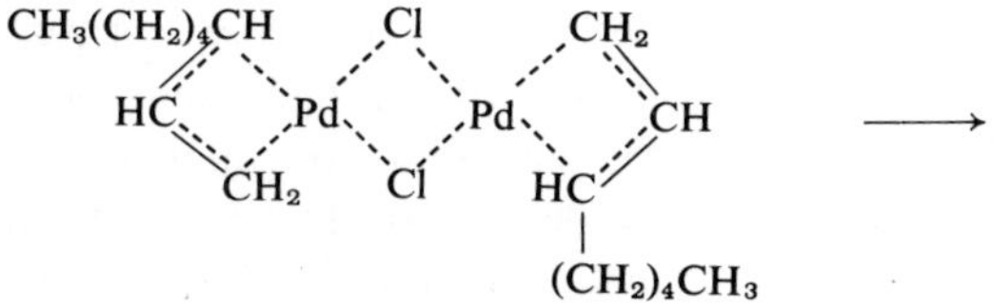

$$CH_3(CH_2)_2 CH=CHCH=CHCH_3 + CH_3(CH_2)_3CH=CHCH=CH_2 + [HCl + Pd]$$

Cyclohexene is dehydrogenated to benzene in the presence of two moles of palladium acetate and sodium acetate in acetic acid solution (155). Presumably this reaction begins with the formation of a π-cyclohexenylpalladium acetate complex. This intermediate then probably undergoes elimination to form 1,3-cyclohexadiene. Reformation of a π-allyl-type complex with loss of acetic acid with the 1,3-cyclohexadiene and another elimination would produce benzene. The reaction is complicated, however, by the fact that

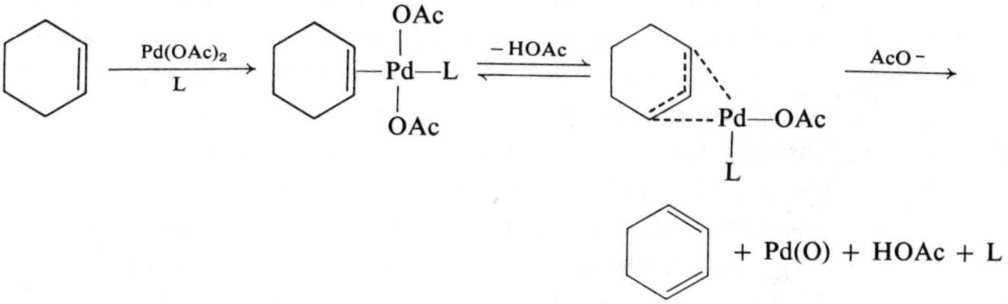

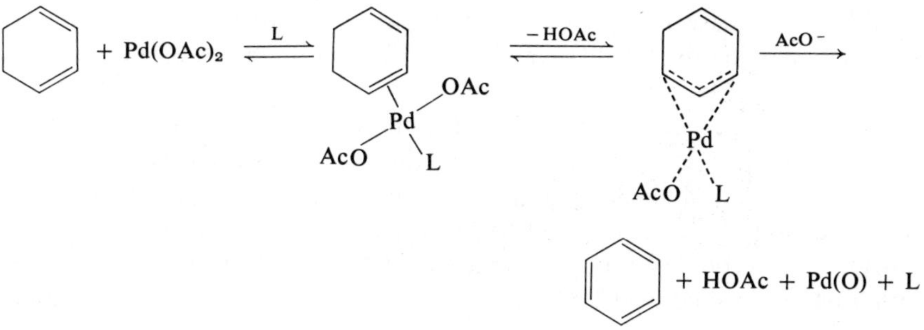

palladium metal produced in the reaction is also a catalyst for the dehydrogenation.

The catalytic dehydrogenation of alcohols to carbonyl compounds can be achieved catalytically with certain homogeneous transition metal catalysts. Rhodium trichloride in the presence of stannous chloride at reflux temperature converts isopropanol into acetone and hydrogen. The accumulation of acetone as product decreases the reaction rate probably because the reaction is reversible (156). A rhodium alkoxide is probably an intermediate and it probably undergoes a β-hydride elimination with formation of ketone and a hydridorhodium species. The last step probably is a reaction of hydrogen chloride with the metal hydride forming hydrogen and the catalyst again. Without stannous chloride the reaction is stoichiometric in rhodium trichloride. Thus, the stannous chloride appears to be accelerating the reaction of hydrogen chloride with the metal hydride. It is known that stannous chloride forms the trichlorotin ligand with transition metal chlorides and this ligand, which exerts a strong trans effect, probably makes the hydride group more reactive toward protons (probably by an oxidative addition mechanism).

$$RhCl_3L_3 + SnCl_2 \rightleftharpoons RhCl_2(SnCl_3)L_3$$

$$RhCl_2(SnCl_3)L_3 + (CH_3)_2CHOH \rightleftharpoons (CH_3)_2CHORhCl(SnCl_3)L_3 + HCl$$

$$(CH_3)_2CHORhCl(SnCl_3)L_3 \rightleftharpoons (CH_3)_2CO + HRhCl(SnCl_3)L_3$$

$$HRhCl(SnCl_3)L_3 + HCl \longrightarrow H_2 + RhCl_2(SnCl_3)L_3$$

A very interesting reaction which combines a dehydrogenation of an alcohol with an aldol condensation and hydrogen transfer reaction occurs when primary alcohols are heated with a sodium alkoxide, a catalytic amount of chlorodicarbonylrhodium(I) dimer, and eight equivalents of triethylphosphine. n-Butanol, for example, produces the commercially important 2-ethyl-1-hexanol in 90% yield by this reaction if the water formed is continuously removed (157). The true catalyst is probably chlorocarbonylbistriethylphosphinerhodium(I). On reaction with sodium alkoxide, an alkoxyrhodium complex would be formed which could eliminate hydride and form the aldehyde. A base-catalyzed aldol condensation of two aldehyde molecules with water elimination would then give the α,β-unsaturated aldehyde condensation product. One of the two rhodium hydride molecules formed in the alcohol oxidation now could add to the unsaturated aldehyde forming a π-oxyallylrhodium complex. Reaction of this complex with n-butanol could produce the reduced aldehyde and reform the starting alkoxyrhodium complex. Reaction of the reduced aldehyde with the second of the two rhodium hydride species could give an alkoxyrhodium intermediate which on exchange with n-butanol would give 2-ethyl-1-hexanol and the starting alkoxyrhodium

$$\text{(structure)} + 4P(C_2H_5)_3 \longrightarrow 2CO + 2RhCl(CO)[P(C_2H_5)_3]_2$$

$$RhCl(CO)[P(C_2H_5)_3]_2 + NaOCH_2CH_2CH_2CH_3 \xrightarrow{-NaCl}$$

$$CH_3CH_2CH_2CH_2ORh(CO)[P(C_2H_5)_3]_2 \xrightarrow{L} \rightleftharpoons$$

$$CH_3CH_2CH_2CHO + HRh(CO)[P(C_2H_5)_3]_2L$$

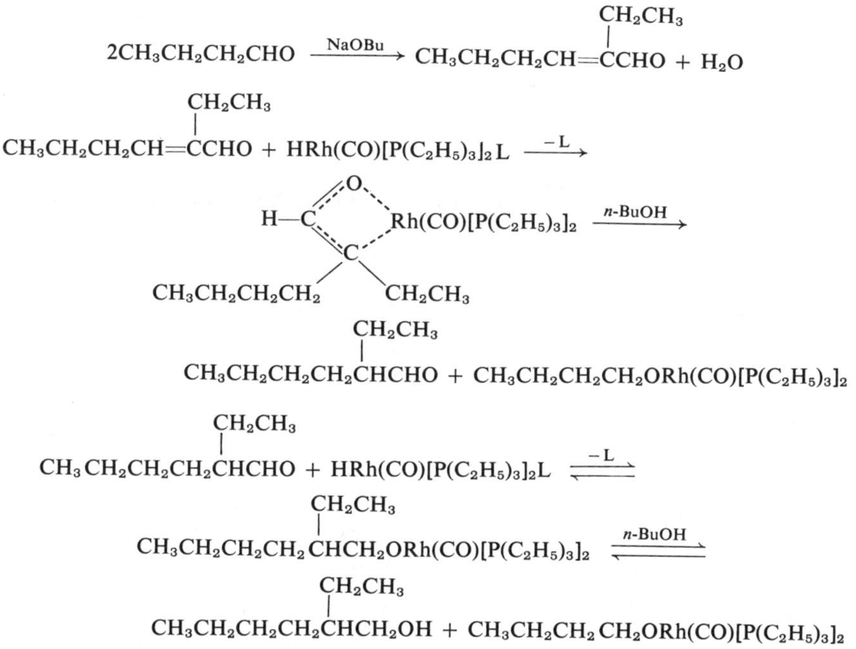

$$2CH_3CH_2CH_2CHO \xrightarrow{NaOBu} CH_3CH_2CH_2CH{=}\overset{\overset{\displaystyle CH_2CH_3}{|}}{C}CHO + H_2O$$

$$CH_3CH_2CH_2CH{=}\overset{\overset{\displaystyle CH_2CH_3}{|}}{C}CHO + HRh(CO)[P(C_2H_5)_3]_2L \xrightarrow{-L}$$

$$CH_3CH_2CH_2CH_2\overset{\overset{\displaystyle CH_2CH_3}{|}}{C}HCHO + CH_3CH_2CH_2CH_2ORh(CO)[P(C_2H_5)_3]_2$$

$$CH_3CH_2CH_2CH_2\overset{\overset{\displaystyle CH_2CH_3}{|}}{C}HCHO + HRh(CO)[P(C_2H_5)_3]_2L \rightleftharpoons \xrightarrow{-L}$$

$$CH_3CH_2CH_2CH_2\overset{\overset{\displaystyle CH_2CH_3}{|}}{C}HCH_2ORh(CO)[P(C_2H_5)_3]_2 \underset{\xleftarrow{}}{\xrightarrow{n-BuOH}}$$

$$CH_3CH_2CH_2CH_2\overset{\overset{\displaystyle CH_2CH_3}{|}}{C}HCH_2OH + CH_3CH_2CH_2CH_2ORh(CO)[P(C_2H_5)_3]_2$$

compound again. Excess n-butanol is necessary to drive the various equilibria in the reaction to completion and form largely 2-ethyl-1-hexanol rather than a mixture of the alcohol and the related saturated aldehyde.

FURTHER READING

A. J. Birch and D. H. Williamson, Hydrogenation with soluble catalysts. *Org. React.* (1974) (in press).

J. Halpern, Homogeneous reduction. *Annu. Rev. Phys. Chem.* **16**, 103 (1965).

J. Kwiatek, Hydrogenation and dehydrogenation. "Transition Metals in Homogeneous Catalysis," Chapter 2. Dekker, New York, 1971.

J. D. Morrison, Asymmetric hydrogenation. *Surv. Progr. Chem.* **3**, 147 (1955).

W. Strohmeier, Homogeneous hydrogenation. *Fortschr. Chem. Forsch.* **25**, 71 (1972).

R. Ugo, Selectivity in hydrogenation. "Aspects of Homogeneous Catalysis." Manfredi, 1970.

Chapter V

Addition and Elimination Reactions of Olefinic Compounds

A. OLEFIN ISOMERIZATION

1. 1:2 Addition–Elimination Reactions of Transition Metal Hydrides

Double-bond isomerization is a very commonly observed reaction. The isomerization often accompanies other transition metal reactions giving less reactive isomerized olefins which, in turn, may give isomeric reactions products. In most instances the mechanism of isomerization is a simple metal hydride addition–elimination sequence. The isomerization requires that the hydride elimination take place with a different hydrogen than the one that was added. In the usual example, olefin coordination with the hydride appears to precede the addition, while the elimination probably initially produces a hydride π complex of the rearranged olefin (see Scheme V-1). Olefin equilibration can be achieved with many catalysts, although equilibration often will not go to completion in reaction mixtures containing reagents which react with the intermediate metal alkyls. The degree of isomerization observed obviously depends on the relative rates of the various reactions involved. Examples are known where double bonds have moved by eight or more carbon atoms during transition metal-catalyzed reactions.

In the usual mechanism as indicated above, a cis addition and a cis elimination of metal hydride is expected and this result is consistent with the facts in the very few instances where data have been obtained (158, 159). Trans addition of the elements of a metal hydride very probably can occur also, particularly when relatively stable, coordinately saturated olefin π complexes react with protons or metal hydrides. If the olefin π complexes are coordinately

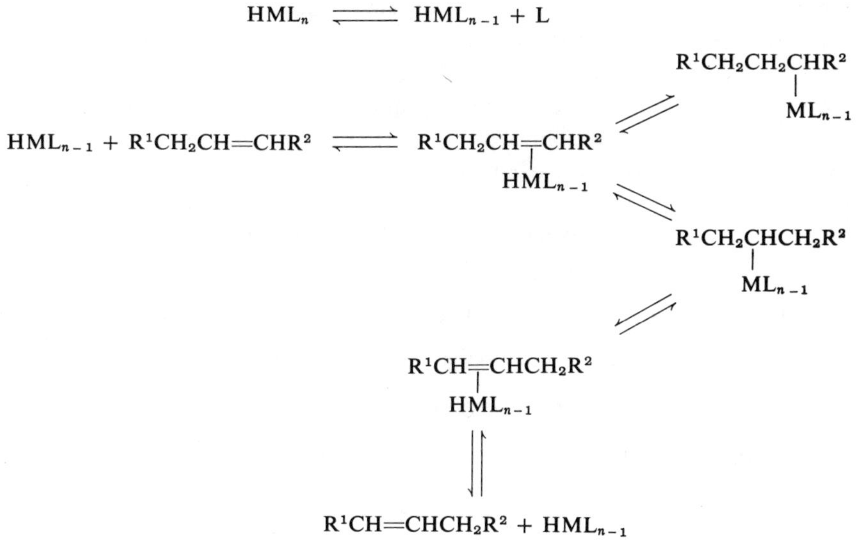

$$HML_n \rightleftarrows HML_{n-1} + L$$

$$
\begin{array}{c}
R^1CH_2CH_2CHR^2 \\
| \\
ML_{n-1}
\end{array}
$$

$$HML_{n-1} + R^1CH_2CH{=}CHR^2 \rightleftarrows R^1CH_2CH{=}CHR^2$$
$$\underset{HML_{n-1}}{|}$$

$$
\begin{array}{c}
R^1CH_2CHCH_2R^2 \\
| \\
ML_{n-1}
\end{array}
$$

$$R^1CH{=}CHCH_2R^2$$
$$\underset{HML_{n-1}}{|}$$

$$R^1CH{=}CHCH_2R^2 + HML_{n-1}$$

Scheme V-1.

unsaturated or dissociate easily, protonation of the metal may occur first and in such cases produce cis addition (cf. Chapter VI, Section D).

The direction of addition and elimination of transition metal hydrides is of obvious importance in predicting initial isomerization products or final products if complete equilibration does not occur. Covalent metal hydride additions appear to be controlled by a combination of steric and electronic effects that may vary substantially from metal to metal and from complex to complex. The almost complete lack of data on directions of additions does not allow any generalizations to be made. In most reactions the direction of addition is assumed from the isomeric composition of the products obtained, either olefins or more complicated insertion products. Without more information about relative stabilities and reactivities of the intermediate species, however, such data may give incorrect estimates of the isomeric composition of the initial adducts formed.

The initial direction of addition has been determined in the gas phase reaction of hydridotetracarbonylcobalt(I) with propylene (see Scheme V-2). This addition could put cobalt on the terminal carbon or the middle carbon. Simply looking at the composition of the isomeric mixture of cobalt compounds produced may not answer the question of how the initial addition occurred since equilibration of isomers may have taken place by the time the products are isolated. To show that equilibration did not occur, the reaction was carried out with a large excess of completely deuterated propylene and metal hydride containing normal hydrogen. Since the organocobalt intermediates are very unstable, the first formed monoprotio and monodeutero elimination products were identified. Correcting for statistical effects and assuming no isotope effects (a possibly quite inaccurate assumption), it was deduced that 70% addition of the cobalt occurred on the middle carbon and 30% on the terminal carbon (160).

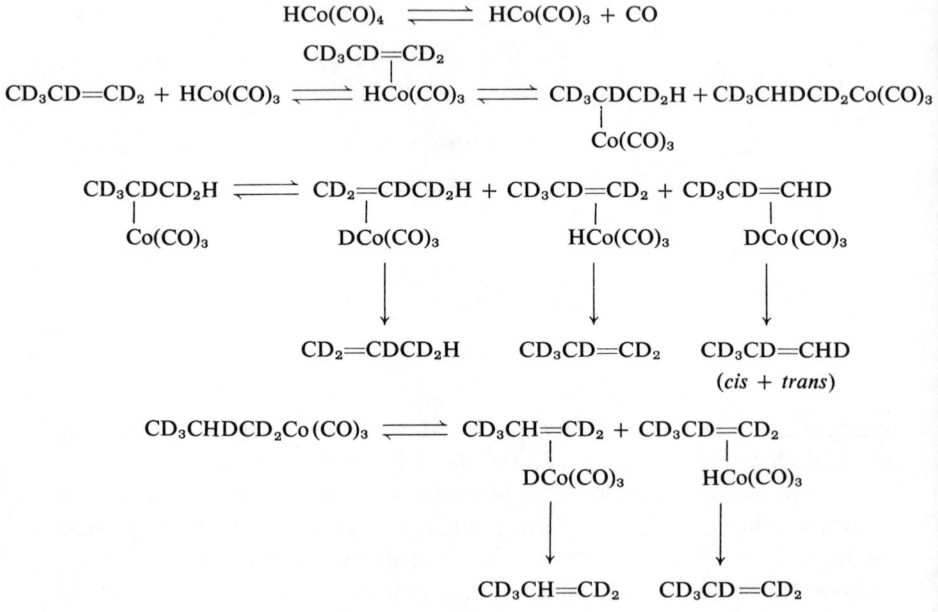

$$HCo(CO)_4 \rightleftharpoons HCo(CO)_3 + CO$$

$$CD_3CD=CD_2$$

$$CD_3CD=CD_2 + HCo(CO)_3 \rightleftharpoons HCo(CO)_3 \rightleftharpoons CD_3CDCD_2H + CD_3CHDCD_2Co(CO)_3$$
$$\underset{Co(CO)_3}{|}$$

$$CD_3CDCD_2H \rightleftharpoons CD_2=CDCD_2H + CD_3CD=CD_2 + CD_3CD=CHD$$
$$\underset{Co(CO)_3}{|} \qquad \underset{DCo(CO)_3}{|} \qquad \underset{HCo(CO)_3}{|} \qquad \underset{DCo(CO)_3}{|}$$

$$\downarrow \qquad\qquad \downarrow \qquad\qquad \downarrow$$

$$CD_2=CDCD_2H \qquad CD_3CD=CD_2 \qquad CD_3CD=CHD$$
$$(cis + trans)$$

$$CD_3CHDCD_2Co(CO)_3 \rightleftharpoons CD_3CH=CD_2 + CD_3CD=CD_2$$
$$\underset{DCo(CO)_3}{|} \qquad \underset{HCo(CO)_3}{|}$$

$$\downarrow \qquad\qquad \downarrow$$

$$CD_3CH=CD_2 \qquad CD_3CD=CD_2$$

Scheme V-2.

Another, well-studied example is a rhodium-catalyzed olefin isomerization (161). The initial direction of addition was not determined, but the overall mechanism was found to be of the hydride addition–elimination type. The hydridorhodium catalyst in this example was formed in the reaction mixture by the oxidative addition of hydrogen chloride to chlorobisethylenerhodium(I) dimer in alcohol solution. It is presumed that the reacting olefin replaces ethylene in the complex and then addition and elimination occur. Since a variety of possible ligands are present in the reaction mixture (alcohol, olefins, and chloride ion) the exact ligands on the reaction intermediate were uncertain and may not matter as far as the addition–elimination reactions are concerned. The proposed mechanism of isomerization is shown in Scheme V-3 with 1-butene as the olefin.

Numerous similar isomerization catalysts are known (162).

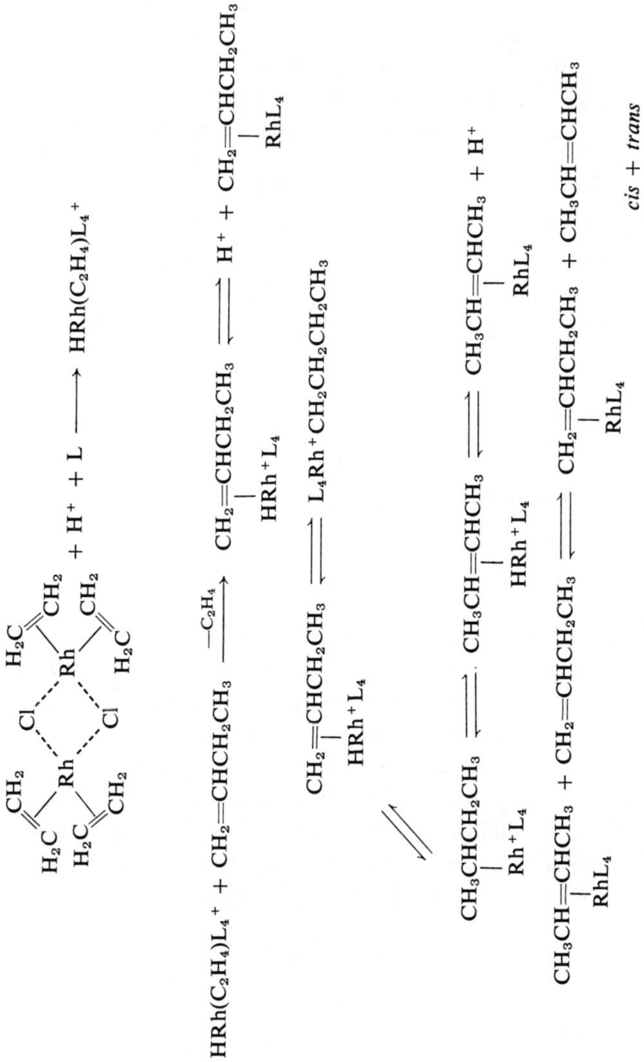

$$\text{HRh(C}_2\text{H}_4\text{)L}_4^+ + \text{H}^+ + \text{L} \longrightarrow \text{HRh(C}_2\text{H}_4\text{)L}_4^+$$

$$\text{HRh(C}_2\text{H}_4\text{)L}_4^+ + \text{CH}_2{=}\text{CHCH}_2\text{CH}_3 \xrightarrow{-\text{C}_2\text{H}_4} \underset{\overset{|}{\text{HRh}^+\text{L}_4}}{\text{CH}_2{=}\text{CHCH}_2\text{CH}_3} \rightleftharpoons \text{H}^+ + \underset{\overset{|}{\text{RhL}_4}}{\text{CH}_2{=}\text{CHCH}_2\text{CH}_3}$$

$$\underset{\overset{|}{\text{HRh}^+\text{L}_4}}{\text{CH}_2{=}\text{CHCH}_2\text{CH}_3} \rightleftharpoons \text{L}_4\text{Rh}^+\text{CH}_2\text{CH}_2\text{CH}_2\text{CH}_3$$

$$\underset{\overset{|}{\text{Rh}^+\text{L}_4}}{\text{CH}_3\text{CHCH}_2\text{CH}_3} \rightleftharpoons \underset{\overset{|}{\text{HRh}^+\text{L}_4}}{\text{CH}_3\text{CH}{=}\text{CHCH}_3} \rightleftharpoons \underset{\overset{|}{\text{RhL}_4}}{\text{CH}_3\text{CH}{=}\text{CHCH}_3} + \text{H}^+$$

$$\underset{\overset{|}{\text{RhL}_4}}{\text{CH}_3\text{CH}{=}\text{CHCH}_3} + \text{CH}_2{=}\text{CHCH}_2\text{CH}_3 \rightleftharpoons \underset{\overset{|}{\text{RhL}_4}}{\text{CH}_2{=}\text{CHCH}_2\text{CH}_3} + \text{CH}_3\text{CH}{=}\text{CHCH}_3$$

cis + *trans*

Scheme V-3.

The factors influencing the direction of elimination of metal hydride once a metal alkyl is formed have not been specifically investigated, but indications are that steric effects can be very important. The few results available suggest that elimination tends to produce the metal hydride π complex with the olefin which has the fewest substituents. The more substituted olefins apparently form less stable complexes for steric reasons (see Table V-1). For the same reason *cis*-olefin π complexes would be preferred over the corresponding trans ones. A preference for formation of *cis*-olefins, initially during isomerization reactions has been observed in several cases (e.g., refs. 163, 164).

2. Isomerization by 1:3 Hydrogen Shifts

Olefin isomerization may occur by a second mechanism when allylic hydrogens are activated and/or free metal hydride is not present in the reaction mixture. This mechanism involves a 1:3 shift of a hydrogen from an allylic position to a double bond carbon by way of a π-allyl metal hydride intermediate.

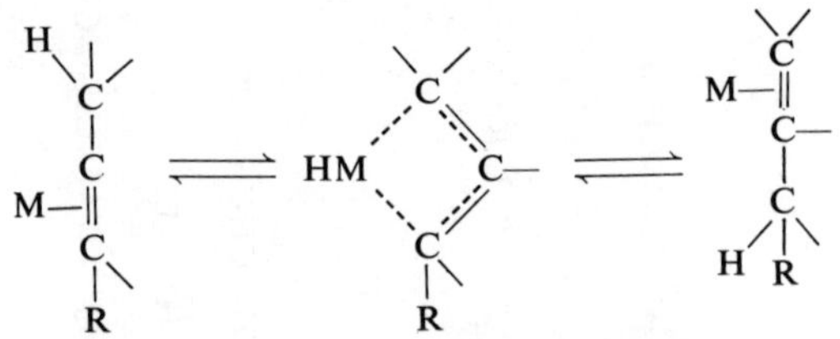

Typical of many reactions which may be of this type is the sodium tetrachloroplatinate isomerization of mesityl oxide (4-methyl-3-penten-2-one) at room temperature to a platinum complex of 4-methyl-4-penten-2-one. The reaction goes to completion because of the formation of a stable chelated platinum complex. It is perhaps surprising that this carbonyl–olefin chelated complex is stable enough to overcome the energy requirement of moving the double bond out of conjugation with the carbonyl group. The complexed olefin produced may be displaced from the complex with tri-*n*-butylphosphine and the isomerized olefin isolated in about 90% yield (165).

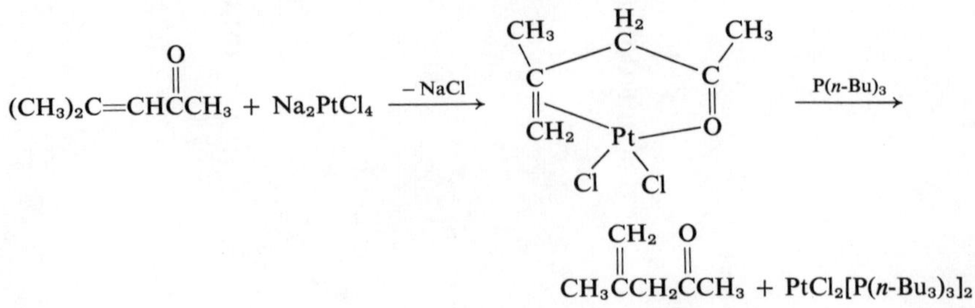

The low temperature and the neutral reaction medium used in the above reaction suggests that metal hydride probably is not involved and a 1:3 hydride shift mechanism is more probable.

An unambiguous example of the 1:3 hydride shift mechanism has been found in the reactions of a highly reactive nickel complex at low temperature. Bromotrifluorophosphine-π-allylnickel(II) reacts with sodium trimethylboro-hydride at $-135°$ to form hydridotrifluorophosphine-π-allylnickel(II), insoluble sodium bromide, and volatile trimethylboron. The pure nickel complex, on warming to $-40°$, undergoes an internal hydrogen shift forming trifluorophosphine-π-propylenenickel(0), a two-coordinate nickel(0) complex. The propylene can be liberated from this complex by reaction with triphenyl phosphite. The most interesting feature of the dicoordinate complex is that on cooling below about $-50°$ it is reconverted into the hydrido-π-allyl complex clearly illustrating the 1:3 hydride shift mechanism (166).

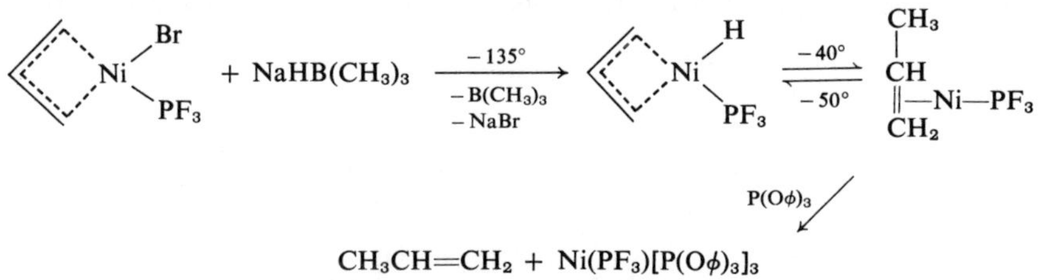

$$CH_3CH=CH_2 + Ni(PF_3)[P(O\phi)_3]_3$$

A similar migration of hydrogen from rhodium to a coordinated π-allyl group has been reported (167).

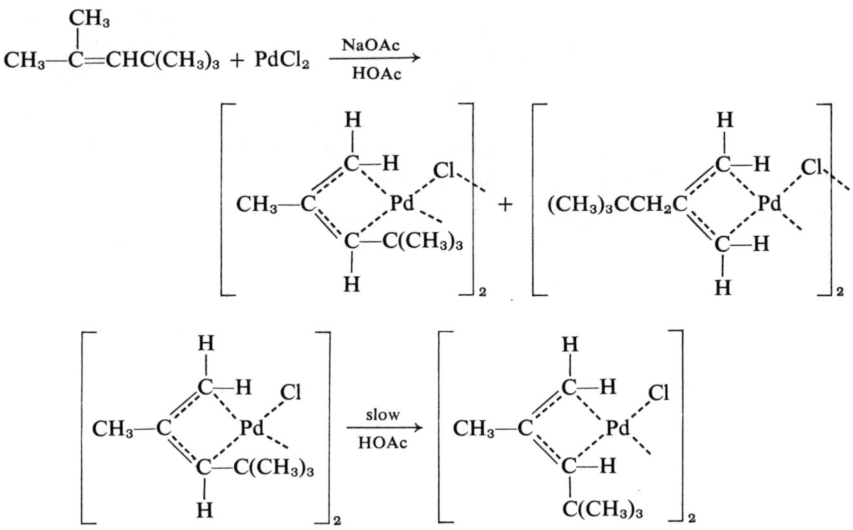

An apparent ionic analog of the covalent 1:3 hydrogen shift may occur when π-olefin complexes of palladium are converted into π-allylpalladium salts with sodium acetate in acetic acid (168). It is not certain whether the hydrogen is removed directly from a methylene group or from a π-allylic hydride isomer present in low concentration in the reaction mixture.

The initially formed major products are the anti isomers, but these slowly isomerize to syn isomers in the reaction mixture, either by a reversal of the reaction, by protonation, or perhaps more probably, by formation of a σ-allyl intermediate which then closes to the *syn-π-allyl* complex.

B. HYDRIDE TRANSFER REACTIONS

A consequence of the facile addition–elimination reactions that many transition metal hydrides undergo is the frequent, easy transfer of metal hydride from one metal alkyl to an olefin with formation of another metal alkyl. These are equilibrium reactions and the concentrations of the various

$$R^1\text{—}\underset{\underset{|}{|}}{\overset{\overset{H}{|}}{C}}\text{—}\underset{|}{\overset{|}{C}}\text{—}M \;\rightleftharpoons\; \underset{H}{\overset{R^1}{{}}}\!\!\diagup\!\!C\!\!=\!\!C\!\!\diagup \; + \; MH$$

$$MH \; + \; \underset{\diagup}{\overset{R^2}{\diagdown}}C\!\!=\!\!C\!\!\underset{\diagdown}{\overset{\diagup}{}} \;\rightleftharpoons\; R^2\text{—}\underset{\underset{|}{|}}{\overset{\overset{H}{|}}{C}}\text{—}\underset{|}{\overset{|}{C}}\text{—}M$$

components present depend on the olefins, metal, and ligands involved. An interesting and potentially useful application of this exchange reaction is the conversion of an olefin into a Grignard reagent. If a Grignard reagent is stirred with an olefin in the presence of a catalytic amount of titanium tetrachloride, an equilibrium mixture of olefins and Grignard reagents is formed. The Grignard reagent apparently alkylates the titanium tetrachloride. The alkyl then eliminates hydridotrichlorotitanium forming an olefin. The hydride then adds to the other olefin present forming a new alkyltitanium compound. A final exchange of this titanium alkyl with magnesium halide produces the new Grignard reagent and regenerates the titanium tetrahalide catalyst. The use of *n*-propylmagnesium bromide as the starting Grignard reagent allows the formation of volatile propylene in the exchange. If the

$$CH_3CH_2CH_2MgX + TiX_4 \;\rightleftharpoons\; [CH_3CH_2CH_2TiX_3] + MgX_2$$

$$[CH_3CH_2CH_2TiX_3] \;\rightleftharpoons\; [HTiX_3] + CH_3CH\!\!=\!\!CH_2$$

$$[HTiX_3] + RCH\!\!=\!\!CH_2 \;\rightleftharpoons\; [RCH_2CH_2TiX_3]$$

$$[RCH_2CH_2TiX_3] + MgX_2 \;\rightleftharpoons\; RCH_2CH_2MgX + TiX_4$$

$$X = Br \text{ or } Cl$$

propylene is allowed to evaporate from the reaction mixture the equilibrium is shifted to the formation of the Grignard reagent of the olefin added. The reagent may then be used for conventional Grignard reactions (169).

A completely analogous reaction occurs when nickel(II) chloride is used in place of titanium tetrachloride (170).

$$CH_3CH_2CH_2MgBr + C_6H_5CH{=}CH_2 \xrightarrow{NiCl_2} C_6H_5CH_2CH_2MgBr + CH_3CH{=}CH_2$$

Another useful example of a hydride exchange reaction is the platinum(II)-catalyzed addition of silicon hydride derivatives to olefins (Scheme V-4). The silicon hydride group initially adds oxidatively to the platinum(II)

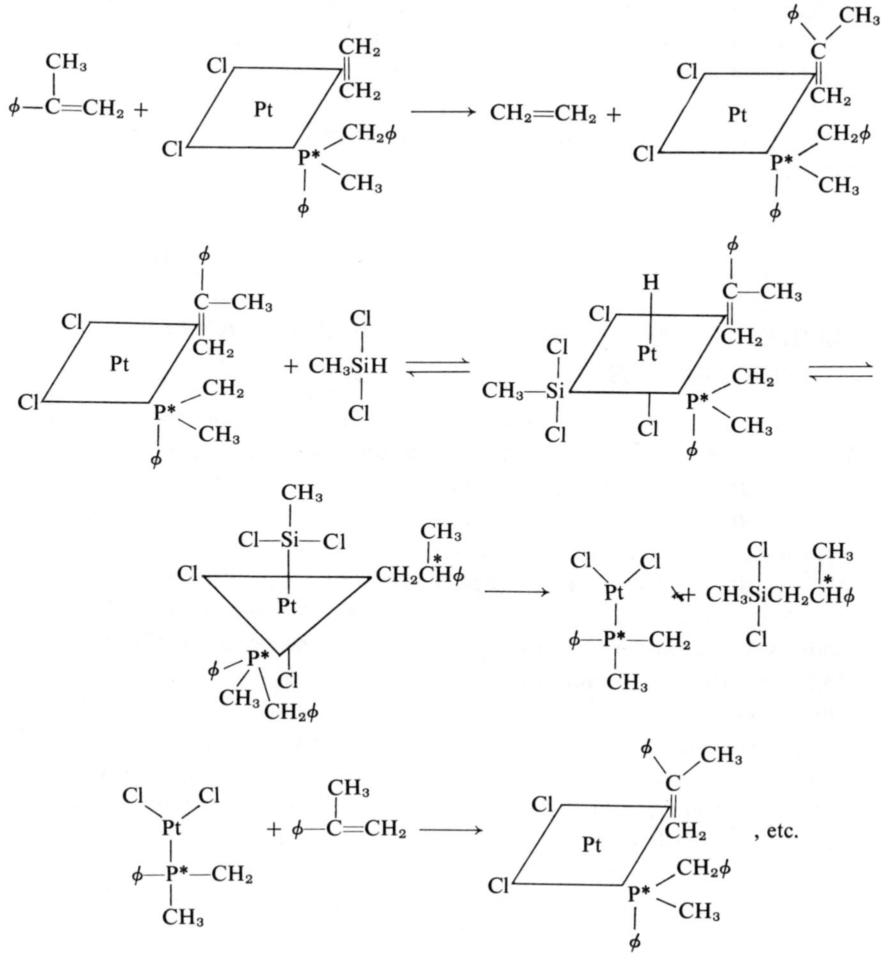

Scheme V-4.

catalyst or if a platinum(IV) salt is employed a reduction by the hydride probably occurs before the addition. The platinum(IV) hydride then adds to the olefin probably by way of the olefin π complex. The alkyl then undergoes reductive elimination of alkylsilane. A readdition of olefin and the silicon hydride to the platinum(II) product would start the cycle again. The reaction is illustrated by the addition of dichloromethylsilane to 2-phenylpropene catalyzed by cis- dichloro[(R)-benzylmethylphenylphosphine]-π-ethylene-platinum(II). A chiral phosphine ligand was used in an effort to induce optical activity in the product silane. After 40 hr at 40° a 43% yield of product was obtained which, however, was only about 5% optically pure (171). Only terminal silation is observed because the platinum hydride adds only in that direction, but with other Group VIII metal catalysts this is not the case and the reason for terminal addition must be that only the terminal isomer reacts [see $Co_2(CO)_8$-catalyzed reactions described in Chapter III, Section B).

In a similar reaction employing an optically active silicon hydride derivative, the product was formed with complete retention of the configuration about the silicon atom (172).

$$CH_3(CH_2)_5CH{=}CH_2 + (\alpha\text{-}C_{10}H_7)(C_6H_5)CH_3\,Si^*H \xrightarrow{\quad H_2PtCl_6 \quad}$$
$$CH_3(CH_2)_7Si^*(CH_3)(C_6H_5)(\alpha\text{-}C_{10}H_7$$

C. OLEFIN DIMERIZATION, OLIGOMERIZATION, AND POLYMERIZATION

Numerous catalysts are known which will convert olefins into dimers, oligomers (very low molecular weight polymers composed usually of trimers, tetramers, pentamers, etc.), and high molecular weight polymers. Dimers are generally formed by a metal hydride adding reversibly to the olefin and then occasionally the metal alkyl formed adds (via an olefin π complex) to another olefin molecule. The latter complex then eliminates metal hydride liberating the olefin dimer, probably by way of the hydride–dimer π complex. Catalyst adducts which undergo metal hydride elimination more slowly than they add to another olefin molecule, produce higher molecular weight oligomers. Oligomers may also be formed if dimers are produced initially and then these react further either with themselves or with more monomeric olefin. High polymers are formed when the elimination is very much slower than the addition to another olefin unit (Scheme V-5). Obviously, hydrogen substituents must be present in the olefin in order for the appropriate elimination to occur. This is always the case since for all practical purposes, only ethylene and monosubstituted olefins and a few very special (usually strained) types of disubstituted olefins will undergo the transition metal-catalyzed dimerization or polymerization reactions. Steric problems concern-

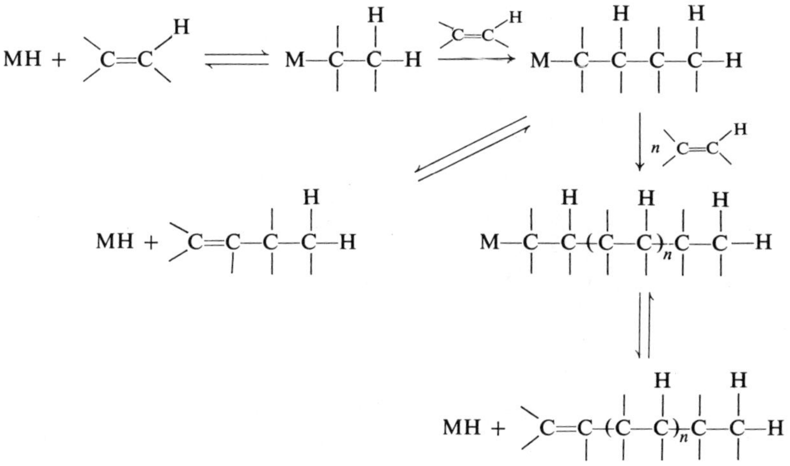

$$MH + \underset{/}{\overset{\backslash}{C}}=\underset{\backslash}{\overset{/}{C}}\overset{H}{} \rightleftharpoons M\!-\!\overset{\overset{H}{|}}{\underset{|}{C}}\!-\!\overset{\overset{H}{|}}{\underset{|}{C}}\!-\!H \xrightarrow{\overset{\backslash}{C}=\overset{/}{C}\overset{H}{}} M\!-\!\overset{\overset{H}{|}}{\underset{|}{C}}\!-\!\overset{\overset{H}{|}}{\underset{|}{C}}\!-\!\overset{\overset{H}{|}}{\underset{|}{C}}\!-\!\overset{\overset{H}{|}}{\underset{|}{C}}\!-\!H$$

$$MH + \underset{/}{\overset{\backslash}{C}}=C\!-\!\overset{\overset{H}{|}}{\underset{|}{C}}\!-\!\overset{|}{\underset{|}{C}}\!-\!H \qquad M\!-\!\overset{\overset{H}{|}}{\underset{|}{C}}\!-\!\overset{|}{\underset{|}{C}}\!\left(\overset{\overset{H}{|}}{\underset{|}{C}}\!-\!\overset{|}{\underset{|}{C}}\right)_{\!n}\!\overset{\overset{H}{|}}{\underset{|}{C}}\!-\!\overset{|}{\underset{|}{C}}\!-\!H$$

$$MH + \underset{/}{\overset{\backslash}{C}}=C\!-\!\!\left(\overset{\overset{H}{|}}{\underset{|}{C}}\!-\!\overset{|}{\underset{|}{C}}\right)_{\!n}\!\overset{\overset{H}{|}}{\underset{|}{C}}\!-\!\overset{|}{\underset{|}{C}}\!-\!H$$

Scheme V-5.

ing formation of the olefin π complexes probably are responsible for the difficulty of polymerizing the more highly substituted olefins. Since metal hydrides may add in both possible directions to unsymmetrical olefins, dimerizations or polymerizations may produce mixtures of isomeric products. The metal alkyl intermediates in the polymerizations also may add both ways to the next olefin unit and produce isomers. Metal hydride addition–elimination sequences likewise may occur and produce double-bond isomers as well. In spite of all these potential complications the reactions often turn out to be quite specific.

Olefin π-complexing ability is most likely an important factor in determining which olefins undergo the dimerization, polymerization, or, in fact, practically any other transition metal reaction. The other important factor is the rate of the shift of the reacting substituent on the metal to the π-complexed olefin; however, very little is known about this second factor. Some data are available on the stability of π complexes, however. An instructive series of equilibrium constants indicating the ability of an olefin to replace ethylene in acetylacetonatobisethylenerhodium(I) has been obtained (173, 174). The data are given in Table V-1.

$$AcacRh(CH_2{=}CH_2)_2 + olefin \overset{K}{\rightleftharpoons} AcacRh(olefin)(CH_2{=}CH_2) + CH_2{=}CH_2$$

Qualitatively the trends will probably turn out to be similar for other complexes with different ligands and with complexes of other transition metals. Clearly, increasing the number of alkyl substituents on ethylene strongly decreases the stability of the π complex formed, in line with the

TABLE V-1
Equilibrium Constants for the
Replacement of Ethylene in
Acetylacetonatobisethylenerhodium(I) by
Another Olefin[a]

Olefin or CO	K (at 25°)
CO	>100.0
$CHF{=}CF_2$	88.0
$CH_2{=}CHCN$	>50.0
cis-$CHF{=}CHF$	1.6
trans-$CHF{=}CHF$	1.2
($CH_2{=}CH_2$	1.0)
$CH_2{=}CHF$	0.32
$CH_2{=}CF_2$	0.1
$CH_2{=}CHCl$	0.17
$CH_2{=}CHCH_3$	0.08
$CH_2{=}CHCH_2CH_3$	0.09
$CH_2{=}CHOCH_3$	0.018
cis-$CH_3CH{=}CHCH_3$	0.004
trans-$CH_3CH{=}CHCH_3$	0.002
$CH_2{=}C(CH_3)_2$	0.00035
$CH_2{=}CCl_2$	No reaction
$Cl_2C{=}CCl_2$	No reaction

[a] From Cramer (173).

observed tendencies of the more substituted olefins not to polymerize or dimerize. Sterically small, strongly electronegative substituents, on the other hand, increase the stability of the π complexes possibly because removing electron density from the olefin allows stronger back bonding from the metal.

Catalysts containing Group VIII metals generally produce dimers, whereas transition metals from Groups IV, V, and VI often produce higher polymers reflecting a change in the relative rates of olefin addition compared with rates of metal hydride elimination. Changing ligands or oxidation state no doubt also influences these rates to some extent, but the major factor seems to be the metal itself.

1. Dimerization

The conversion of ethylene into butene with a rhodium catalyst is typical of the many olefin dimerization reactions known. The mechanism of this reaction has been investigated in some detail (175). Chlorobisethylenerhodium(I) dimer was the catalyst employed and it was used in the presence of at least two moles of hydrogen chloride for optimum activity. The bridged chloride dimer is believed to be converted into the mononuclear anion with chloride ion. This species then reacts with hydrogen chloride by oxidative addition, either forming a metal hydride initially or perhaps by direct protonation of the coordinated ethylene, to produce a solvated ethylrhodium complex. This complex apparently may reversibly lose coordinated ethylene. The ethylene-dissociated ethylrhodium(III) complex can be isolated as the monohydrated cesium salt. The ethylene coordinated form, however, is required for the addition step which occurs next to form an n-butylrhodium intermediate. A rhodium hydride elimination followed by reductive elimination of hydrogen chloride would give solvated dichloro-π-(1-butene)rhodium anion. A final replacement of 1-butene by ethylene would complete the catalytic cycle. The initially formed 1-butene isomerizes rapidly under the reaction conditions to 2-butene by a hydride addition–elimination sequence (Scheme V-6). Metal hydrides are not shown in the mechanism because they

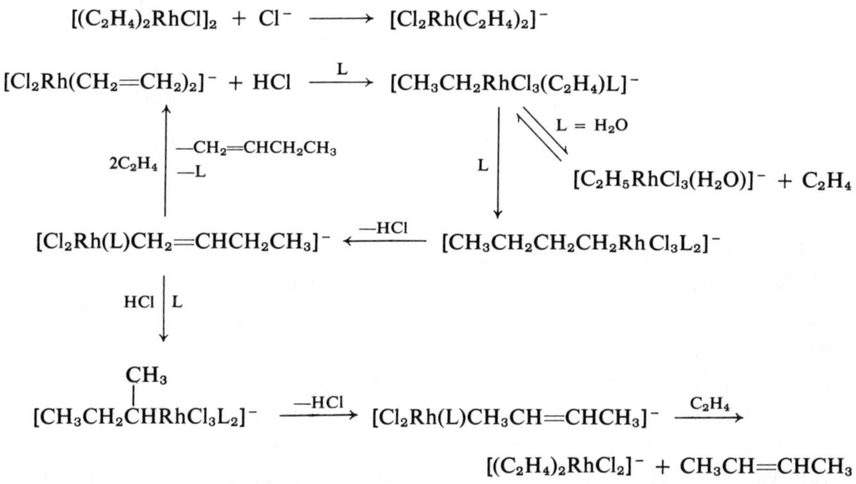

$$[(C_2H_4)_2RhCl]_2 \; + \; Cl^- \; \longrightarrow \; [Cl_2Rh(C_2H_4)_2]^-$$

$$[Cl_2Rh(CH_2{=}CH_2)_2]^- \; + \; HCl \; \xrightarrow{\;L\;} \; [CH_3CH_2RhCl_3(C_2H_4)L]^-$$

$$2C_2H_4 \Big\uparrow \begin{array}{c} -CH_2{=}CHCH_2CH_3 \\ -L \end{array}$$

$$L \Bigg\downarrow \quad \begin{array}{c} L = H_2O \\ [C_2H_5RhCl_3(H_2O)]^- \; + \; C_2H_4 \end{array}$$

$$[Cl_2Rh(L)CH_2{=}CHCH_2CH_3]^- \; \xleftarrow{\;-HCl\;} \; [CH_3CH_2CH_2CH_2RhCl_3L_2]^-$$

$$HCl \Big\downarrow L$$

$$\underset{[CH_3CH_2\overset{\displaystyle CH_3}{\overset{|}{C}HRhCl_3L_2]^-}}{} \; \xrightarrow{\;-HCl\;} \; [Cl_2Rh(L)CH_3CH{=}CHCH_3]^- \; \xrightarrow{\;C_2H_4\;}$$

$$[(C_2H_4)_2RhCl_2]^- \; + \; CH_3CH{=}CHCH_3$$

Scheme V-6.

were not detected, but nevertheless were very probably involved in the addition and elimination steps.

Olefin dimerization is quite a widely applicable reaction. Even olefins containing some functional groups can be dimerized. Reactions of potential commercial interest are the dimerization of acrylonitrile and methyl acrylate to difunctional products (176, 177).

$$CH_2{=}CHCOOCH_3 \; \xrightarrow[CH_3OH]{RhCl_3} \; CH_3OCOCH{=}CHCH_2CH_2COOCH_3$$

With monosubstituted olefins the possibility of producing mixtures of isomeric dimers exists. A study of the propylene dimerization reaction catalyzed by a nickel complex has yielded useful information about the way steric effects influence the isomeric composition of the dimer mixtures formed. The catalyst employed was a bromotriorganophosphine-π-allylnickel(II) complex in combination with aluminum bromide. The aluminum bromide presumably reacts with the nickel complex by coordinating with an unshared pair of electrons on the bromine atom. The complexing either loosens or completely removes that ligand so that a position becomes available for reactants to enter the coordination sphere of the nickel. A close analogy is the reaction of bromodicarbonyl-π-cyclopentadienyliron(I) with aluminum bromide and ethylene. The cationic π-ethylene complex can be isolated from the reaction mixture as the hexafluorophosphate salt (178).

In the nickel-catalyzed dimerization of propylene, the isomeric composition of the dimers was found to be quite dependent on the triorganophosphine ligand in the catalyst. The major effects were apparently steric in origin (179). The true catalyst in the dimerization is most likely not the π-allyl complex but a hydridotriorganophosphine-π-propylenenickel(II) tetrabromoaluminate salt formed by reaction of the π-allyl group with propylene, probably forming 1,4-hexadiene, the 1,5-isomer, or 2-methyl-1,4-pentadiene. The hydride catalyst can then undergo addition to the coordinated propylene group in either direction giving an *n*-propyl- or an isopropylnickel derivative. Coordination of either of these complexes with another propylene molecule followed by a second addition, again in either of two possible directions, would lead to four isomeric hexylnickel derivatives. β-Elimination of nickel hydride in either of two directions with two of the alkyls and in the only way possible with the other two and dissociation would give a total of six possible isomeric, six-carbon olefins excluding cis-trans isomers. The reactions thought to be involved are summarized in Scheme V-7. Some minor olefin isomerization apparently occurs also under the reaction conditions. From an analysis of the various olefins formed the favored reaction paths can be determined. The products obtained depend strongly on the size of the triorganophosphine group in the complex. With triphenylphosphine as the ligand 74% of the product consists of 2-methylpentenes, 22% n-hexenes, and 4% 2,3-dimethyl-butene, whereas if di-t-butylethylphosphine is the ligand the yields are 22, 1, and 77%, respectively. An analysis of the isomeric olefins produced indicated a preference for the initial nickel hydride addition to put nickel on the internal carbon forming the isopropylnickel intermediate no matter which phosphine ligand was present. The relatively small size of the hydride group apparently keeps steric effects minor in this step. The steric effect appears mainly in the addition of the isopropylnickel group to the propylene. The sterically largest ligand, di-t-butylethylphosphine, causes the nickel to go mainly to the terminal carbon and isopropyl to the second carbon of the propylene. This direction of addition gives a primary nickel alkyl which is probably less sterically crowded at the nickel atom than it would be if the addition had occurred in the reverse direction as is preferred if the phosphine ligand is small. Presumably such steric effects will prove to be general, but few other similar examples have been studied.

This transition metal-catalyzed dimerization contrasts with the ionic dimerization caused by alkali metals. Propylene is dimerized at 150° by a potassium graphite catalyst, KC_8, in more than 80% yield to 2-methyl-4-pentene, an isomer not seen in the nickel hydride dimerization. The potassium graphite catalyst probably reacts with the propylene, first forming allyl-potassium. Addition of allylpotassium to propylene to produce the more stable primary carbanion would give 2-methyl-4-pentenylpotassium. Since the latter compound would be a stronger base than allylpotassium, hydrogen transfer from propylene occurs forming 2-methyl-4-pentene and allylpotas-sium and a new cycle can begin (180).

$$CH_3CH{=}CH_2 + 2K \rightleftharpoons KH + K^+C^-H_2CH{=}CH_2.$$

$$K^+C^-H_2CH{=}CH_2 + CH_2{=}CHCH_3 \longrightarrow CH_2{=}CHCH_2\overset{\overset{\displaystyle CH_3}{|}}{C}HCH_2{}^-K^+$$

$$CH_2{=}CHCH_2\overset{\overset{\displaystyle CH_3}{|}}{C}HCH_2{}^-K^+ + CH_3CH{=}CH_2 \longrightarrow$$

$$CH_2{=}CHCH_2\overset{\overset{\displaystyle CH_3}{|}}{C}HCH_3 + K^+C^-H_2CH{=}CH_2$$

2. Oligomerization

The basic mechanisms of addition and elimination are the same whether dimers, oligomers, or polymers are formed, but the structures of the products may vary depending on the catalyst employed. The differences arise from differences in the direction of addition of the metal hydrides and/or alkyls

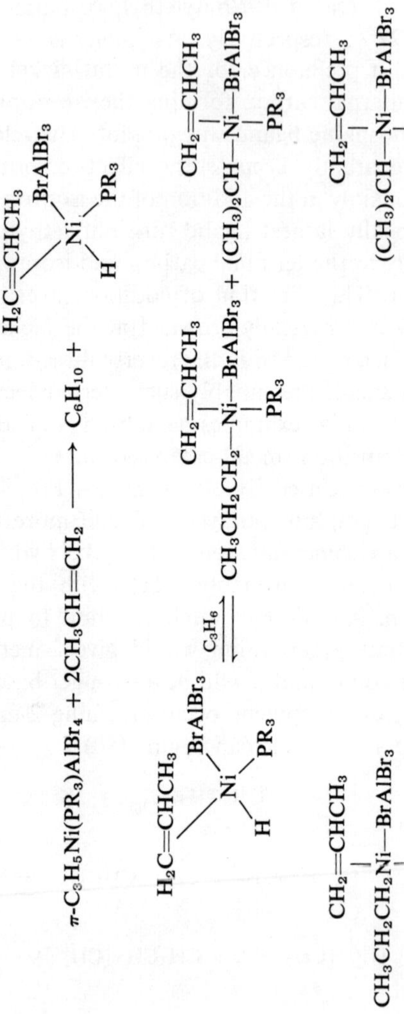

$$\pi\text{-}C_8H_5Ni(PR_3)AlBr_4 + 2CH_3CH{=}CH_2 \longrightarrow C_6H_{10} +$$

$$H_2C{=}CHCH_3 \quad BrAlBr_3$$
$$\underset{\underset{H}{|}}{Ni}$$
$$PR_3$$

$$\begin{array}{c} H_2C{=}CHCH_3 \quad BrAlBr_3 \\ \backslash\;/ \\ Ni \\ /\;\backslash \\ H \qquad PR_3 \end{array} \underset{C_3H_6}{\rightleftharpoons}$$

$$\begin{array}{c} CH_2{=}CHCH_3 \\ | \\ CH_3CH_2CH_2{-}Ni{-}BrAlBr_3 \\ | \\ PR_3 \end{array}$$

$$\begin{array}{c} CH_2{=}CHCH_3 \\ | \\ CH_3CH_2CH_2{-}Ni{-}BrAlBr_3 + (CH_3)_2CH{-}Ni{-}BrAlBr_3 \\ | \qquad\qquad\qquad\qquad\qquad | \\ PR_3 \qquad\qquad\qquad\qquad\qquad PR_3 \end{array}$$

$$\begin{array}{c} CH_2{=}CHCH_3 \\ | \\ (CH_3)_2CH{-}Ni{-}BrAlBr_3 \\ | \\ PR_3 \end{array}$$

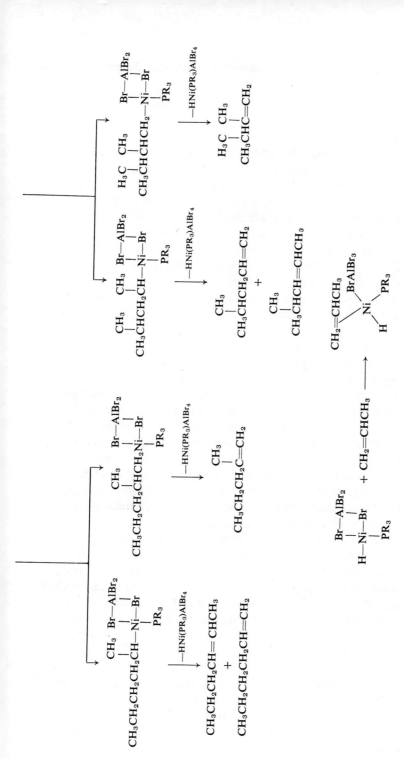

Scheme V-7.

in the reactions. Ethylene, for example, is converted into a mixture of 14% 1-butene, 30% 2-ethyl-1-butene, 17% 2-ethyl-1-hexene, and higher, presumably, 2-ethyl-1-olefins with even numbers of carbons, at −70° by a catalyst prepared by reacting titanium tetrachloride with dimethylaluminum chloride dimer (181). The products likely to be formed from titanium tetrachloride and dimethylaluminum chloride are coordinately unsaturated and therefore highly reactive. The catalyst is likely formed in a series of reactions. An exchange of methyl for chlorine and abstraction of a second chlorine by the methylaluminum dichloride formed would produce a methyltitanium methyltrichloroaluminate complex. This complex could π-complex with ethylene, add the methyl to the ethylene, and eliminate hydride, producing propylene in an amount equal to the catalyst concentration and form, by coordinating with another ethylene, the likely true catalyst, hydridodichloro-π-ethylene-titanium(IV) methyltrichloroaluminate. Whether the methyltrichloroaluminate group remains weakly coordinated to the titanium during the reactions is not known, but if not it is certainly close enough to enter the coordination sphere anytime a vacancy appears and thereby stabilize the reaction intermediates. As shown in the nickel-catalyzed dimerization above, the haloaluminate anions can probably function either as mono- or bidentate ligands. In the titanium-catalyzed oligomerization the hydride catalyst can easily shift hydrogen and form the ethyltitanium complex. An addition of this complex to ethylene followed by a β-hydride elimination and replacement of 1-butene by ethylene would regenerate the catalyst by a mechanism identical with that of the nickel and rhodium examples already described. The tendency for titanium hydride elimination to occur relative to ethylene insertion must be less than with nickel and rhodium since some 1-hexene is formed and some is used in subsequent reactions with the titanium catalyst. The major difference between these catalysts is apparently in their relative reactivity toward 1-olefins versus ethylene. The rhodium and nickel catalysts show very little tendency to form higher olefins by further reactions with the initially formed dimers, whereas a large fraction of the titanium product arises by this route. Clearly, the preferred direction of addition of the ethyltitanium complex to 1-olefins is to place titanium on the terminal position and the ethyl group on the second carbon. Elimination of hydride would then give the observed 2-ethyl-1-olefins. Olefin isomerization is relatively minor under the reaction conditions with the titanium catalyst (see Scheme V-8). A possible alternative mechanism for forming branched chain olefins would be for titanium hydride addition to occur with the 1-olefins, bonding titanium to the second carbon and then inserting ethylene. This reaction, however, would produce 3-methyl-1-olefins and these are not observed.

The dimerization of propylene with the titanium catalyst has not been reported, but 1-butene reacts as would be predicted from the results obtained

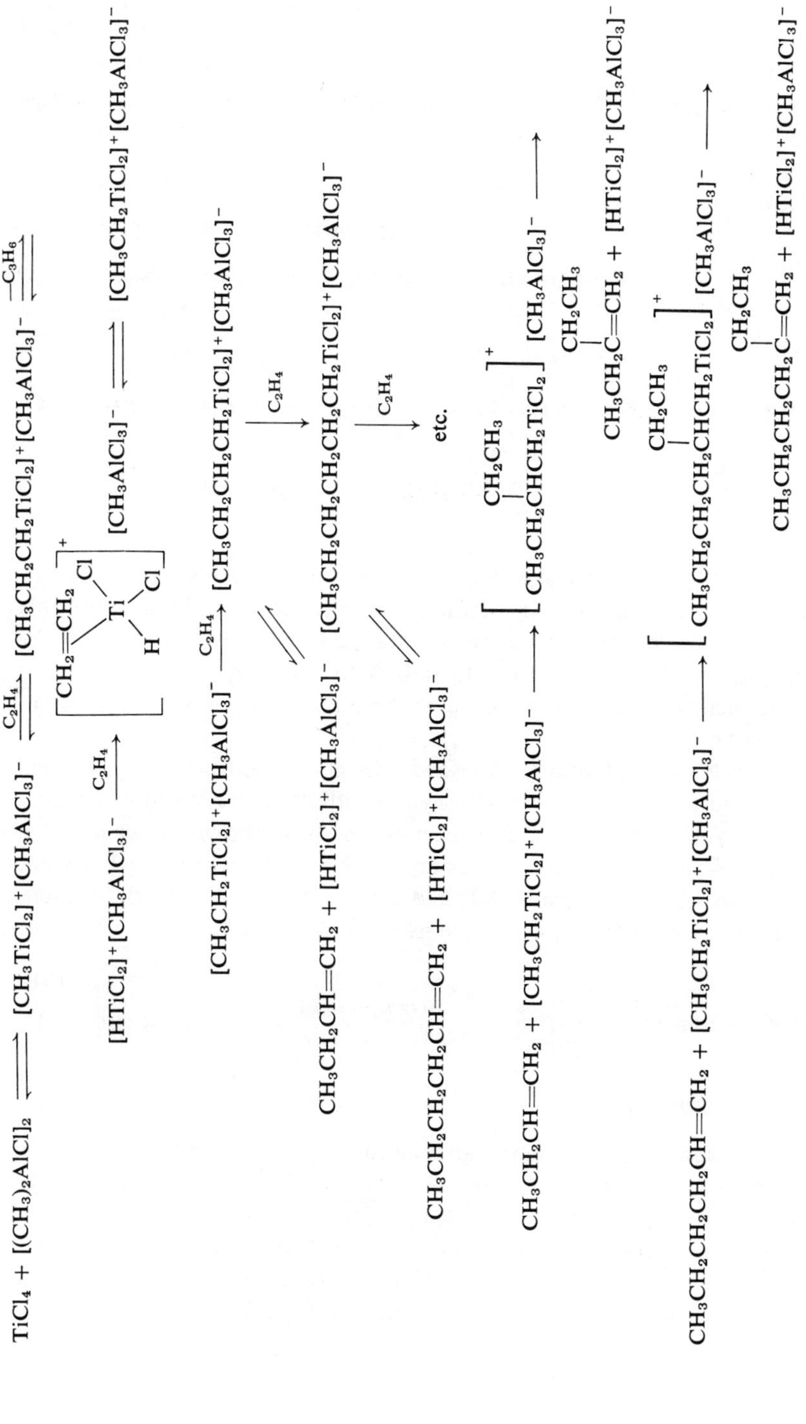

$$TiCl_4 + [(CH_3)_2AlCl]_2 \rightleftharpoons [CH_3TiCl_2]^+[CH_3AlCl_3]^- \xrightarrow{C_2H_4} [CH_3CH_2CH_2TiCl_2]^+[CH_3AlCl_3]^- \xrightarrow{-C_3H_6}$$

$$[HTiCl_2]^+[CH_3AlCl_3]^- \xrightarrow{C_2H_4} \left[\begin{array}{c} CH_2{=}CH_2 \ \ Cl \\ Ti \\ H \ \ Cl \end{array} \right]^+ [CH_3AlCl_3]^- \rightleftharpoons [CH_3CH_2TiCl_2]^+[CH_3AlCl_3]^-$$

$$[CH_3CH_2TiCl_2]^+[CH_3AlCl_3]^- \xrightarrow{C_2H_4} [CH_3CH_2CH_2CH_2TiCl_2]^+[CH_3AlCl_3]^-$$

$$\rightleftharpoons CH_3CH_2CH{=}CH_2 + [HTiCl_2]^+[CH_3AlCl_3]^-$$

$$[CH_3CH_2CH_2CH_2TiCl_2]^+[CH_3AlCl_3]^- \xrightarrow{C_2H_4} [CH_3CH_2CH_2CH_2CH_2CH_2TiCl_2]^+[CH_3AlCl_3]^-$$

$$\rightleftharpoons CH_3CH_2CH_2CH_2CH{=}CH_2 + [HTiCl_2]^+[CH_3AlCl_3]^-$$

etc.

$$\rightarrow \left[\begin{array}{c} CH_2CH_3 \\ | \\ CH_3CH_2CHCH_2TiCl_2 \end{array} \right]^+ [CH_3AlCl_3]^-$$

$$\longrightarrow CH_3CH_2C{=}CH_2 + [HTiCl_2]^+[CH_3AlCl_3]^-$$
$$\qquad\qquad | \\ \qquad\qquad CH_2CH_3$$

$$\rightarrow \left[\begin{array}{c} CH_2CH_3 \\ | \\ CH_3CH_2CH_2CH_2CHCH_2TiCl_2 \end{array} \right]^+ [CH_3AlCl_3]^-$$

$$\longrightarrow CH_3CH_2CH_2CH_2C{=}CH_2 + [HTiCl_2]^+[CH_3AlCl_3]^-$$
$$\qquad\qquad\qquad\qquad | \\ \qquad\qquad\qquad\qquad CH_2CH_3$$

Scheme V-8.

with ethylene. The products are mainly "head-to-tail" dimers and trimers (alkyltitanium adding to place titanium on the terminal position of the double bond) with smaller amounts of head-to-head products being formed. Both the 2-alkyl-1-olefins and internal olefins produced in these oligomerizations appear to be stable to further reaction under the reaction conditions.

$$CH_3CH_2CH=CH_2 \xrightarrow[-70°]{[HTiCl_2]^+[CH_3AlCl_3]^-}$$

$$\underset{37\%}{\overset{\overset{\displaystyle CH_2CH_3}{\displaystyle |}}{CH_3CH_2CH_2CH_2C=CH_2}} + \left\{ \begin{array}{l} CH_3CH_2CH_2CH_2CH=CHCH_2CH_3 \\ + \\ CH_3CH_2CH_2CH_2CH_2CH=CHCH_3 \end{array} \right.$$
$$ 3\%$$

$$+ \underset{13\%}{\overset{\overset{\displaystyle CH_2CH_3CH_2CH_3}{\displaystyle |}}{CH_3CH_2CH_2CH_2CHCH_2C=CH_2}} + \left\{ \begin{array}{l} \overset{\overset{\displaystyle CH_2CH_3}{\displaystyle |}}{CH_3CH_2CH_2CH_2CHCH_2CH_2CH=CHCH_3} + \text{etc.} \\ + \overset{\displaystyle CH_2CH_3}{\displaystyle |} \\ CH_3CH_2CH_2CH_2CHCH_2CH=CHCH_2CH_3 \end{array} \right.$$
$$ 6\%$$

Thus, the titanium catalyst would be expected to produce mainly 2-methyl-1-pentene and 1- and 2-hexene as dimers from propylene. These were only minor products in the above nickel-catalyzed reactions with either the sterically small or large phosphine ligands. Why the titanium hydride prefers to add to olefins in the reverse manner to the nickel hydride in these reactions is not known.

The results in the titanium-catalyzed ethylene dimerization suggest that reaction of any terminal olefin with ethylene with this catalyst should produce the 2-ethyl-substituted olefin. The reaction has been shown to occur and it provides a useful synthetic route to 2-ethyl-1-olefins. In one example 1-pentene was reacted with ethylene at $-80°$ and a 91.5% yield of 2-ethyl-1-pentene was formed along with 8% n-heptene isomers (182).

$$CH_3CH_2CH_2CH=CH_2 + CH_2=CH_2 \xrightarrow[-80°]{[HTiCl_2]^+[CH_3AlCl_3]^-} \underset{}{\overset{\overset{\displaystyle CH_2CH_3}{\displaystyle |}}{CH_3CH_2CH_2C=CH_2}}$$

3. Polymerization

The conversion of olefins into high molecular weight polymers has received a tremendous amount of attention in recent years because of the value of the polymers for producing low cost injection molded products, films, and fibers. The discovery was made in the laboratories of K. Ziegler in the early 1950's that ethylene could be converted into a solid, high molecular weight polymer at low pressure and ordinary temperatures with a combination of a

Group IV, V, or VI transition element compound and triethylaluminum. Since that time practically every major chemical company in the world has carried out research on the reaction. It is probably safe to say that no other reaction in the history of chemistry has been so much studied as this one has. The vast majority of catalysts investigated were heterogeneous. Despite considerable effort, however, little has been learned about the reaction mechanism with the heterogeneous catalysts. A study of a soluble catalyst prepared from dichlorobis-π-cyclopentadienyltitanium(IV) and diethylaluminum chloride dimer has been much more successful in this respect and a reasonable mechanism has been proposed (53). The mechanism is very similar to the subsequently discovered titanium-catalyzed oligomerization discussed above. The change in ligands, however, has made the titanium hydride elimination much less favorable with respect to the ethylene insertion reaction. The higher number of electrons available from the cyclopentadienyl ligands probably makes the titanium alkyls less likely to undergo β-hydride elimination in order to obtain (share) the two additional electrons from the olefin that would be formed. Unfortunately, this cyclopentadienyltitanium catalyst

$$Cp_2TiCl_2 + [(C_2H_5)_2AlCl]_2 \longrightarrow [Cp_2TiCH_2CH_3]^+[C_2H_5AlCl_3]^-$$

$$[Cp_2TiCH_2CH_3]^+[C_2H_5AlCl_3]^- + nCH_2{=}CH_2 \longrightarrow$$
$$[Cp_2Ti(CH_2CH_2)_nCH_2CH_3]^+[C_2H_5AlCl_3]^-$$

$$[Cp_2Ti(CH_2CH_2)_nCH_2CH_3]^+[C_2H_5AlCl_3]^- \longrightarrow$$
$$[Cp_2TiH]^+[C_2H_5AlCl_3]^- + CH_2{=}CH(CH_2CH_2)_{n-1}CH_2CH_3$$

$$[Cp_2TiH]^+[C_2H_5AlCl_3]^- + CH_2{=}CH_2 \longrightarrow [Cp_2TiCH_2CH_3]^+[C_2H_5AlCl_3]^-, \text{ etc.}$$

is only able to polymerize ethylene; many heterogeneous catalysts will also polymerize a wide variety of other terminal olefins.

The selectivity of some heterogeneous catalysts is remarkable since the asymmetric carbon produced in the polymerization of terminal olefins can be formed with specific stereochemistry with respect to the previous asymmetric carbons in the polymer chain. If the configuration about the asymmetric carbons alternate, the polymer is said to be syndiotactic; if they are all the same in at least long segments of the chain it is called isotactic. In either case no net optical activity results because in the whole polymer the same number of d and l carbons are formed. The isotactic specificity is probably caused by ligands forming an asymmetric catalyst, but it is not clear how syndiotactic structures arise.

4. Cyclic Dimerizations

Cyclobutane derivatives have been formed from olefins in a few instances with transition metal catalysts. The mechanism appears to involve formation of a bisolefin π complex, ligand coupling to form a metalacyclopentane, and then reductive elimination of the cyclobutane. The conversion of nor-bornadiene into the cyclic exo-trans-exo dimer is an example of the reaction. The proposed intermediates have not been detected (183). Monocoordination

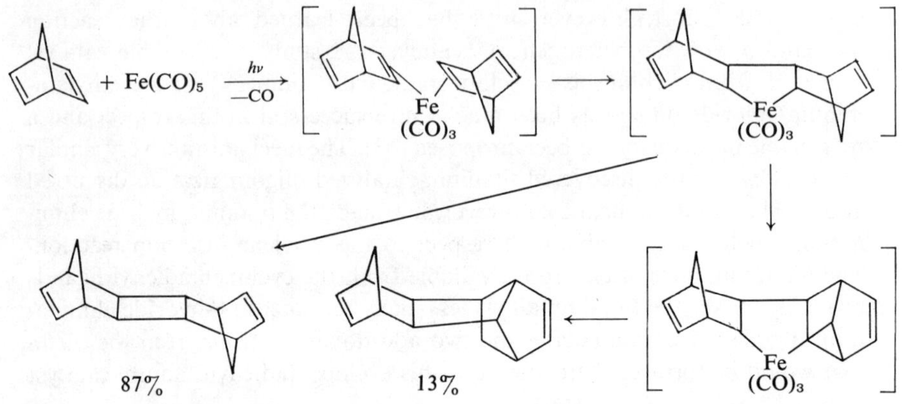

apparently occurs on the less hindered, exo side of the norbornadiene ring since exo products are formed. A small amount of a rearranged isomer is also produced perhaps by way of a Wagner–Meerwein-type rearrangement in the probable iron(II) intermediate.

An iridacyclopentane derivative of undetermined molecular weight related to the proposed intermediate in the iron carbonyl-catalyzed norbornadiene dimerization has been isolated. On reaction with excess triphenylphosphine the *exo-trans-exo*-norbornadiene dimer is obtained (184).

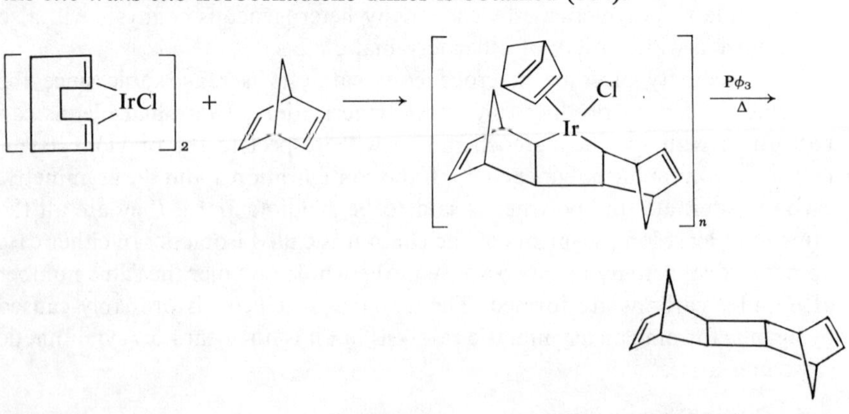

Similar cyclic dimerizations have been observed with some highly fluorinated olefins. In the case of tetrafluoroethylene the intermediate ferraoctafluoro-cyclopentane derivative can be isolated. Heating this complex to 160° causes reductive elimination to the cyclobutane compound, but two fluorine atoms are also lost, no doubt to help satisfy the coordinately unsaturated iron carbonyl fragment left in the pyrolysis (185, 186).

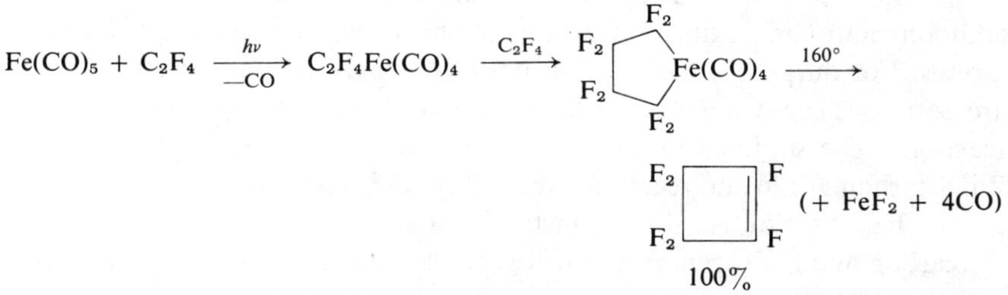

Another kind of cyclic dimerization can take place when two double bonds occur in the same molecule. Heating diallyl ether with methanolic rhodium(III) chloride, for example, produces 3-methylene-4-methyltetrahydrofuran (187).

$$RhCl_3 + CH_2{=}CHCH_2OCH_2CH{=}CH_2 + CH_3OH \longrightarrow$$
$$[(CH_2{=}CHCH_2OCH_2CH{=}CH_2)RhCl]_2 + HCl + H_2C(OCH_3)_2$$

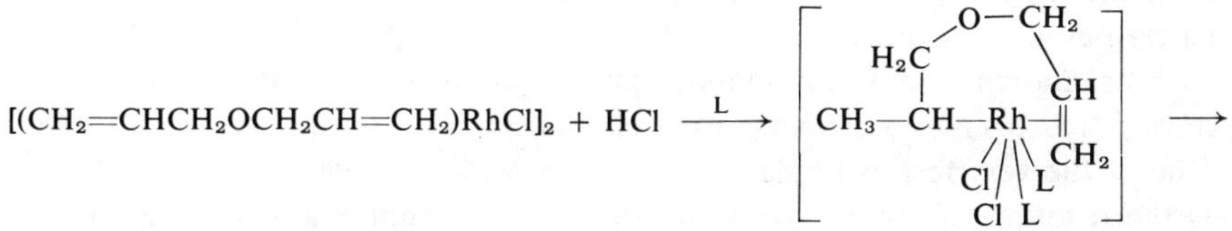

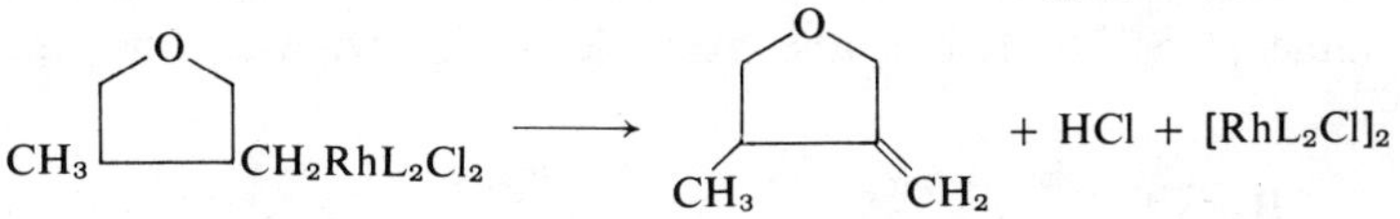

$$[RhL_2Cl]_2 + CH_2{=}CHCH_2OCH_2CH{=}CH_2 \xrightarrow{-L}$$
$$[[CH_2{=}CHCH_2OCH_2CH{=}CH_2]RhCl]_2$$

D. VINYLIC AND ALLYLIC SUBSTITUTION REACTIONS

The net result of the addition of a metal compound to an olefin followed by elimination of a different metal compound from the adduct is a vinylic substitution, or an allylic substitution with a double-bond shift. The reactions can be summarized as follows.

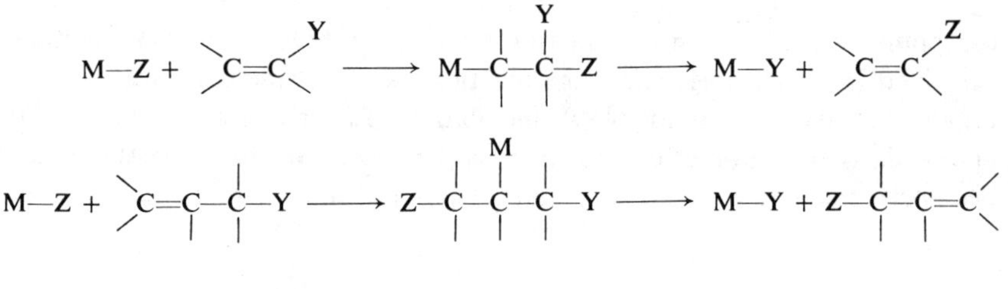

Numerous different Z and Y groups can be used in the reaction, the scope
of which is still being explored. If Z and Y are both hydrogen the reaction is
simply the olefin isomerization already discussed. Both ionic and covalent
additions and eliminations have been observed depending on the Z and Y
groups. For purposes of discussion both the vinylic and allylic substitution
are best considered together since many compounds undergo both types of
reactions. The subject can be conveniently separated into reactions where
Z is a carbon group and reactions where Z is other than a carbon or hydrogen
group. This division is, in most instances, a separation by mechanism since
the carbon and hydrogen groups will generally add by a covalent mechanism
and the other groups by ionic processes.

1. Substitution by Carbon Groups

The substitution of vinylic or allylic hydrogen by an organic group (Z = an
organic group and Y = H) is a useful synthetic reaction. The intermediate
metal alkyl adducts are generally very unstable and usually cannot be detected.
Even the initial organometallic reagents are often made in the reaction mixture
since they may be unstable as well. The best examples of this reaction are
with organopalladium compounds. The organometallic reagent should not
have β-hydrogen substituents; then elimination will occur before addition.
One of the very few examples which begins with an isolable organometallic
is the reaction of styrene with chloro-o-dimethylaminomethylpalladium(II)
dimer (188). Numerous other examples are known where the organopalladium

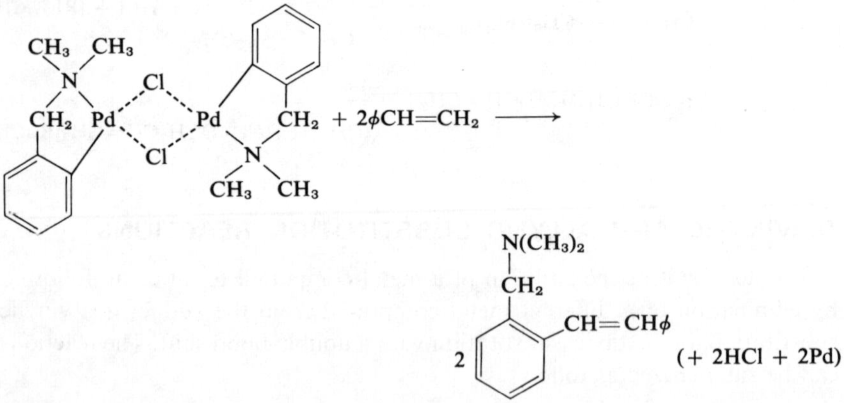

compound is prepared in the reaction mixture by three of the usual methods
described in Chapter II: exchange reactions with organolead, organotin, or
organomercury compounds (189), metalation of aromatic compounds (190),
and oxidative addition of organic halides (191, 192). Mechanistic studies have
been performed under mild conditions where product isomerization does not

occur, using the exchange reaction of an arylmercuric acetate with palladium acetate to prepare the organometallic reagent in the presence of the reacting olefin. Since the more substituted an olefin is the slower it usually reacts, the reaction with excess olefin stops cleanly with only monosubstitution. Table V-2 gives the products of reaction of propylene with p-anisyl-, phenyl-, and p-carbomethoxyphenylpalladium acetates (193). The ligands in the organo-palladium acetate complexes are not known with certainty since these materials cannot be isolated, but probably the complexes are four-coordinate with two solvent molecules completing the coordination sphere. The data in Table V-2 show that changes in the electronic effects in the organopalladium

TABLE V-2
Reactions of Various "Arylpalladium Salts" with Propylene[a]

Aryl group	Total yield of product	trans- 1-Aryl- 1-propene	cis- 1-Aryl- 1-propene	1-Aryl- 2-propene	2-Aryl- 2-propene
p-CH$_3$OC$_6$H$_4$	74	50	3	4	43
C$_6$H$_5$	100	57	5	12	26
p-CH$_3$OCOC$_6$H$_4$	95	58	—	18	24

[a] Reactions were carried out at 30° with 30 psi of propylene in acetonitrile solvent for 1 hr. All data given as percent. From ref. 193.

reagent have a relatively minor influence on the direction of addition of the reagent to propylene. The following equations summarize the reactions involved. Probable intermediate π complexes are not shown.

$$\text{ArHgOAc} + \text{Pd(OAc)}_2 \xrightarrow{\text{L}} [\text{ArPdL}_2\text{OAc}] + \text{Hg(OAc)}_2$$

$$[\text{ArPdL}_2\text{OAc}] + \text{CH}_2\!\!=\!\!\text{CHCH}_3 \longrightarrow \left[\begin{array}{c}\text{CH}_3\\|\\\text{ArCH}_2\text{CPdL}_2\text{OAc}\\|\\\text{H}\end{array}\right] + \left[\begin{array}{c}\text{CH}_3\\|\\\text{AcOPdCH}_2\text{CAr}\\|\\\text{H}\end{array}\right]$$

$$\left[\begin{array}{c}\text{CH}_3\\|\\\text{ArCH}_2\text{CPdL}_2\text{OAc}\\|\\\text{H}\end{array}\right] \xrightarrow{-[\text{HPdL}_2\text{OAc}]} \begin{array}{c}\text{H}\quad\text{CH}_3\\\text{C}\!\!=\!\!\text{C}\\\text{Ar}\quad\text{H}\end{array} + \begin{array}{c}\text{Ar}\quad\text{CH}_3\\\text{C}\!\!=\!\!\text{C}\\\text{H}\quad\text{H}\end{array} + \text{ArCH}_2\text{CH}\!\!=\!\!\text{CH}_2$$

$$\left[\begin{array}{c}\text{CH}_3\\|\\\text{AcOPdCH}_2\text{CAr}\\|\\\text{H}\end{array}\right] \xrightarrow{-[\text{HPdL}_2\text{OAc}]} \begin{array}{c}\text{CH}_3\\|\\\text{CH}_2\!\!=\!\!\text{CAr}\end{array}$$

The preferred direction of addition in all examples is for the palladium to add to the more substituted carbon of the double bond with the organic group going to the less substituted carbon. The change from the strongly electron-supplying *p*-methoxyphenyl group to phenyl, to the electron-withdrawing *p*-carbomethoxyphenyl causes a decrease in the amount of addition of the organic group to the more substituted position of the double bond, from 43 to 26 to 24%. The direction of the change is the expected one based on the idea that withdrawing electrons from the ring would make the aryl group less anionic and add more easily to the more negative terminal carbon of the double bond. Since the electronic effects do not appear to be controlling the direction of addition, steric effects probably are. The organic group behaves as the sterically larger unit and the metal as the smaller. The metal atom, of course, is larger than the carbon of the organic group, but the carbon–palladium bond length of about 2 Å is considerably longer than the carbon–carbon bond length of 1.5 Å and the square-planar geometry about the palladium also would decrease its effective steric size. When electron-withdrawing substituents are present on the olefinic double bond, the organo-palladium addition occurs exclusively to place palladium on the carbon atom alpha to the substituent (193).

The stereochemistry of the addition of the organopalladium reagent to an olefin has been determined from a study of the products obtained from the reaction of phenylpalladium acetate, generated *in situ* by exchange, with *trans*- and *cis*-1-phenyl-1-propene (194). The reaction is quite specific at 25°, but it is not at 100° (192). The hydridopalladium–π-olefin complexes produced in the last stage of the reaction at 100° apparently are involved in the partial equilibration of isomers by the usual internal addition–elimination sequences. The reaction products at 25° are mainly the isomeric 1,2-diphenyl-1-propenes expected from a cis addition of the phenylpalladium acetate followed by a cis elimination of hydridopalladium acetate. The *cis*-1-phenyl-1-propene also produces a minor amount of isomerized olefins. The interrelationships between the intermediates and products are shown in Scheme V-9.

The direction of elimination observed with various alkylpalladium acetates is influenced by (1) the tendency to give the least substituted olefin π complex, (2) the preferred rotational isomer formed since the *cis*-hydrogen closest to the palladium is most likely to be lost, and (3) the rate of the elimination relative to rotation about the single bond formed from the double bond in the addition. Rotation about that single bond must occur before a *cis*-vinylic hydrogen can be available for loss with the palladium group. Another factor which may be important in specific cases is a chelation effect. If, in the adduct, a substituent is favorably placed so that it can coordinate with the palladium to form a five-membered chelating ring, then coordination probably would direct the hydride elimination exocyclically with respect to the ring since

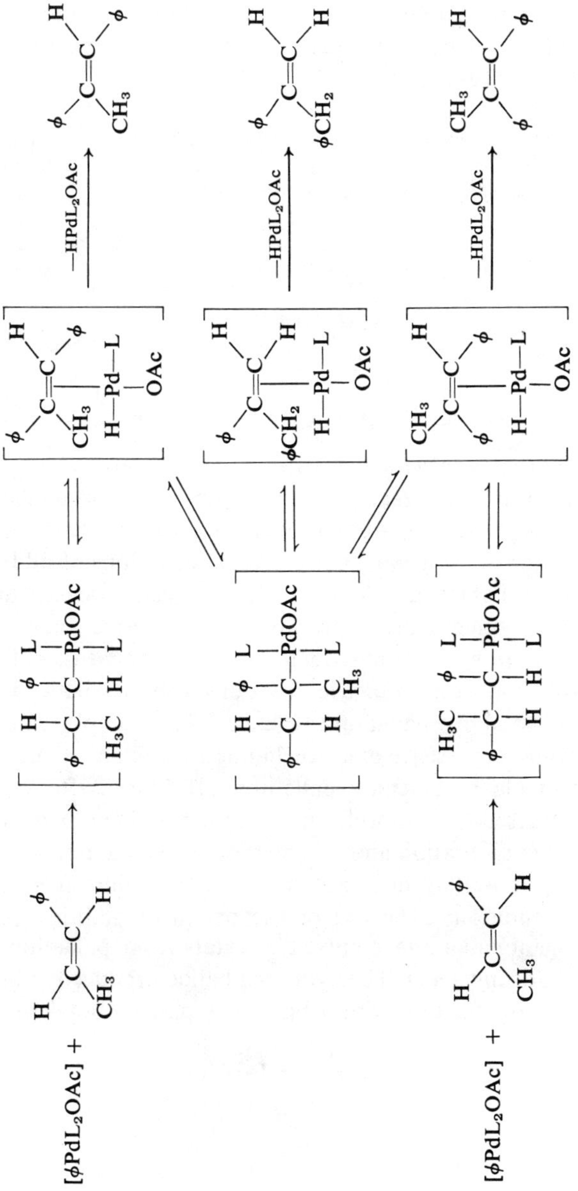

Scheme V-9.

an endocyclic elimination would produce a more strained, π-complexed product. An example may be the addition of carbomethoxypalladium acetate, prepared by an exchange reaction of carbomethoxymercuric acetate with palladium acetate, to 1-pentene where almost 60% of the product is the nonconjugated ester (194).

$$CH_3OCOHgOAc + Pd(OAc)_2 \xrightarrow{L} [CH_3OCOPdL_2OAc] + Hg(OAc)_2$$

$$[CH_3OCOPdL_2OAc] + CH_2{=}CH(CH_2)_2CH_3 \longrightarrow$$

$$\left[\begin{array}{c} \overset{\displaystyle OAc}{\underset{\displaystyle}{|}} \\ O{\sim}Pd{-}L \\ CH_3OC{\diagdown}{\diagup}CHCH_2CH_2CH_3 \\ CH_2 \end{array} \right] \xrightarrow{-HPdL_2OAc}$$

$$CH_3OCOCH{=}CHCH_2CH_2CH_3 + CH_3OCOCH_2CH{=}CHCH_2CH_3$$

Other transition metal organic derivatives such as those of ruthenium and rhodium will undergo similar reactions with olefins. Probably all the transition metals could be used under some conditions. Stereochemistry would probably remain the same for other metals, but preferred directions of addition could change. On the more practical side, the use of rare and expensive metals such as palladium stoichiometrically is prohibitively expensive for any reactions on larger than laboratory scale unless metal recovery is carried out. Fortunately many precious metal stoichiometric reactions can be made catalytic by various means. The above palladium reactions, for example, can be carried out catalytically in the presence of a stoichiometric amount of cupric chloride which reoxidizes palladium metal to palladium(II) chloride after each reaction cycle (189). The presence of chloride anion in the reaction mixtures, however, does cause olefin equilibration and this method of reoxidation is not suitable in all cases. Cupric acetate does not reoxidize palladium metal under the usual reaction conditions. The use of the oxidative addition reactions to produce the organopalladium complex is catalytic in palladium since the halide is the reoxidizing agent. The hydrogen halide produced in this reaction must be removed by reaction with a base such as tri-n-butylamine because

$$ArI + Pd \xrightarrow{L} [ArPdL_2I]$$

$$[ArPdL_2I] + \overset{}{\underset{H}{\diagup}}C{=}C\overset{}{\underset{}{\diagdown}} \longrightarrow \overset{}{\underset{Ar}{\diagdown}}C{=}C\overset{}{\underset{}{\diagup}} + [HPdL_2I]$$

$$[HPdL_2I] \longrightarrow HI + 2L + Pd$$

$$HI + n\text{-}Bu_3N \longrightarrow n\text{-}Bu_3NH^+I^-$$

at $100°$ the temperature necessary to effect the oxidative addition, the hydrogen halide reacts with the intermediate organopalladium compounds to produce hydrocarbon and a palladium halide if it is not neutralized (191, 192).

2. Substitution of Hydrogen by Groups Not Bonding through Carbon

Several inorganic palladium compounds will react with olefins by the addition–elimination sequence producing substituted olefins. Palladium acetate in acetic acid is typical. It converts olefins into enol or allylic acetates. 1-Butene reacts to give 80% 2-acetoxy-1-butene, 9% crotyl acetate, and 9% 1-acetoxy-1-butene (94d). The addition probably occurs by external attack of

$$CH_3CH_2CH{=}CH_2 + Pd(OAc)_2 \longrightarrow$$

$$\left[\begin{array}{c} OAc \\ | \\ CH_3CH_2CHCH_2PdL_2OAc \end{array} \right] + \left[\begin{array}{c} PdL_2OAc \\ | \\ CH_3CH_2CHCH_2OAc \end{array} \right]$$

$$\left[\begin{array}{c} OAc \\ | \\ CH_3CH_2CHCH_2PdL_2OAc \end{array} \right] \longrightarrow \begin{array}{c} OAc \\ | \\ CH_3CH_2C{=}CH_2 \end{array} + HPdL_2OAc$$

$$\left[\begin{array}{c} PdL_2OAc \\ | \\ CH_3CH_2CHCH_2OAc \end{array} \right] \longrightarrow$$

$$CH_3CH{=}CHCH_2OAc + CH_3CH_2CH{=}CHOAc + HPdL_2OAc$$

(20:80, *cis : trans*)

acetic acid on the palladium acetate π complex (195). The addition mainly places the metal on the less substituted carbon of the double bond, the opposite of what is observed in the organopalladium complex addition. In 70% dimethyl sulfoxide–30% acetic acid, however, the direction of palladium acetate addition is reversed and 73% of the product is crotyl acetate. The probable explanation for this reversal is that the state of aggregation of the palladium acetate changes from the known trimer in acetic acid to a sterically much smaller, monomeric species in dimethyl sulfoxide. The addition is probably sterically controlled and the palladium group simply changes relative size compared with the acetate group when the solvent is changed.

The vapor phase catalytic oxidation of ethylene to vinyl acetate with air and acetic acid with a heterogeneous palladium acetate–copper acetate catalyst is carried out commercially. The reaction mechanism is certainly basically the same as in the homogeneous reaction.

Palladium(II) cyanide will react in a similar manner in dimethylformamide solvent at $210°$. Propylene, in a stoichiometric reaction, forms mainly methacrylonitrile (196).

$$CH_3CH{=}CH_2 + Pd(CN)_2 \xrightarrow{L} \left[\overset{\overset{\textstyle CN}{\textstyle |}}{CH_3CHCH_2PdL_2CN} \right] \longrightarrow$$

$$\overset{\overset{\textstyle CN}{\textstyle |}}{CH_3C}{=}CH_2 + [HPdL_2CN]$$

Palladium(II) chloride at 100° in anhydrous formamide reacts analogously with ethylene producing vinyl chloride in 80% yield (197). Even though palladium chloride elimination is favored over hydride elimination, the addition recurs until the palladium is eliminated irreversibly as the hydride.

$$CH_2{=}CH_2 + PdCl_2 \underset{}{\overset{L}{\rightleftharpoons}} [ClCH_2CH_2PdL_2Cl] \longrightarrow ClCH{=}CH_2 + [HPdL_2Cl]$$

The palladium(II) chloride reaction with olefins takes a different course if water is present. Carbonyl compounds rather than chlorides are formed. The reaction with ethylene produces acetaldehyde and it can be done catalytically in the presence of oxygen and cupric chloride. The catalytic reaction is known as the Wacker Process since it was discovered at the German company, Wacker Chemie G.m.b.H. in 1956 by J. Smidt and co-workers. This process or its newer heterogeneous modifications for the preparation of acetaldehyde is one of the most important commercially useful reactions employing transition metal catalysts. The process involves three reactions: (1) the palladium(II) olefin oxidation, (2) the cupric chloride reoxidation of the palladium metal formed in the first step, and (3) the oxygen oxidation of the cuprous chloride formed in the second step back to the cupric state (198). The equations, using tetrachloropalladate ion as the catalysts, are the following:

$$CH_2{=}CH_2 + PdCl_4{}^{2-} + H_2O \longrightarrow CH_3CHO + Pd + 2HCl + 2Cl^-$$

$$Pd + 2CuCl_2 \longrightarrow PdCl_2 + 2CuCl$$

$$2CuCl + \tfrac{1}{2}O_2 + 2HCl \longrightarrow 2CuCl_2 + H_2O$$

$$\overline{CH_2{=}CH_2 + \tfrac{1}{2}O_2 \longrightarrow CH_3CHO}$$

The overall result is that one mole of ethylene plus a half a mole of oxygen (supplied as air) produces one mole of acetaldehyde. The reaction proceeds in very high yield.

Numerous mechanistic studies have been performed on this important reaction. Kinetic studies in aqueous solution have shown that the rate of product formation depends in a first-order fashion on the ethylene concentration and the tetrachloropalladate ion while there is an inverse dependence on the hydrogen ion concentration and an inverse square dependence on the chloride ion concentration (199). The kinetic data are consistent with

$$d(CH_3CHO)/dt = k[C_2H_4][PdCl_4{}^{2-}]/[H^+][Cl^-]^2$$

steps requiring replacement of two chloride ligands on the palladium, one by ethylene and the other by water. The reversibility of these two steps accounts for the inverse square dependence on the chloride ion concentration. The third step appears to be loss of a proton from the coordinated water molecule. Coordinated water is always more acidic than noncoordinated water. This reaction accounts for the inhibiting effect of acid on the oxidation. The rate determining transfer of hydroxyl from palladium to the coordinated olefin then occurs, necessarily from a cis intermediate. The 2-hydroxyethyl-palladium species formed next loses palladium hydride probably producing a π-vinyl alcohol–hydridopalladium complex. When carried out in deuterium oxide instead of water, no deuterium is found in the acetaldehyde; thus the π-vinyl alcohol complex cannot simply dissociate (200). The liberated vinyl alcohol would have gone to acetaldehyde with deuterium incorporation. Probably the π-vinyl alcohol–hydridopalladium intermediate undergoes a series of addition–elimination steps ending in the loss of the β-hydroxylic hydrogen with the palladium. This mechanism leaves all the original ethylenic hydrogens in the acetaldehyde product (see Scheme V-10). An unstable π-vinyl alcohol complex with tetracarbonyliron(0) has been reported (201) as have similar complexes with platinum (202, 203). An X-ray structure has been obtained for one of these complexes by F. A. Cotton, J. N. Francis, B. A. Freng, and M. Tsutsui, *J. Amer. Chem. Soc.* **95**, 2483 (1973).

The palladium(II) aqueous oxidation of monosubstituted olefins produces mainly ketones rather than aldehydes indicating a preference for palladium to become attached to the less substituted terminal carbon of the double bond. The direction is probably determined electronically in this example with the electronegative hydroxyl group going to the more positive carbon of the double bond. A simple, preparatively useful procedure for converting terminal olefins into 2-carbonyl derivatives based on the palladium(II) oxidation has been reported (204).

$$CH_3(CH_2)_9CH{=}CH_2 + \tfrac{1}{2}O_2 \xrightarrow{\text{PdCl}_2,\ \text{CuCl}_2} CH_3(CH_2)_9\overset{\displaystyle O}{\overset{\displaystyle \|}{C}}CH_3$$
$$85\%$$

The reaction of palladium(II) chloride with olefins in alcoholic solution results in the formation of vinyl ethers, very probably by a mechanism similar to that of the aqueous oxidation (205, 206). The alkoxide group is probably added by alcohol attacking the olefin π-complex externally rather than by a shift of alkoxide from palladium.

$$CH_2{=}CH_2 + CH_3OH + PdCl_2 \longrightarrow CH_3OCH{=}CH_2 + Pd + HCl$$

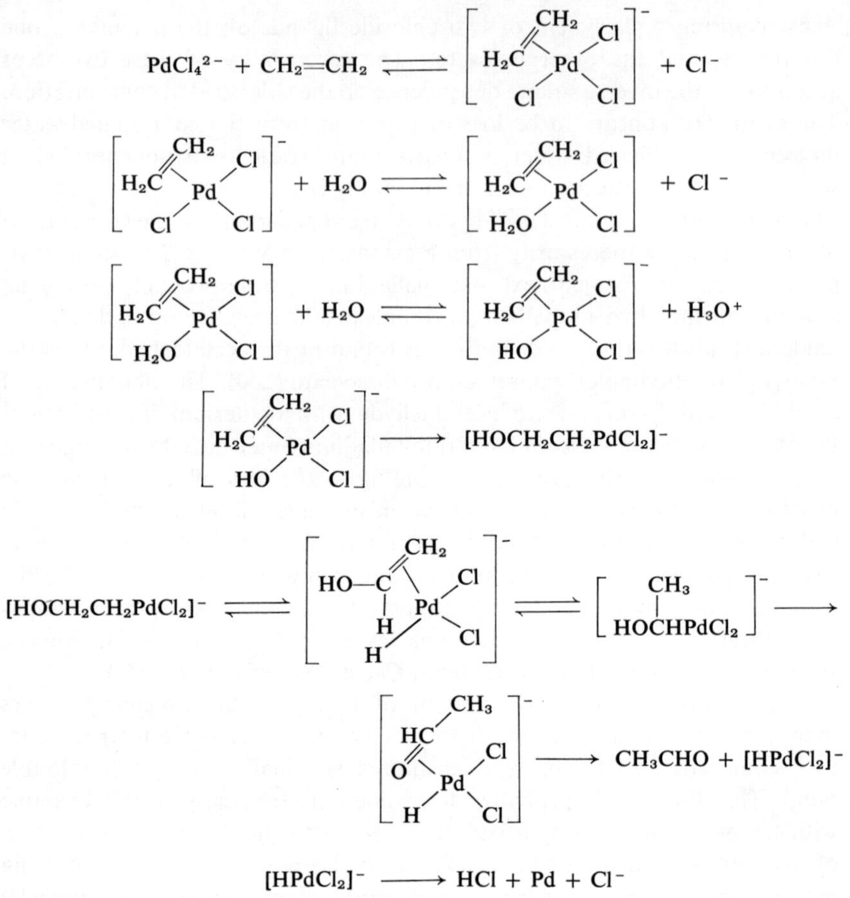

$$PdCl_4^{2-} + CH_2{=}CH_2 \rightleftharpoons \left[\underset{Cl}{\overset{CH_2}{H_2C{=}}}Pd\overset{Cl}{\underset{Cl}{}} \right]^- + Cl^-$$

$$\left[\underset{Cl}{\overset{CH_2}{H_2C{=}}}Pd\overset{Cl}{\underset{Cl}{}} \right]^- + H_2O \rightleftharpoons \left[\underset{H_2O}{\overset{CH_2}{H_2C{=}}}Pd\overset{Cl}{\underset{Cl}{}} \right] + Cl^-$$

$$\left[\underset{H_2O}{\overset{CH_2}{H_2C{=}}}Pd\overset{Cl}{\underset{Cl}{}} \right] + H_2O \rightleftharpoons \left[\underset{HO}{\overset{CH_2}{H_2C{=}}}Pd\overset{Cl}{\underset{Cl}{}} \right]^- + H_3O^+$$

$$\left[\underset{HO}{\overset{CH_2}{H_2C{=}}}Pd\overset{Cl}{\underset{Cl}{}} \right]^- \longrightarrow [HOCH_2CH_2PdCl_2]^-$$

$$[HOCH_2CH_2PdCl_2]^- \rightleftharpoons \left[\overset{CH_2}{\underset{H}{HO{-}C}}\underset{H}{\overset{|}{{}}}Pd\overset{Cl}{\underset{Cl}{}} \right]^- \rightleftharpoons \left[\overset{CH_3}{\underset{HOCHPdCl_2}{|}} \right]^- \longrightarrow$$

$$\left[\underset{H}{\overset{CH_3}{HC{=}}}\overset{}{\underset{O}{}}Pd\overset{Cl}{\underset{Cl}{}} \right]^- \longrightarrow CH_3CHO + [HPdCl_2]^-$$

$$[HPdCl_2]^- \longrightarrow HCl + Pd + Cl^-$$

Scheme V-10.

Several alkoxypalladation products of chelating diolefins are isolable and are known to have the trans stereochemistry expected from an external addition mechanism. If the reaction is carried out catalytically with cupric chloride as reoxidant and ethylene glycol as the alcohol, cyclic acetals are obtained in high yield (207, 208). Unsaturated alcohols may react with

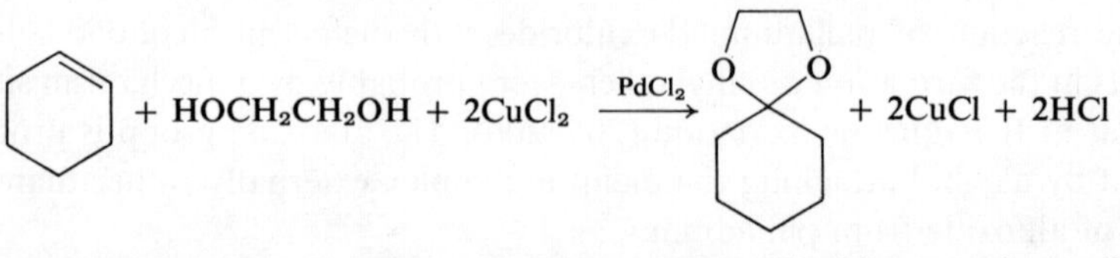

themselves. Methallyl alcohol and rhodium(III) chloride react in methanol solution to form a cyclic ether complex probably as follows (209).

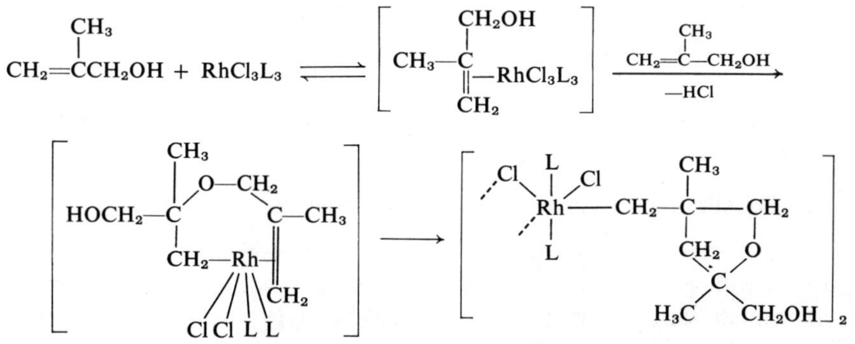

$$
\begin{array}{c}
\underset{\substack{| \\ CH_2=CCH_2OH}}{\overset{CH_3}{}} + RhCl_3L_3 \rightleftharpoons \left[\underset{\substack{\| \\ CH_2}}{\overset{CH_2OH}{\underset{|}{CH_3-C-RhCl_3L_3}}} \right] \xrightarrow[-HCl]{} \underset{\substack{| \\ CH_2=C-CH_2OH}}{\overset{CH_3}{}}
\end{array}
$$

Another related reaction occurs when palladium(II) chloride–olefin complexes are reacted with primary amines in tetrahydrofuran solution and imines are formed in low yield (210). The mechanism has not been investigated, but it is probably similar to the above alcohol reaction with an additional double-bond isomerization of the initially formed vinylamine to the imine.

$$[(CH_2=CH_2)PdCl_2]_2 + n\text{-BuNH}_2 \longrightarrow CH_3CH=NBu + Pd + 2HCl$$

Platinum(II)–olefin complexes react with amines to give isolable adducts. An example giving stereochemical information as well is the reaction of cis-dichloro-(S)-1-butene-(S)-1-phenylethylamineplatinum(II) with diethylamine. Since the π-bonded olefin is asymmetric and reaction is at the 2-carbon of the 1-butene, it can be determined which side the olefin is attacked from the absolute configuration of the amine product formed. The results showed that a trans (external attack) addition of the amine had occurred (211).

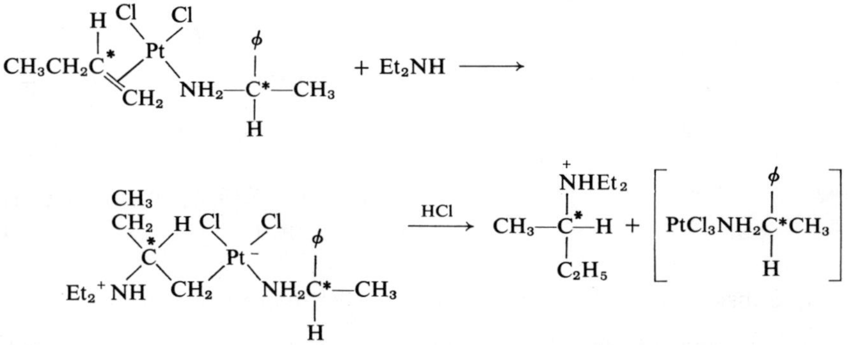

A similar but catalytic reaction seems to occur between ethylene and secondary amines with rhodium(III) chloride as catalyst. The intermediate complex, at the elevated temperatures employed, transfers the proton from the nitrogen to the carbon bonded to the metal and a saturated tertiary amine

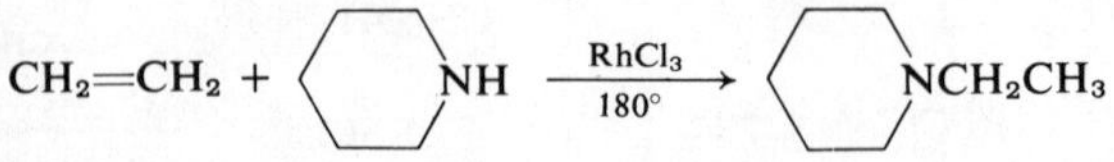

$$CH_2{=}CH_2 + \left[\underset{}{\bigcirc}NH\right] \xrightarrow[180°]{RhCl_3} \left[\underset{}{\bigcirc}NCH_2CH_3\right]$$

is formed (212). This is an example of a catalyzed 1 : 2 addition. More examples will be discussed in Section E.

3. Substitution of Vinylic Substituents Other than Hydrogen

Vinylic ester and halide substituents are readily exchanged for other ester groups with palladium(II) catalysts. 2-Chloropropene, for example, reacts with acetate salts in acetic acid using palladium(II) acetate as a catalyst to form isopropenyl acetate (102).

$$\underset{\underset{CH_2=CCH_3}{|}}{Cl} + OAc^- \xrightarrow[HOAc]{Pd(OAc)_2} \underset{\underset{CH_2=CCH_3}{|}}{OAc} + Cl^-$$

The palladium-catalyzed exchange of acetate for trideuteroacetate in *cis*- and *trans*-propenyl acetate shows each exchange occurs with inversion of configuration, a result consistent with a trans addition–trans elimination process (213).

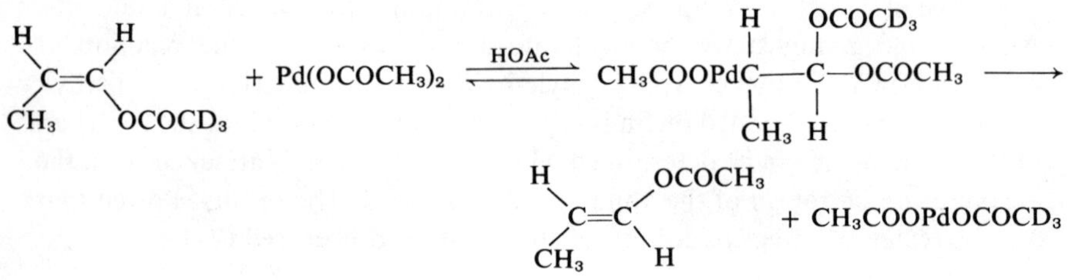

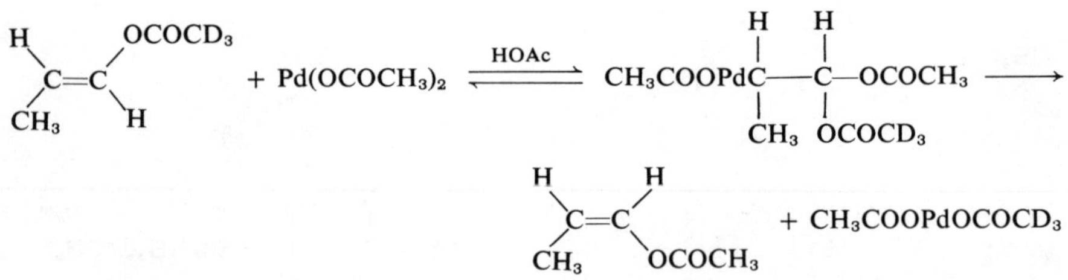

$$\underset{CH_3}{\overset{H}{\diagdown}}C{=}C\underset{OCOCD_3}{\overset{H}{\diagup}} + Pd(OCOCH_3)_2 \underset{}{\overset{HOAc}{\rightleftharpoons}} CH_3COOPdC\underset{CH_3 \; H}{\overset{H \quad OCOCD_3}{\underset{|\quad\;|}{\overset{|\quad\;|}{{-}C{-}OCOCH_3}}}} \longrightarrow$$

$$\underset{CH_3}{\overset{H}{\diagdown}}C{=}C\underset{H}{\overset{OCOCH_3}{\diagup}} \quad + CH_3COOPdOCOCD_3$$

$$\underset{CH_3}{\overset{H}{\diagdown}}C{=}C\underset{H}{\overset{OCOCD_3}{\diagup}} + Pd(OCOCH_3)_2 \underset{}{\overset{HOAc}{\rightleftharpoons}} CH_3COOPdC\underset{CH_3 \; OCOCD_3}{\overset{H \quad H}{\underset{|\quad\;|}{\overset{|\quad\;|}{{-}C{-}OCOCH_3}}}} \longrightarrow$$

$$\underset{CH_3}{\overset{H}{\diagdown}}C{=}C\underset{OCOCH_3}{\overset{H}{\diagup}} \quad + CH_3COOPdOCOCD_3$$

4. Substitution of Allylic Substituents Other than Hydrogen

The exchange of allylic substituents with allylic rearrangement catalyzed by palladium(II) salts is a very general reaction.

Substitution of allylic chlorine by carbon groups occurs when organo-palladium complexes are reacted with allylic chlorides. The electron-withdrawing effect of the halogen causes addition of the palladium group to the carbon beta to the halogen-bearing carbon. The elimination of the chloro-palladium group is preferred over the alternative palladium hydride elimination and allylically substituted olefins are produced (214). The organopalladium complexes were prepared *in situ* by an exchange between palladium(II) chloride and an arylmercuric chloride. The anionic allylic exchange usually

$$PdL_2Cl_2 + ArHgCl \longrightarrow [ArPdL_2Cl] + HgCl_2$$

$$[ArPdL_2Cl] + \underset{/}{\overset{\backslash}{C}}=C-\underset{|}{\overset{|}{C}}-Cl \longrightarrow \left[Ar-\underset{|}{\overset{|}{C}}-\underset{|}{\overset{\overset{\displaystyle PdL_2Cl}{|}}{C}}-\underset{|}{\overset{|}{C}}-Cl \right] \longrightarrow$$

$$Ar-\underset{|}{\overset{|}{C}}-\underset{|}{\overset{|}{C}}=C\overset{\diagup}{\diagdown} + PdL_2Cl_2$$

proceeds most readily when the protonated form of the anion is relatively acidic. Then the reaction usually proceeds when the reactants are warmed with a catalytic quantity of any of a variety of palladium complexes. Allylic ethers, esters, alcohols, chlorides, and amines have all been exchanged with other amines, alcohols, acids, dimethyl malonate, acetylacetone, methyl acetoacetate, etc. (215–217).

$$\underset{|}{\overset{|}{C}}=\underset{|}{\overset{|}{C}}-\underset{|}{\overset{|}{C}}-Y + HZ \xrightarrow{\text{Pd}} Z-\underset{|}{\overset{|}{C}}-\underset{|}{\overset{|}{C}}=C\overset{\diagup}{\diagdown} + HY$$

E. CATALYZED 1:2 ADDITIONS TO OLEFINS

An example of this type of reaction has been mentioned above where piperidine was added to ethylene forming *N*-ethylpiperidine with rhodium(III) chloride as catalyst. A few other examples are also known.

The addition of hydrogen cyanide to olefins and dienes can be catalyzed by tetrakis(triphenyl phosphite)palladium(0) (218). The mechanism probably involves oxidative addition of hydrogen cyanide to the palladium(0) complex to form hydridocyanobis(triphenyl phosphite)palladium(II). This hydride then may add to the olefin to form an alkyl. The last step would be reductive elimination of the alkyl nitrile with reformation of the palladium(0) catalyst.

$$Pd[P(O\phi)_3]_4 + HCN \rightleftharpoons HPd[P(O\phi)_3]_2CN + 2P(O\phi)_3$$

$$HPd[P(O\phi)_3]_2CN + CH_2=CH_2 \longrightarrow CH_3CH_2Pd[P(O\phi)_3]_2CN \xrightarrow{P(O\phi)_3}$$

$$CH_3CH_2CN + Pd[P(O\phi)_3]_4$$

A similar reaction occurs with related nickel(0) complexes in the presence of zinc chloride (219).

The oxidation of olefins with palladium choride in aqueous solution in the presence of a high concentration of cupric chloride, in contrast to when it is present at low concentration in the Wacker process, produces chlorohydrins rather than carbonyl compounds. The intermediate 2-hydroxyethylpalladium complexes apparently react with the cupric chloride to replace the palladium by chlorine. The mechanism of the replacement is not clear, but a 1:2 shift of chlorine from palladium to carbon in chlorine-bridged copper complexes seems to be a reasonable possibility (220).

$$PdCl_4{}^{2-} + H_2O + CH_2{=}CH_2 + CuCl_2 \longrightarrow \left[\begin{array}{c} ClCuCl_2 \\ | \\ HOCH_2CH_2PdClCuCl_2 \\ | \\ Cl \end{array} \right]^{2-} \longrightarrow$$

$$HOCH_2CH_2Cl + PdCl_4{}^{2-} + 2CuCl$$

A similar effect of a high concentration of cupric chloride is observed in the reaction of organopalladium complexes with olefins. Thus, phenylpalladium chloride prepared by exchange from palladium(II) chloride and phenylmercuric chloride reacts with ethylene in 10% aqueous acetic acid containing a 2 M concentration of cupric chloride to form almost exclusively 2-phenylethyl chloride rather than the styrene formed when a significantly lower concentration of cupric chloride is present (221).

$$[C_6H_5PdL_2Cl] + CH_2{=}CH_2 + 2CuCl_2 \longrightarrow C_6H_5CH_2CH_2Cl + 2CuCl + PdL_2Cl_2$$

The inclusion of lead tetraacetate in the reaction of organopalladium complexes with olefins also results in a catalyzed 1:2 addition, this time of the organic group and an acetate group. It seems quite possible that there is an actual transfer of the organic group from palladium to lead in this case and that a lead (IV) alkyl triacetate is formed which decomposes by reductive elimination to the product acetate and lead(II) acetate. The reaction is catalytic in palladium since the lead(IV) reoxidizes it in each cycle. Yields are not high in this addition since competing palladium hydride elimination is serious and the lead(IV) acetate may attack the olefin directly (222).

$$[C_6H_5PdL_2OAc] + Pb(OAc)_4 + CH_2{=}CH_2 \longrightarrow$$

$$C_6H_5CH_2CH_2OAc + Pb(OAc)_2 + PdL_2(OAc)_2$$

FURTHER READING

D. G. H. Ballard, Polymerization of olefins and vinyl monomers. *Advan. Catal.* **23**, 263 (1973).

M. N. Berger, G. Boocock, and R. N. Haward, Olefin polymerization. *Advan. Catal.* **19**, 211 (1969).

C. W. Bird, Oligomerization of olefins. "Transition Metal Intermediates in Organic Syntheses," Chapter 2. Academic Press, New York, 1967.

R. F. Heck, Addition-Elimination reactions of palladium compounds with olefins. *Fortschr. Chem. Forsch.* **16,** 221 (1971).

P. M. Henry, Pd(II) catalyzed exchange and isomerization reactions. *Accounts Chem. Res.* **6,** 16 (1973).

F. D. Mango and J. H. Schachtschneider, Catalysis of symmetry forbidden reactions. "Transition Metals in Homogeneous Catalysis," Chapter 6. Dekker, New York, 1971.

H. W. Quinn and J. H. Tsai, Olefin complexes of the transition metals. *Advan. Inorg. Chem. Radiochem.* **12,** 217 (1969).

F. G. A. Stone, Reactions of fluorinated olefins. *Pure Appl. Chem.* **30,** 551 (1972).

M. Tsutsui, M. Levy, A. Nakamura, M. Ichikaura, and K. Mori, "Introduction to Metal π-Complex Chemistry." Plenum, New York, 1970.

Chapter VI

Reactions of Dienes, Trienes, and Tetraenes with Transition Metal Compounds

A. DIENE π COMPLEXES AND THEIR REARRANGEMENTS

Dienes may π-complex through one double bond as a simple olefin would do, but more often, if the double bonds are favorably situated, both double bonds coordinate. A typical example is tricarbonyl-π-butadieneiron(0) which is obtained by heating pentacarbonyliron(0) with butadiene (223).

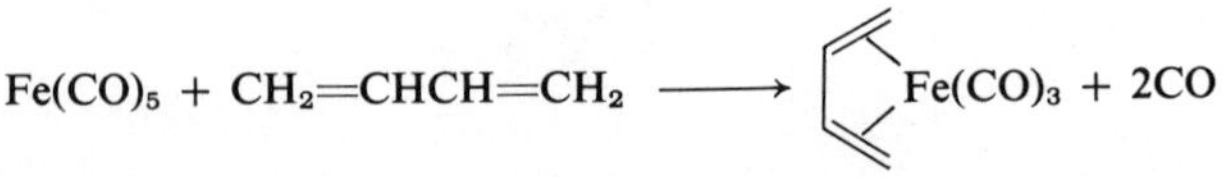

$$Fe(CO)_5 + CH_2\!=\!CHCH\!=\!CH_2 \longrightarrow \quad Fe(CO)_3 + 2CO$$

Recently, butadiene π complexes have been prepared by a direct reaction of the diene with metal atoms, prepared by condensation of the vaporized metal in high vacuum. The metal atoms are condensed on the diene at liquid nitrogen temperatures and then the mixture is allowed to warm up and react. Nickel, for example, is believed to form a bisbutadienenickel(0) complex (224) and iron, bisbutadienecarbonyliron(0), after reaction with CO (225).

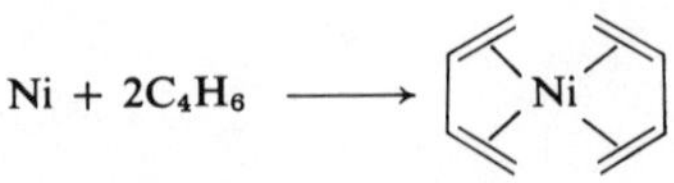

$$Ni + 2C_4H_6 \longrightarrow \quad Ni$$

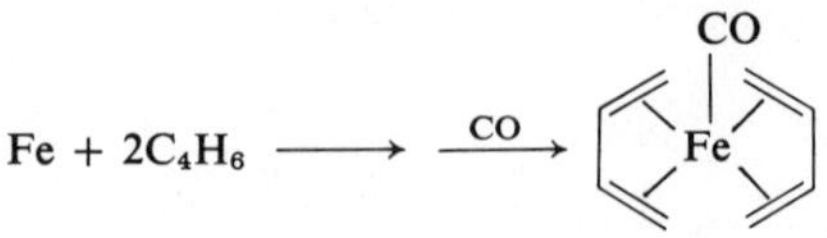

$$Fe + 2C_4H_6 \longrightarrow \xrightarrow{CO} \quad Fe$$

The stability of the conjugated diene–iron carbonyl group is high enough so that some vinyl aromatic compounds will form complexes with loss of the aromaticity of the benzene ring. For example, heating p-divinylbenzene with dodecacarbonyltriiron produces an orange complex containing two π-diene-iron systems with one isolated double bond left in the original benzene ring. An X-ray structure showed the isolated double bond to be 1.31 Å long, while the coordinated ones were 1.40 Å (226). In another example, heating 1,2-

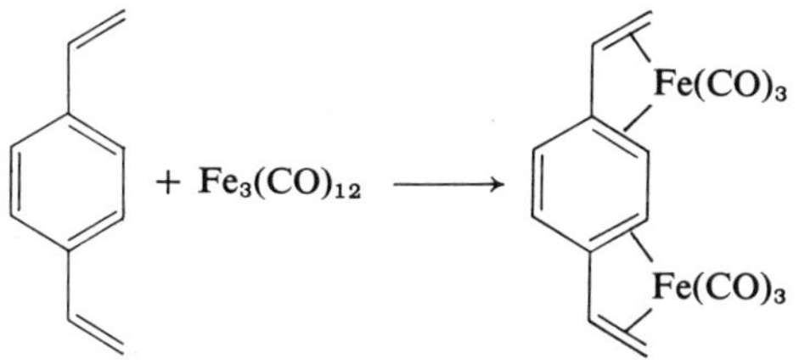

bis(bromomethyl)benzene with nonacarbonyldiiron gave a complex with two uncoordinated double bonds and a π-dieneiron system. Pyrolysis of the complex at 510° produced benzocyclobutene (227).

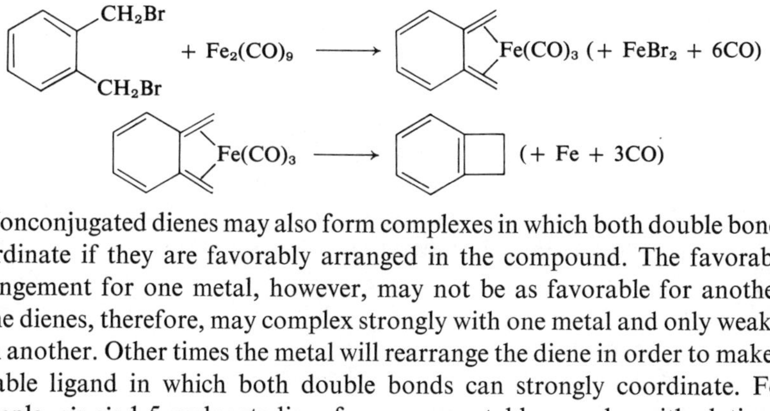

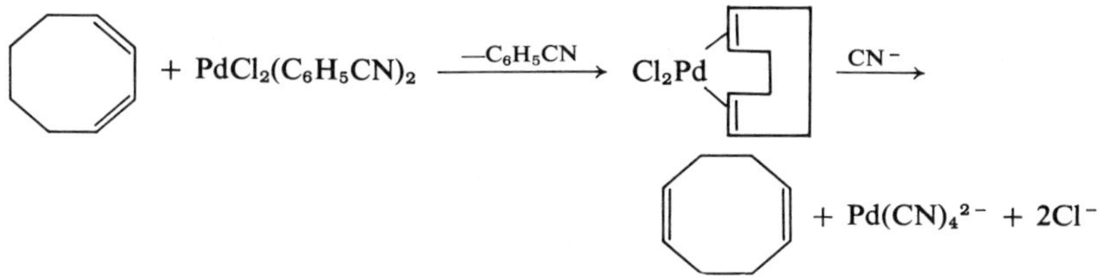

Nonconjugated dienes may also form complexes in which both double bonds coordinate if they are favorably arranged in the compound. The favorable arrangement for one metal, however, may not be as favorable for another. Some dienes, therefore, may complex strongly with one metal and only weakly with another. Other times the metal will rearrange the diene in order to make a suitable ligand in which both double bonds can strongly coordinate. For example, cis,cis-1,5-cyclooctadiene forms a very stable complex with platinum or palladium chloride, but the 1,3-diene does not. If dichlorobisbenzonitrilepalladium(II) is reacted with 1,3-cyclooctadiene at room temperature, the π-cis-1,5-cyclooctadiene complex is formed very rapidly. The rearranged diene may be liberated from the complex with cyanide ion (228). On the other hand, iron(0) prefers to coordinate with 1,3- rather than 1,5-dienes. Consequently, when *cis,cis*-1,5-cyclooctadiene (1,5-COD) is heated with pentacarbonyliron(0) a catalytic isomerization occurs by way of a 1,3-dieneiron complex producing 1,3-cyclooctadiene (229). Probably both palladium and

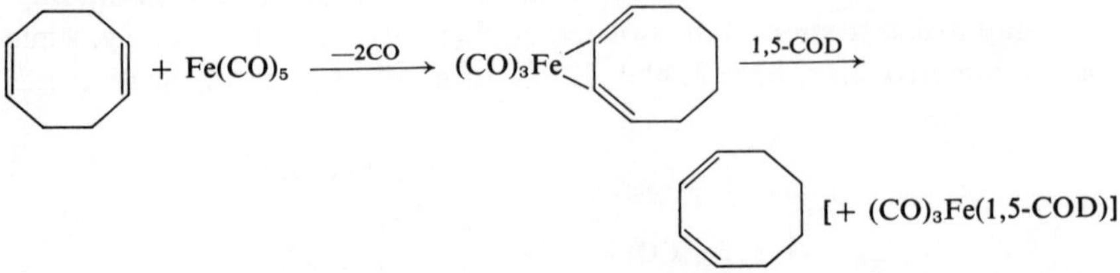

iron produce isomerization in the cyclooctadiene ring by the π-allylic hydride mechanism. The reason for the diene preferences is not clear, since both palladium and iron have 90° bond angles available with similar metal–carbon bond lengths. Iron in trigonal-bipyramidal complexes would have 120° equatorial bond angles also, but these are too large for a 1,3-diene.

The cyclodecadienes behave differently toward these catalysts. Dichlorobisbenzonitrilepalladium(II) forms a stable complex with *cis,cis*-1,6-cyclodecadiene, while *cis,trans*-1,5-cyclodecadiene reacts with the same reagent at 0° to produce the palladium chloride complex of *cis*-1,2-divinylcyclohexane. A carbon–carbon bond is apparently broken at the weakest point in the ring where two allylic fragments can be formed. It is believed that the arrangement goes by way of a bis-π-allylic palladium intermediate, but its structure is uncertain (230). The overall reaction is actually a metal-catalyzed Cope rearrangement. Pentacarbonyliron(0) at 115° causes rearrangement of *cis,trans*-

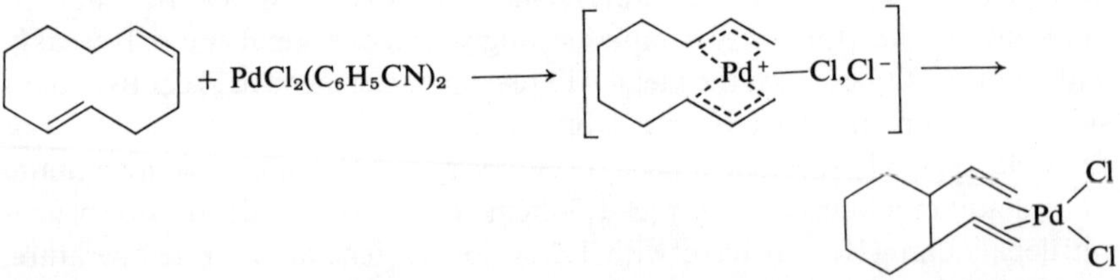

1,5-cyclodecadiene to the same product, whereas at 20°–50° with exposure to ultraviolet light to dissociate coordinated carbon monoxide, the diene rearranges to a mixture of *cis,cis*- and *trans,trans*-1,6-cyclodecadiene (231). The reactions may go by way of π-allylic hydrides.

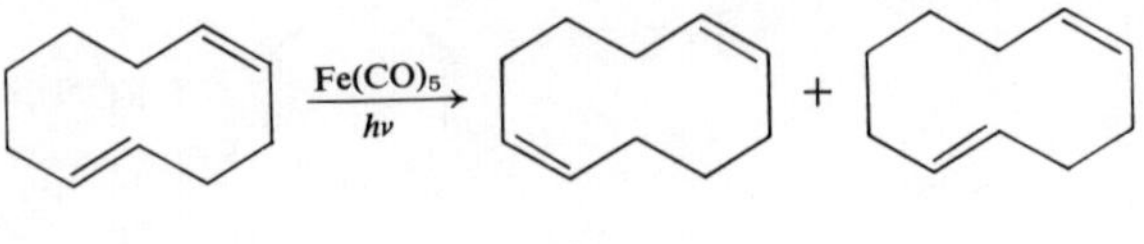

Ring enlargement also can occur in appropriately substituted dienes. 4-Vinylcyclohexene and dichlorobisbenzonitrilepalladium(II) react to form the *cis,cis*-1,5-cyclooctadiene complex, again probably by way of a bis-π-allylic complex (232). If the structure does not have a weak bond between two

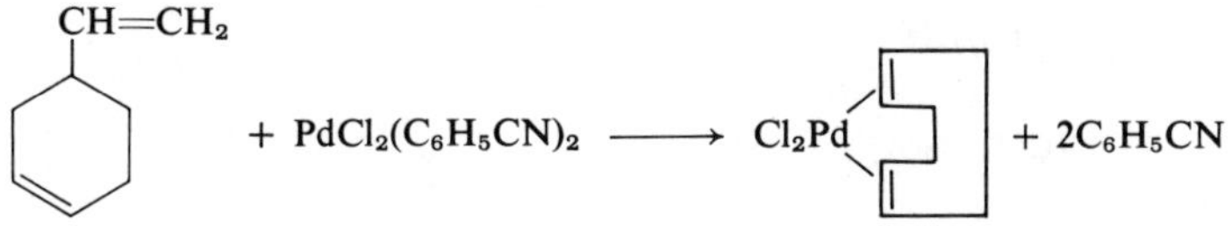

$$\text{(4-vinylcyclohexene)} + PdCl_2(C_6H_5CN)_2 \longrightarrow Cl_2Pd\text{(complex)} + 2C_6H_5CN$$

allylic carbons, the isomerization does not occur, at least under the same mild conditions. 2-Vinylcyclohexene, for example, does not rearrange under the same conditions (233).

More deep-seated rearrangements may occur with other catalysts under these or more vigorous conditions. The stability of the tridentate π-complexes of aromatic compounds with chromium(0) is probably the driving force for the conversion of *cis,cis*-1,5-cyclooctadiene with hexacarbonylchromium(0) into *o*-xylene and tricarbonyl-π-*o*-xylenechromium(0) in boiling di-*n*-butyl ether. The yield, however, was only 2% and the remaining material was not identified. At lower temperatures the reaction forms tetracarbonyl-*cis,cis*-1,5-cyclooctadienechromium(0) which does not yield the aromatic complex on heating (234). The mechanism of this rearrangement is not known.

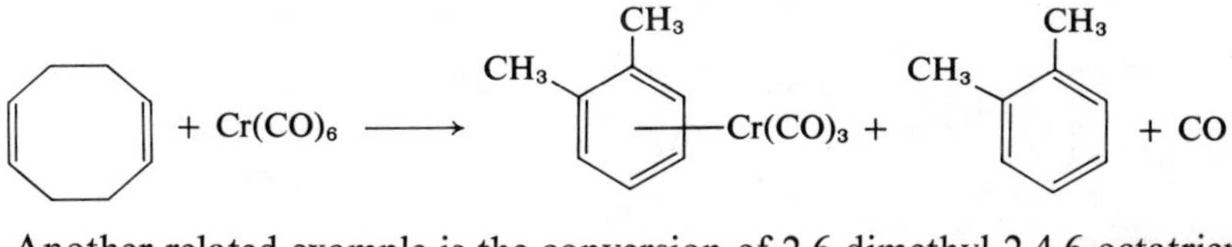

$$\text{(cyclooctadiene)} + Cr(CO)_6 \longrightarrow \text{(xylene)}Cr(CO)_3 + \text{(xylene)} + CO$$

Another related example is the conversion of 2,6-dimethyl-2,4,6-octatriene with hexacarbonylchromium at 200° into tricarbonyl-π-(1,2,3-trimethyl-benzene) chromium(0) in 40% yield (235). Two carbons are lost in the reaction, probably in the form of ethane. Rhenium(I), in contrast to chromium(0),

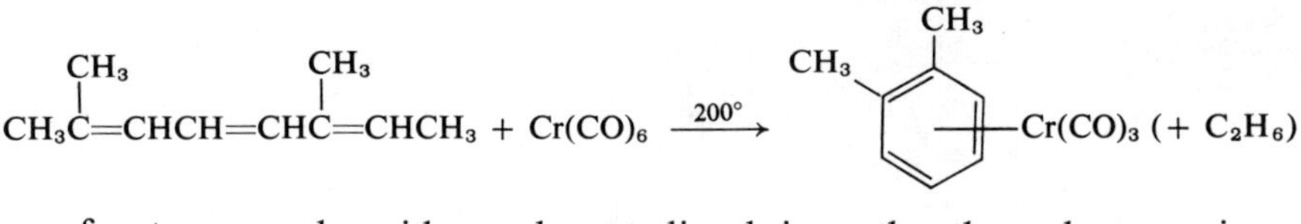

$$CH_3C\!=\!CHCH\!=\!CHC\!=\!CHCH_3 + Cr(CO)_6 \xrightarrow{200°} \text{(trimethylbenzene)}Cr(CO)_3 (+ C_2H_6)$$

prefers to π-complex with a cyclopentadienyl ring rather than a benzene ring. Decacarbonyldirhenium reacts at 250° with *cis,cis*-1,5-cyclooctadiene to form tricarbonyl-1,2-cyclopentenocyclopentadienylrhenium(I) in 5% yield (236).

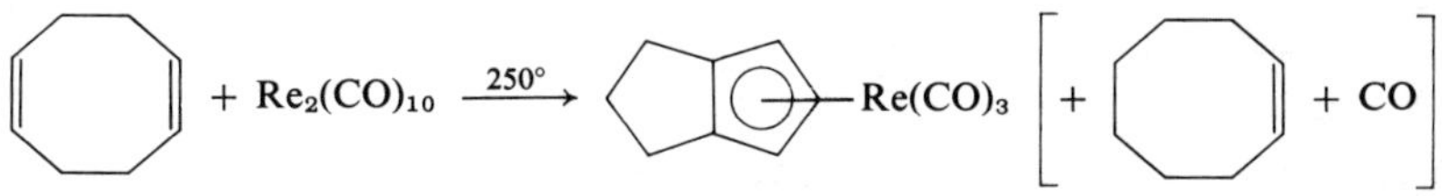

$$\text{(cyclooctadiene)} + Re_2(CO)_{10} \xrightarrow{250°} \left[\text{(complex)}Re(CO)_3 + \text{(cyclooctadiene)} + CO \right]$$

A related formation of the bicyclo[3.3.0]octadienyl ligand occurs in the reaction of 1,5,9-cyclododecatriene with *cis*-bis(trimethylgermyl)tetracarbonylruthenium where probably butadiene is lost (237).

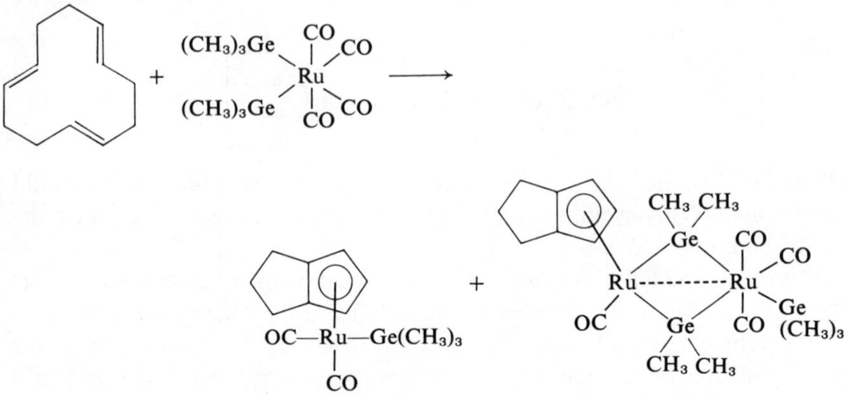

$(CH_3)_3Ge$
CO
CO
Ru
$(CH_3)_3Ge$
CO
CO
+
OC—Ru—Ge$(CH_3)_3$
CO
+
CH_3 CH_3
Ge
CO
CO
Ru
Ru
Ge
OC
Ge
CO
$(CH_3)_3$
CH_3 CH_3

Another example of cyclopentadienyl complex formation was mentioned in Chapter I, Section A, where titanium tetrachloride was reacted with an olefin (any olefin) and trichloropentamethylcyclopentadienyltitanium(IV) was formed (6). The mechanisms of these very interesting reactions have not been studied.

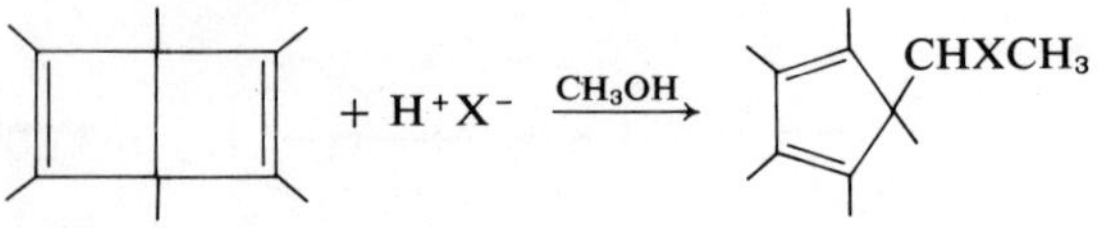

+ H^+X^-
CH_3OH
$CHXCH_3$

X = Cl, Br, CH_3O

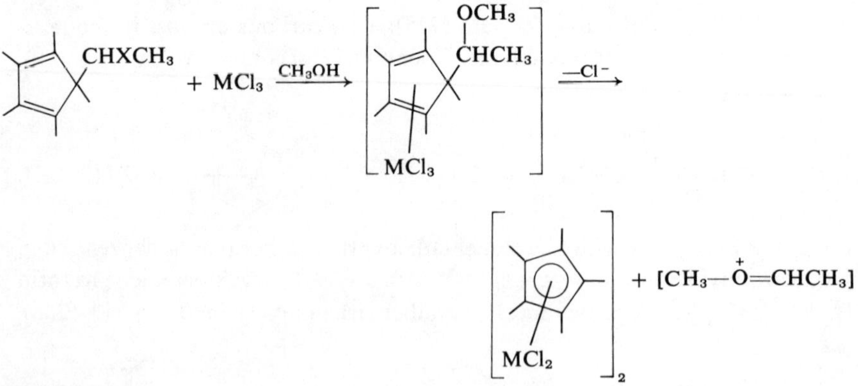

$CHXCH_3$
+ MCl_3
CH_3OH
OCH_3
$CHCH_3$
MCl_3
—Cl^-
MCl_2
2
+ $[CH_3—\overset{+}{O}=CHCH_3]$

M = Rh, Ir

$$CH_3\overset{+}{O}=CHCH_3 + CH_3OH \longrightarrow (CH_3O)_2CHCH_3$$

The apparent preference for formation of cyclopentadienyl π complexes is also observed in the reactions of hexamethyl(Dewar benzene) with rhodium and iridium trichlorides. In these cases, however, it is clear that the metal ions are not responsible for the ring contraction reaction. Hexamethyl(Dewar benzene) is known to react with protonic acids to form 5-(1-substituted-ethyl)pentamethylcyclopentadienes, and acid is produced from rhodium and iridium trichlorides in methanol under the reaction conditions. In separate reactions it was shown that 5-(1-chloroethyl)pentamethylcyclopentadiene reacted with both metal chlorides in methanol solution to give high yields of the π-pentamethylcyclopentadienyl metal complexes and dimethylacetal (238). The yield of the iridium complex was much higher with the prerearranged starting complex than it was with hexamethyl(Dewar benzene). The metals are apparently necessary in the step where the methoxymethylcarbonium ion is lost to form the cyclopentadienyl complex. In the same reaction in the presence of acid and a trace of tin(II) chloride as catalyst, potassium tetrachloroplatinate and hexamethyl(Dewar benzene) gave only a pentamethylcyclopentadiene π complex (239).

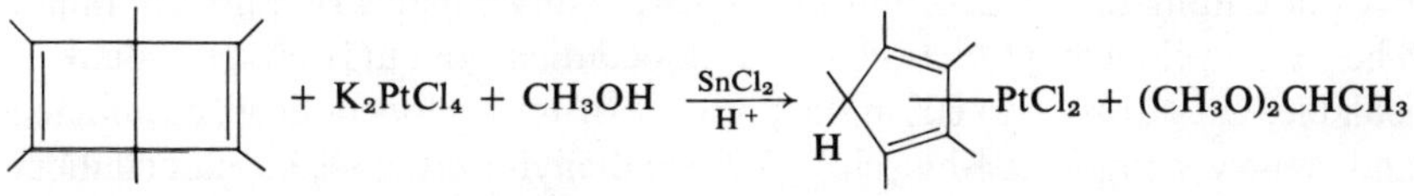

B. HYDRIDE TRANSFER REACTIONS

Hydride or perhaps more correctly hydrogen transfer between ligands on the same metal may occur very easily. An intermediate metal hydride is probably involved. Such hydrogen transfers appear to be important steps in numerous transition metal reactions, particularly when diene-type ligands are present. Two simple examples below illustrate the reaction. The reaction of bis(cis,cis-1,5-cyclooctadiene)nickel(0) with cyclopentadiene forms π-5-cyclooctenyl-π-cyclopentadienylnickel(0) (240). The reaction probably proceeds by a replacement of cyclooctadiene by cyclopentadiene followed by hydrogen transfer.

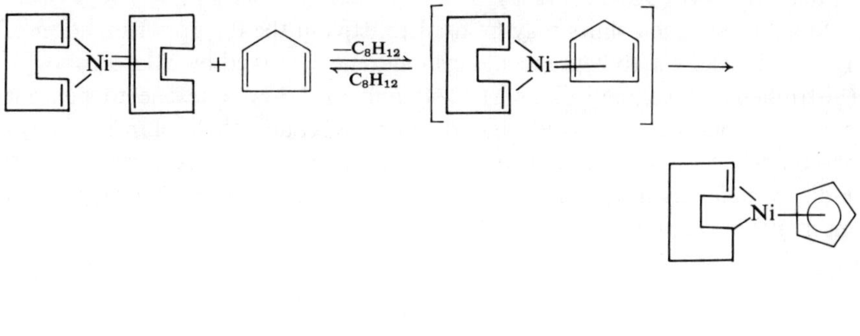

In the second example the same nickel complex was reacted with acetylacetone and a similar hydrogen transfer occurred (241).

An interesting disproportionation of *cis,cis*-1,5-cyclooctadiene apparently involving a series of hydrogen transfers occurs when cobalt(II) acetylacetonate [Co(Acac)₂] is electrolytically reduced in the presence of the diene and a proton source. Initially, a cobalt(I) hydride is thought to be formed which adds to the diene to form a 4-cyclooctenylcobalt derivative which, in turn, undergoes hydride elimination, readdition, etc., until a π-allylcobalt structure is obtained. The π-2-cyclooctenyl-π-*cis,cis*-1,5-cyclooctadienecobalt(I) intermediate is isolable. On warming to 60° in the presence of 1,5-cyclooctadiene, cyclooctene and a new complex, π-bicyclo[3.3.0]octadienyl-π-*cis,cis*-1,5-cyclooctadiene-cobalt(I) are formed (242). The last reaction probably involves a hydrogen shift from the 1,5-cyclooctadiene ligand, forming a 2,5-cyclooctadienyl-π-cyclooctene complex. Replacement of cyclooctene by 1,5-cyclooctadiene, a hydrogen shift, and a ring closure could produce π-1,5-cyclooctadiene-π-h^3-bicyclo[3.3.0]octenylcobalt(I). Two sequential hydrogen shifts from the bicyclo[3.3.0]octenyl ring to the cyclooctadiene ring would form the cyclopentadienyl system and cyclooctene. Replacement of the cyclooctene by the diene would then form the observed product, π-1,5-cyclooctadiene-π-bicyclo[3.3.0]octadienylcobalt(I) (see Scheme VI-1). The thermal reaction of 1,5-cyclooctadienecyclooctenylcobalt(I) with 1,5-cyclooctadiene gave a low yield of 1,5-cyclooctadiene-2,5-cyclooctadienylcobalt(I) thus supporting the above mechanism. Further reaction also gave the same product as obtained in the electrolytic reaction (243).

Similar hydrogen shifts may be used to explain the disproportionation of 1,4-cyclohexadiene to benzene and cyclohexene catalyzed by chlorocarbonyl-(bistriphenylphosphine)iridium(I) (244) and of 1,3-cyclohexene to benzene and cyclohexene catalyzed by dichloro-π-pentamethylcyclopentadienyl-rhodium(III) dimer (245). The true catalysts in these examples must be π-diene–metal hydride complexes in order for the internal hydrogen shift mechanism to apply.

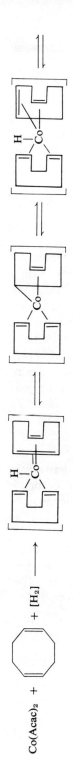
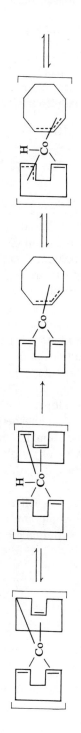
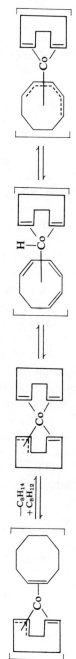
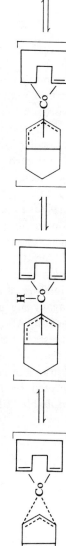
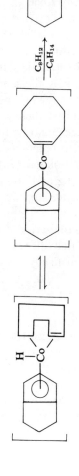

Scheme VI-1.

C. DIENE REACTIONS WITH METAL HYDRIDES

Transition metal hydrides probably add 1:4 to conjugated dienes often subsequently forming π-allyl metal complexes. Typical is the reaction of hydridotetracarbonylcobalt(I) with butadiene at 25° to form a mixture of 35% *syn*- and 65% *anti*-tricarbonyl-π-1-methylallylcobalt(I) (246). This

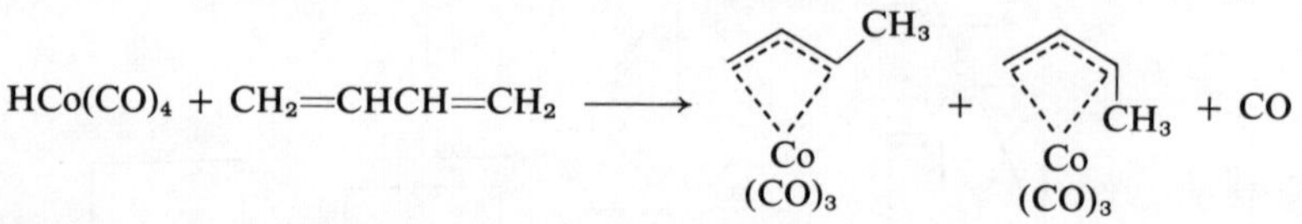

$$HCo(CO)_4 + CH_2{=}CHCH{=}CH_2 \longrightarrow$$

reaction is undoubtedly more complicated than the equation suggests. An initial dissociation of carbon monoxide could occur, followed by coordination with the diene and a hydrogen shift from cobalt to the terminal carbon of the diene system. The π-allyl system, however, is apparently not formed immediately since an intermediate 1:4 addition product can be detected if the reaction is carried out in a carbon monoxide atmosphere under conditions where the π-allyl product would not form the 1:4 intermediate (cf. Chapter IX, Section A,2). It is difficult to comprehend with this mechanism why after the hydride shift has taken place the π allyl formation is not very rapid. More likely the reaction involves a true 1:4 addition of the hydridotetracarbonylcobalt to the diene, either radical or more probably ionic in character. The adduct then would need to lose carbon monoxide before the π-allyl structure could be formed. In the addition of hydridopentacarbonylmanganese(I) to butadiene a mixture of *cis*- and *trans*-crotylmanganese derivatives can be isolated (70), again suggesting that dissociation has not occurred before the addition reaction. Although π-complexing of simple olefins with the hydride appears necessary before addition can occur, this may not be the case with the more reactive conjugated dienes.

A whole series of additions of compounds containing active hydrogens to conjugated dienes catalyzed by transition metal compounds are known. Mechanistically, they probably fall into two groups. In one group the active

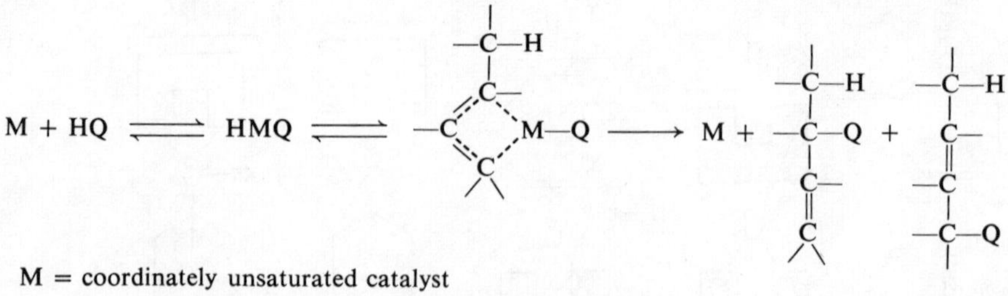

M = coordinately unsaturated catalyst

HQ = HCN, $CH_3COCH_2COCH_3$, $CH_3OCOCH_2COOCH_3$, CH_3COOH, etc.

hydrogen group probably oxidatively adds to the catalyst forming a hydride which then adds to the diene. A reductive elimination of the product with regeneration of the catalyst would be the last step. Two isomeric adducts can be formed depending on the way the intermediate adduct eliminates. If a π-allylic intermediate is formed, as would be expected, the coupling can occur at the first or the third carbon of the diene. Usually mixtures are obtained.

The second group of reactions may involve metal hydrides also. The hydrides could be obtained by one of the reactions of Chapter II, Section A, but not by oxidative addition of the active hydrogen compound. The hydride could then add to the diene to form the π-allyl complex. A nucleophilic attack by the active hydrogen compound on the allylic group would displace metal and then a proton transfer would form product and regenerate the catalyst. Which mechanism is followed depends on the ease of the oxidative

$$\text{MH} + \overset{}{\underset{}{\text{C}}}=\text{C}-\overset{|}{\text{C}}=\text{C} \longrightarrow \overset{|}{\text{C}} \cdots \text{M} \xrightarrow{\text{HX:}} \text{HM} + \cdots$$

MH = metal hydride
HX: = CH_3OH, CH_3CH_2OH, $(CH_3)_2NH$, etc.

addition. The oxidative addition is probably favored when the anion from the molecule undergoing the addition forms a particularly strong bond with the metal.

Examples of the first type may include the $Ni[P(OEt)_3]$-catalyzed addition of HCN to butadiene (247, 248) and the $[\phi_2PCH_2CH_2P\phi_2]PdBr_2$-catalyzed addition of 2,4-pentanedione, ethyl acetoacetate, and dimethyl malonate to butadiene (249). The hydrogen cyanide addition has become the basis for the

$$\text{HCN} + CH_2\text{=}CHCH\text{=}CH_2 \xrightarrow[100°]{Ni[P(OEt)_3]_4} CH_3CH\text{=}CHCH_2CN + CH_2\text{=}CHCHCN \overset{CH_3}{|}$$
$$\textit{cis and trans}$$

$$CH_3O\overset{O}{\overset{||}{C}}CH_2\overset{O}{\overset{||}{C}}OCH_3 + CH_2\text{=}CHCH\text{=}CH_2 \xrightarrow[150°]{[\phi_2PCH_2CH_2P\phi_2]PdBr_2 + NaO\phi}$$
$$CH_2\text{=}CHCHCH(COOCH_3)_2 \overset{CH_3}{|} + CH_3CH\text{=}CHCH_2CH(COOCH_3)_2$$

commercial production of adiponitrile from butadiene. The 3-pentenonitrile obtained as above is isomerized to the terminal olefin and reacted again with hydrogen cyanide. Apparently this can be done in high yield in a simple step with the proper isomerization catalyst present in addition to the nickel catalyst. In one example a borane was used as the isomerization catalyst (250). Presumably, a boron hydride is the true catalyst which by an addition–elimination mechanism forms the much more reactive terminal olefin which then reacts with the hydridonickel cyanide catalyst.

$$CH_2{=}CHCH{=}CH_2 + 2HCN \xrightarrow[BR_3']{Ni[P(OR)_3]_4} NCCH_2CH_2CH_2CH_2CN$$

Examples which probably follow the second mechanism of addition are the $RhCl_3$-catalyzed addition of ethanol to butadiene (251) and the $RhCl_3$-(252) or nickel phosphine-catalyzed (253) addition of amines to butadiene.

$$CH_3CH_2OH + CH_2{=}\overset{\overset{\displaystyle CH_3}{|}}{C}CH{=}CH_2 \xrightarrow[60°]{RhCl_3}$$

$$CH_2{=}CHCH\overset{\overset{\displaystyle CH_3}{|}}{}OCH_2CH_3 + CH_3\overset{\overset{\displaystyle CH_3}{|}}{C}{=}CHCH_2OCH_2CH_3$$

$$(CH_3)_2NH + CH_2{=}CHCH{=}CH_2 \xrightarrow[100°]{RhCl_3}$$

$$CH_2{=}CHCH\overset{\overset{\displaystyle CH_3}{|}}{}N(CH_3)_2 + CH_3CH{=}CHCH_2N(CH_3)_2$$

Simple metal hydrides may cause skeletal rearrangements in nonconjugated dienes with certain structures. The best examples occur with "nickel hydride" catalysts and 1,4-dienes. 1,1-Dideutero-2-methyl-1,4-pentadiene rearranges to 3,3-dideutero-2-methyl-1,4-pentadiene on treatment with a combination of dichlorobistri-*n*-butylphosphinenickel(II) and chlorodiisobutylaluminum dimer at room temperature (254). The actual catalyst is very probably hydridochlorobistri-*n*-butylphosphinenickel(II) formed by elimination of the hydride from an isobutyl derivative which, in turn, was formed by an exchange between the isobutylaluminum compound and the dichloronickel complex. The dichloroisobutylaluminum produced or unreacted chlorodiisobutyl-aluminum acts as a Lewis acid and very likely complexes with the remaining chloro group in the hydride producing a more reactive ionic derivative. The

$$(n\text{-}Bu_3P)_2NiCl_2 + [i\text{-}Bu_2AlCl]_2 \longrightarrow [(n\text{-}Bu_3P)_2NiH]^+[i\text{-}BuAlCl_3]^- + CH_2{=}C(CH_3)_2$$

hydride then apparently adds to the diene, putting nickel on the terminal carbon of the monosubstituted double bond. This is the reverse direction of addition to that observed in the propylene dimerization reaction catalyzed by another nickel hydride complex with smaller ligands (cf. Chapter V, Section C,1). The change in addition direction may just be the result of the larger ligands making the sterically less bulky *n*-alkyl favored over the 2-alkyl derivative. A more likely explanation is that the second double bond is coordinating during the addition and directing the addition to produce the larger ring with the less strained, chelated alkenyl group. The alkenyl complex formed must now undergo an elimination of the 1,1-dideutero-2-methylallyl-nickel group leaving ethylene coordinated to the same nickel atom. This reaction should be facilitated if the second double bond were coordinated to the nickel during the reaction and a π-methylallyl complex were formed directly. The rearrangement now occurs simply by a movement of the π-allylic group so that readdition of the dideuteromethallyl group to the coordinated ethylene occurs at the deuterated end of the π-allylic group. Hydride elimination and exchange with another diene molecule would then complete the catalytic rearrangement. Probably, dissociation of one of the tri-*n*-butylphosphine groups is necessary before the carbon–carbon bond can be broken (Scheme VI-2).

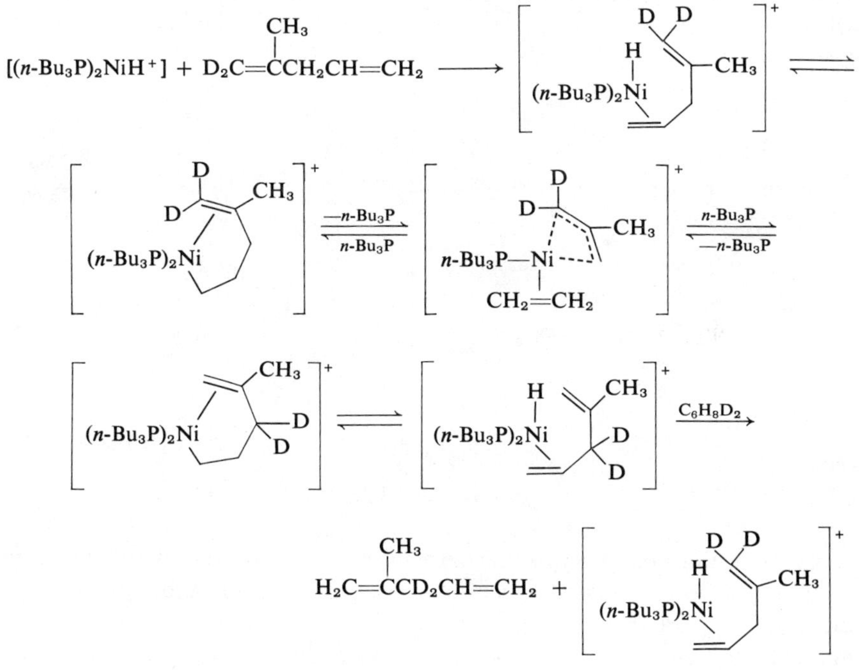

CH3
[(n-Bu3P)2NiH+] + D2C=CCH2CH=CH2
D D
H
CH3
(n-Bu3P)2Ni
D
D
CH3
(n-Bu3P)2Ni
—n-Bu3P
n-Bu3P
D
D
CH3
n-Bu3P—Ni
CH2=CH2
n-Bu3P
—n-Bu3P
CH3
(n-Bu3P)2Ni
D
D
H
CH3
(n-Bu3P)2Ni
D
D
C6H8D2
CH3
H2C=CCD2CH=CH2 +
D D
H
CH3
(n-Bu3P)2Ni

Scheme VI-2.

Treatment of 3-methyl-1,4-pentadiene with the same nickel catalyst as above leads to extensive rearrangement; eight different isomeric products have been identified in the reaction mixture (255). The reaction is complicated

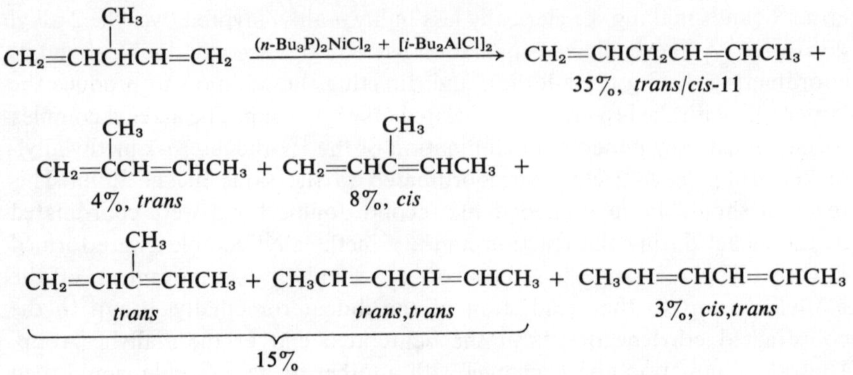

$$\underset{CH_2=CHCHCH=CH_2}{\overset{CH_3}{|}} \xrightarrow{(n\text{-}Bu_3P)_2NiCl_2 \,+\, [i\text{-}Bu_2AlCl]_2} CH_2=CHCH_2CH=CHCH_3 \,+$$
$$35\%, \; trans/cis\text{-}11$$

$$\underset{CH_2=CCH=CHCH_3}{\overset{CH_3}{|}} + \underset{CH_2=CHC=CHCH_3}{\overset{CH_3}{|}} \,+$$
$$4\%, \, trans \qquad\qquad 8\%, \, cis$$

$$\underset{CH_2=CHC=CHCH_3}{\overset{CH_3}{|}} + CH_3CH=CHCH=CHCH_3 + CH_3CH=CHCH=CHCH_3$$
$$\underbrace{\qquad trans \qquad\qquad\qquad trans,trans \qquad\qquad}_{15\%} \qquad 3\%, \, cis,trans$$

by the fact that some of the products are not completely stable under the
reaction conditions. It is clear, however, that at least two different reaction
mechanisms must be involved. The first product listed, 1,4-hexadiene,
probably arises by way of the π-allylic π-olefin mechanism just discussed,
while the other products may arise by any of several other possible routes
(see Scheme VI-3).

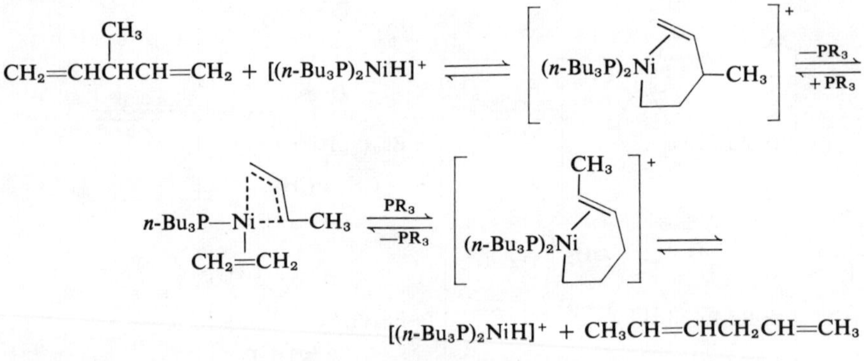

Scheme VI-3.

An interesting possibility for the mechanism of formation of the conjugated
isomers had been proposed which involves nickel hydride addition forming
the 2-nickel alkyl from the product of the above reaction followed by an
internal addition of the metal carbon group to the remaining double bond.
The addition forms a cyclopropane derivative. Reversal of the reaction,
breaking a different cyclopropane carbon–carbon bond, and hydride elimination would produce the second product, 2-methyl-1,3-pentadiene (256) (see
Scheme VI-4) The preferred direction of addition of the hydride to the olefin
is usually not important in these reversible reactions since as long as the

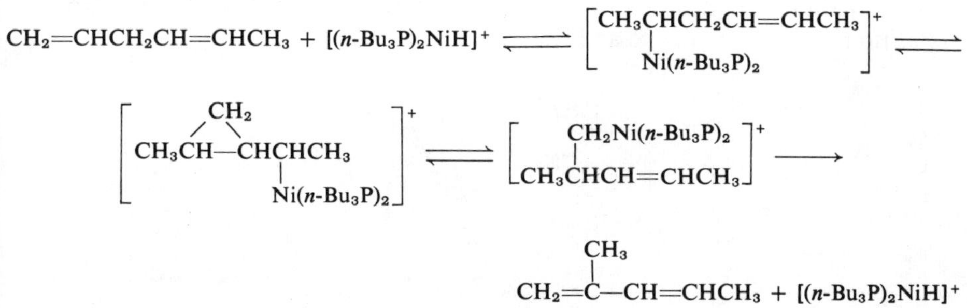

$$CH_2=CHCH_2CH=CHCH_3 + [(n\text{-}Bu_3P)_2NiH]^+ \rightleftharpoons \left[\begin{array}{c} CH_3CHCH_2CH=CHCH_3 \\ | \\ Ni(n\text{-}Bu_3P)_2 \end{array} \right]^+ \rightleftharpoons$$

$$\left[\begin{array}{c} \overset{CH_2}{\diagup \diagdown} \\ CH_3CH-CHCHCH_3 \\ | \\ Ni(n\text{-}Bu_3P)_2 \end{array} \right]^+ \rightleftharpoons \left[\begin{array}{c} CH_2Ni(n\text{-}Bu_3P)_2 \\ | \\ CH_3CHCH=CHCH_3 \end{array} \right]^+ \longrightarrow$$

$$\begin{array}{c} CH_3 \\ | \\ CH_2=C-CH=CHCH_3 \end{array} + [(n\text{-}Bu_3P)_2NiH]^+$$

Scheme VI-4.

hydride is able to add either way to a significant extent the reaction will proceed by way of the adduct which leads to the most stable transition state and, therefore, usually to the most stable products also.

D. PROTON ADDITION AND HYDRIDE ABSTRACTION REACTIONS OF COORDINATED DIENES, TRIENES, AND TETRAENES

Coordinated conjugated dienes often may be protonated with strong acids to form π-allylic complexes. Iron diene complexes, at least, show high selectivity with respect to the isomeric complex produced by protonation. The product produced depends on the reaction media and diene structure (257). The reaction of anhydrous hydrogen chloride in nonpolar media with tricarbonyl-π-butadieneiron(0) produces syn-chloro-π-1-methylallyltricarbonyliron(II), whereas the same reaction with fluoroboric acid in polar media produces the anti cation. Treatment of the syn-chloro complex with silver tetrafluoroborate gives the syn cationic complex (258). The anti isomer would be expected to be formed directly by protonation with a strong acid (in polar media) since the diene ligand is already in a cis configuration. In the

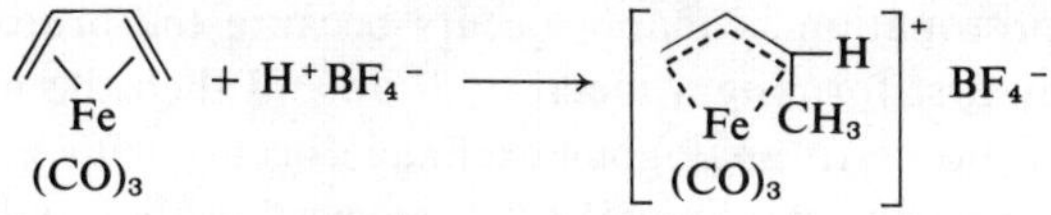

$$\underset{\substack{Fe \\ (CO)_3}}{\diagup\diagdown\diagup\diagdown} + H^+BF_4^- \longrightarrow \left[\underset{\substack{Fe \ CH_3 \\ (CO)_3}}{\diagup\diagdown}H \right]^+ BF_4^-$$

reaction with anhydrous hydrogen chloride, however, a complete addition of the acid to one of the double bonds is assumed to occur and this adduct then, by way of an oxidative addition of the carbon–chlorine group, would form the syn product. If the addition of the chloride ion occurs on the side of the π-complexed diene away from the metal, then the chloro group must be rotated in order to be able to undergo a four-centered oxidative addition.

This rotation would move the methyl group so that it would be in a syn position after the oxidative addition.

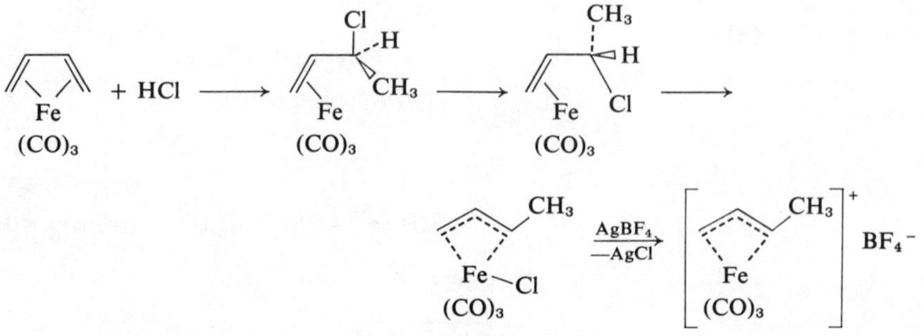

Surprisingly, however, treatment of tricarbonyl-π-1-phenyl-3-methylbuta-dieneiron(0) with deuterium chloride in pentane produces the anti trideutera-ted π-allylic product (259). The fact that deuteration is exclusively at the *anti*-methyl group shows that the above mechanism cannot be operating. A reasonable explanation is that since the chloride in this example would have been tertiary, whereas it was secondary in the above case, it could be completely ionized and oxidative addition would not occur.

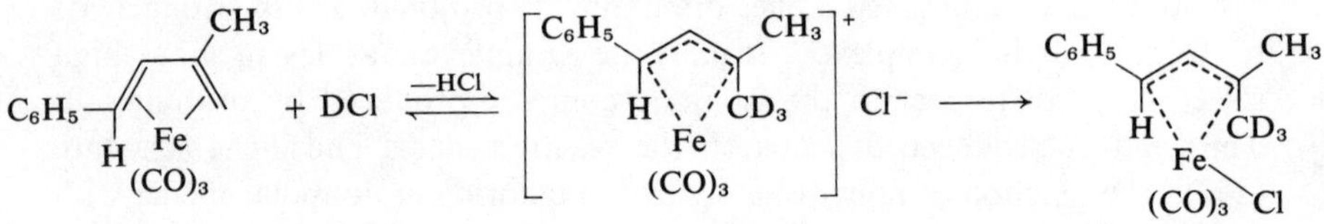

The reversible deuteration of tricarbonyl-π-1,3-cyclohexadieneiron(0) with deuterotrifluoroacetic acid likewise is specific. Only the endo (same side as the metal) protons are exchanged meaning that proton (deuteron) addition and loss occurred only from that side. Since the proton addition and loss can move the double bonds around the cyclohexane ring, complete exchange gives the *endo,endo*-dideuterodiene complex as the final product (259). The specific endo protonation probably occurs because the proton adds to the diene system by first forming a metal hydride and then the hydrogen shifts to the endo position. An analogous exchange occurs with π-1,3-cyclohexa-diene-π-cyclopentadienylrhodium(I) (260) suggesting that endo protonation is a general reaction.

π-Complexes of nonconjugated dienes also often protonate readily. The protonation of *cis,cis*-1,5-cyclooctadiene-π-cyclopentadienylrhodium with trifluoroacetic acid is of interest because the complex rearranges at a measur-able rate to a π-allylic complex and the intermediate complexes can be identified by NMR. In 30 min. at room temperature the 4-cyclooctenyl complex is formed. After 2 days at room temperature the 3-cyclooctenyl species appears and after 55 days the π-2-cyclooctenyl derivative is the sole isomer present. Treatment of the product with triethylamine rapidly converts the cationic allylic compound back into the neutral cis,cis-1,5-cyclooctadiene-containing starting complex. This product is converted rapidly back into the π-allylic form by stronger acids than trifluoroacetic acid (261).

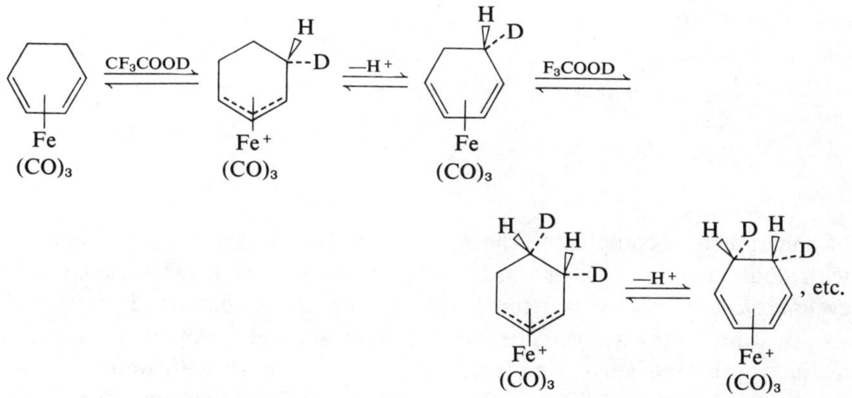

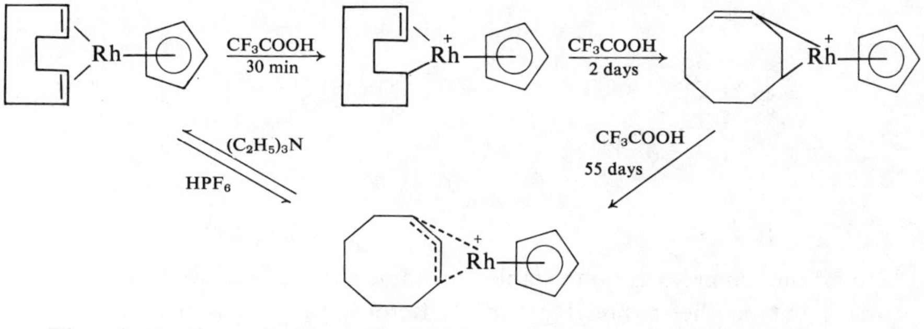

The abstraction of a hydride ion from π-diene complexes often occurs easily as the products are π-pentadienylmetal complexes analogous to the very stable π-cyclopentadienyl complexes. Triphenylcarbonium ion salts with strong acids have most often been used as the hydride-abstracting agents. The abstraction is selective in that exo-hydrogens only are removed, probably for steric reasons. Reaction of the above endo,endo-tricarbonyl-π-dideutero-1,3-cyclohexadieneiron(0) with triphenylmethyl tetrafluoroborate produces a pentadienyl complex with loss of one of the exo-hydrogens (259). This reaction can be reversed with sodium borohydride and the hydrogen enters from the exo side. Thus, borohydride ion appears to attack the ligand directly.

The hydride abstraction reaction has been used to prepare a diene π complex of the keto form of phenol, again demonstrating the high stability of some diene π complexes. The reaction of 1-methoxy-1,4-cyclohexadiene with dodecacarbonyltriiron produces a 3:1 mixture of a π-1-methoxy-1,3-cyclohexadiene and a π-1-methoxy-1,4-cyclohexadiene complex. Reaction of the mixture with triphenylmethyl tetrafluoroborate gives two isomeric methoxycyclohexadienyl complexes, both of which react with water to give the diene π complex of the keto form of phenol (262) (see Scheme VI-5). The

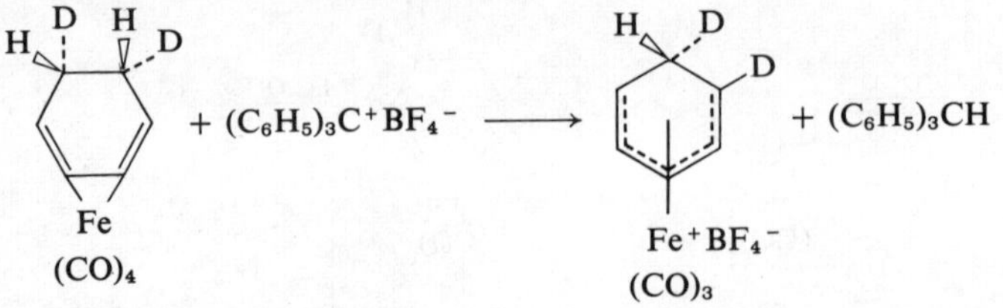

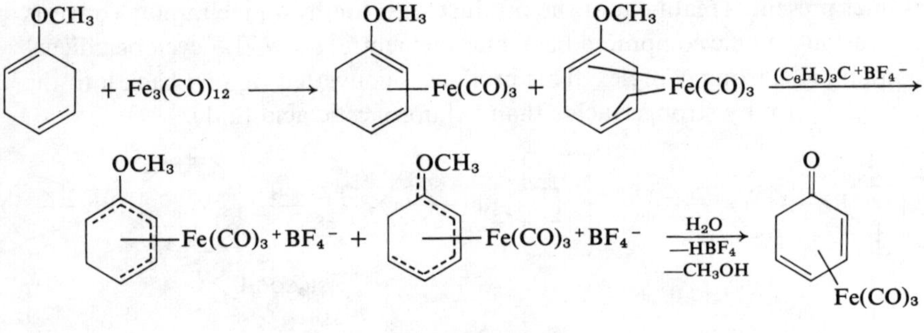

Scheme VI-5.

keto–phenol complex is quite stable and some conventional reactions can be carried out on the carbonyl group. A Reformatsky reaction with methyl bromoacetate produced the expected carbomethoxymethyl hydroxy derivative. Treatment of this product with triphenylmethyl tetrafluoroborate results in loss of the hydroxyl group and a proton forming tricarbonyl(carbomethoxy-methylenecyclohexadiene)iron(0) (263) (see Scheme VI-6).

Iron carbonyl complexes of the enol forms of unsaturated aldehydes and ketones also can be prepared by reaction of the enol acetates with nona-carbonyldiiron, followed by hydrolysis of the acetate group. Direct hydrolysis has not been used, but rather the acetate group is first reacted with methyl-lithium and the reaction products hydrolyzed (264, 265). The strong tendency for iron to complex preferentially with diene or dienyl systems is clearly illustrated in reactions of tricarbonyl-π-cyclooctatetreneiron(0). This complex is fluxional and has only two of the four cyclooctatetraene double bonds coordinated to the iron at any time (266, 267).

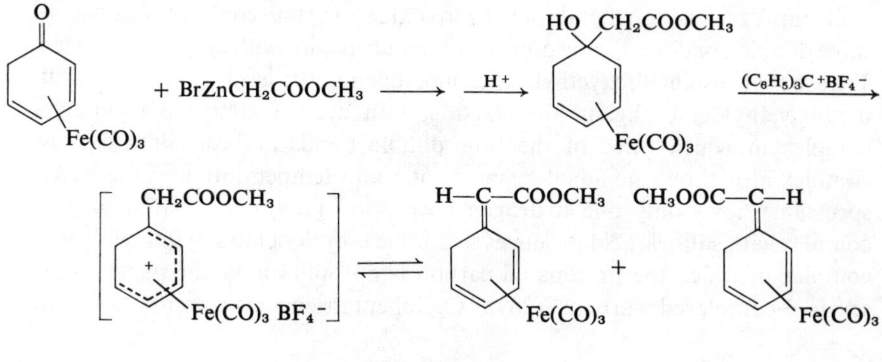

Scheme VI-6.

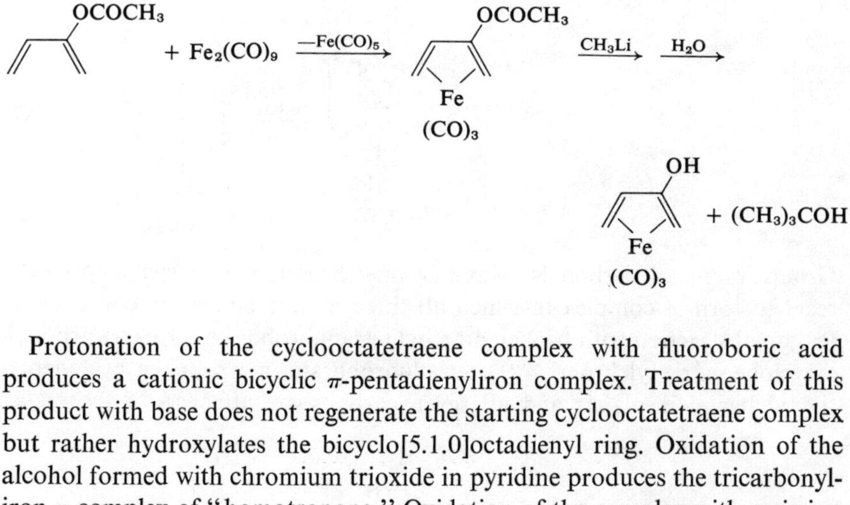

Protonation of the cyclooctatetraene complex with fluoroboric acid produces a cationic bicyclic π-pentadienyliron complex. Treatment of this product with base does not regenerate the starting cyclooctatetraene complex but rather hydroxylates the bicyclo[5.1.0]octadienyl ring. Oxidation of the alcohol formed with chromium trioxide in pyridine produces the tricarbonyliron π complex of "homotropone." Oxidation of the complex with ceric ion liberates "homotropone" providing a convenient synthesis for this very reactive substance (268).

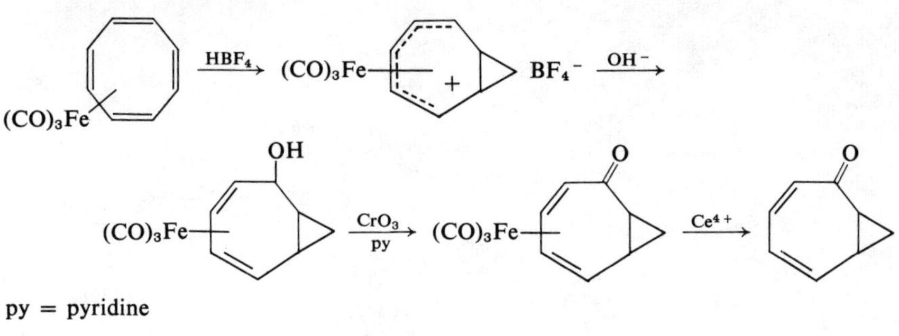

py = pyridine

Group VI transition metals in the zero oxidation state complex readily with three double bonds and show quite a different behavior with cyclooctatetraene. Tricarbonyl(bisdimethoxyethyl ether)molybdenum(0) easily loses the bisdimethoxyethyl ether ligand on reaction with cyclooctatetraene forming a complex in which three of the four double bonds are coordinated. This complex also shows fluxional behavior at room temperature and the NMR spectrum shows only one hydrogen absorption (269). Protonation of this complex with sulfuric acid produces a cationic π-cyclooctatrienylmolybdenum complex in which the protonated carbon is out and above the plane of the seven π-complexed carbons (269). Cycloheptatriene also complexes with

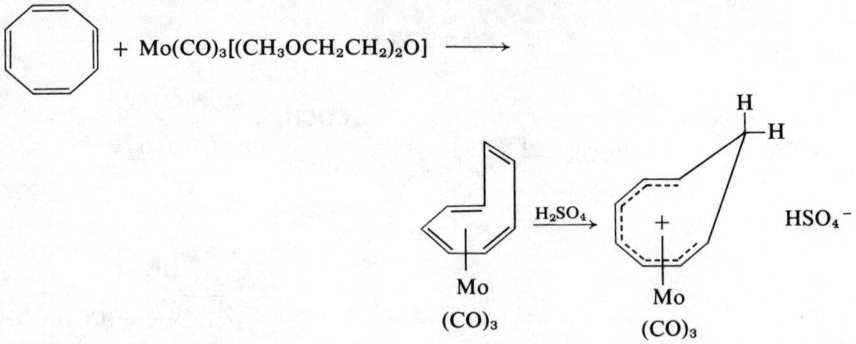

Group VI metal carbonyls. Hexacarbonylchromium and cycloheptatriene react to form a complex in which all three double bonds are coordinated. Proton abstraction with triphenylmethyl tetrafluoroborate forms tricarbonyl-π-cycloheptatrienylchromium(I) tetrafluoroborate in which the cycloheptatrienyl ligand is planar and all seven carbons are attached to the metal.

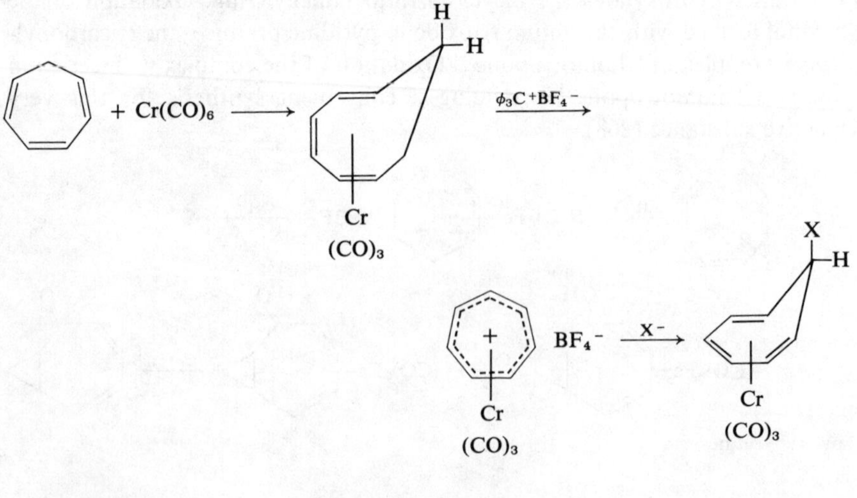

Nucleophilic substitution reactions on this cation give exclusively the cycloheptatriene complex back with the substituent in an exo position (270).

E. REACTIONS OF COORDINATED DIENES WITH NUCLEOPHILES

π-Diene complexes of platinum(II) and palladium(II) readily undergo nucleophilic attack with a variety of reagents. Other transition metal diene complexes apparently either do not undergo the reaction or with one or two exceptions the reactions have not been investigated.

Palladium(II) chloride reacts with conjugated dienes to form 1-chloro-methyl-π-allylpalladium derivatives. These may be covalent additions in nonpolar solvents or they may be ionic with chloride ion attacking the diene π complex. The specificity of these additions is seen in the reactions of the three isomeric 2,4-hexadienes with palladium chloride. All three 2,4-hexadiene isomers at room temperature yield *syn,syn*-1-(1-chloroethyl)-3-methyl-π-allylpalladium chloride dimers, but at $-40°$ the cis,cis and the trans,trans isomers yield one pair of the two possible diastereomers, and the cis,trans isomer the other. Even though in two of the three cases, anti to syn isomerizations occurred, specificity was retained. The diastereomers equilibrate at room temperature. Triphenylphosphine displaces the dienes from the complexes at low temperatures and the trans,trans isomer is obtained from one diastereomer and the cis,trans isomer from the other (271).

$$PdCl_2 + CH_3CH{=}CHCH{=}CHCH_3 \longrightarrow$$

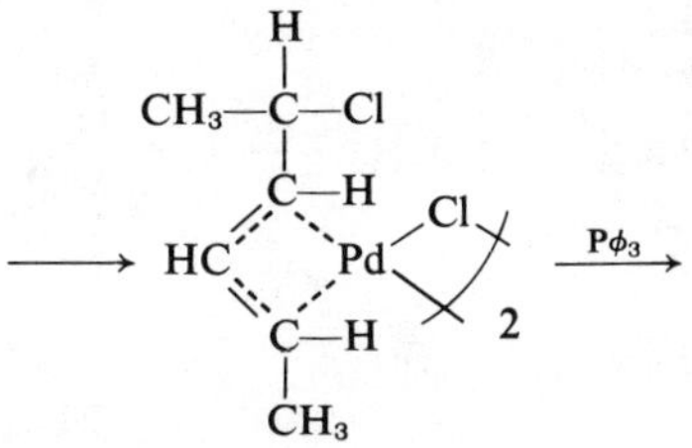

$$CH_3CH{=}CHCH{=}CHCH_3 + PdCl_2[P(C_6H_5)_3]_2$$

In alcoholic solution the conjugated diene–palladium or platinum chloride reaction yields π-(1-alkoxymethyl)allyl derivatives. Sodium chloropalladate and butadiene in methanol solution form chloro-π-(1-methoxymethyl)allyl-palladium(II) dimer as a stable product. The chloro group may be easily replaced by other ligands such as acetylacetonate by reaction with thallium(I) acetylacetonate or cyclopentadienyl by reaction with sodium cyclopenta-dienide (272).

$$CH_2{=}CHCH{=}CH_2 + Na_2PdCl_4 + CH_3OH \longrightarrow$$

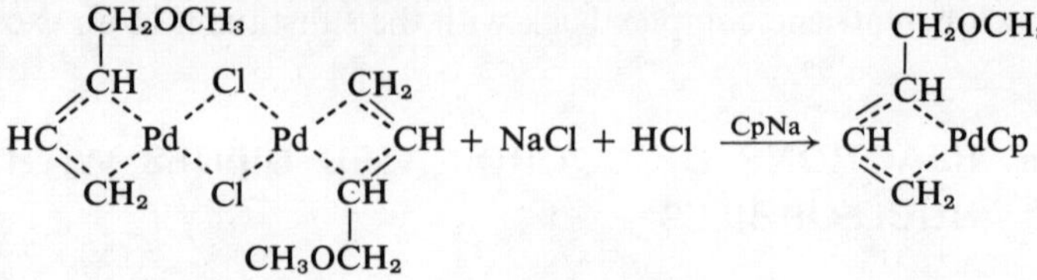

$$Cp = C_5H_5{}^-$$

Nonconjugated dienes which coordinate with the metal (platinum or palladium) through both double bonds also react with nucleophiles. The metal chlorides do not generally add to nonconjugated dienes but only form π complexes. The reaction of norbornadiene with palladium(II) chloride forms a π complex which, in methanol, produces the *exo*-methoxynorbornenyl complex. Treatment of this chlorine-bridged dimer with the bidentate ligand 1,2-bisdiphenylphosphinoethane results in cleavage of the chlorine bridge and a rearrangement which allows the palladium to coordinate with both phosphorus atoms in the ligand. In the rearrangement the palladium–carbon bond adds to the double bond producing an *endo*-nortricyclylpalladium derivative. It is of interest that cleavage of the palladium–carbon bond in the product by bromine proceeds with retention of configuration presumably because the cleavage goes by an oxidative addition of bromine to the palladium followed by a 1:2 shift of bromine from palladium to carbon (a reductive elimination) (273). In the halogen cleavage of coordinately saturated metal–carbon complexes inversions of configuration has been observed in the few cases studied.

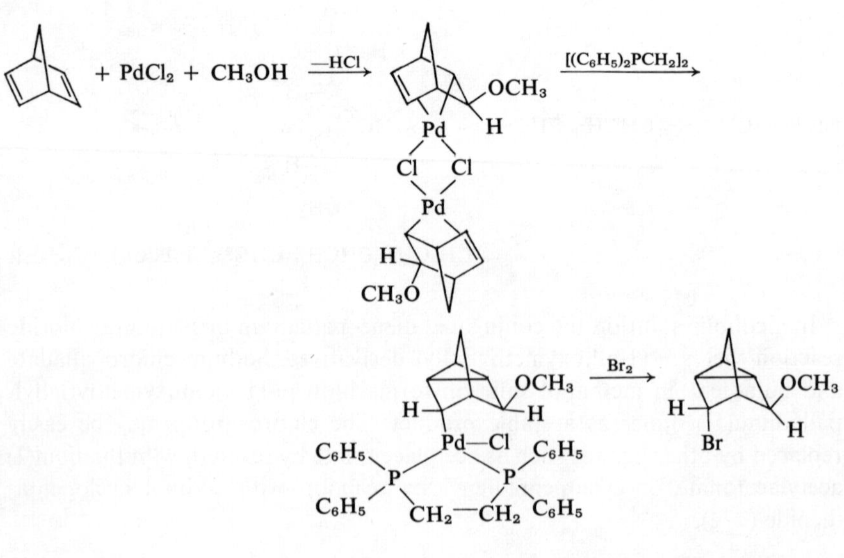

The chloro(*exo*-methoxydicyclopentenyl)platinum(II) dimer from dicyclopentadiene, platinum(II) chloride, and methanol has been resolved by fractional crystallization of the complex formed with *S*-1-phenylethylamine. Treatment of the resolved amine complex with cold hydrochloric acid reforms the optically active, bridged chlorine dimer, whereas hot hydrochloric acid produces optically active dichlorodicyclopentadieneplatinum(II). Optically active dicyclopentadiene was obtained from the complex on treatment with cyanide ion. Racemic alcohols reacted selectively with the optically active dichlorodicyclopentadieneplatinum(II). Excess *dl*-2-butanol, for example, reacted, leaving the unreacted alcohol partially resolved (274).

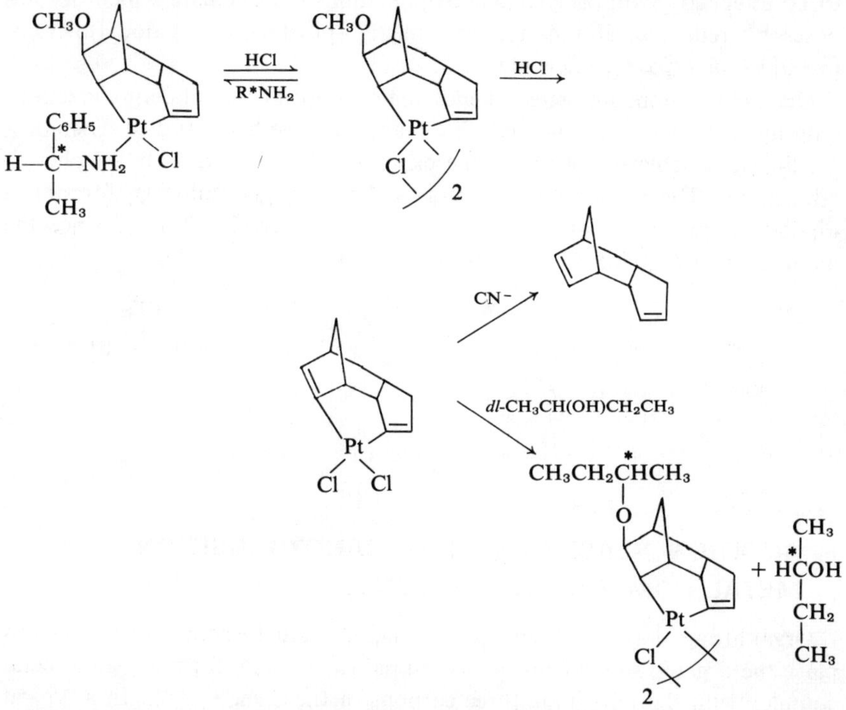

The reactions of other nucleophiles with nonconjugated diene π complexes at the present time are rather unpredictable, as with the chloride above, in that attack may occur either at the metal or at one of the coordinated double bonds. Particularly complicated examples occur with the reaction product of dimethyl malonate anion with dichloro-1,5-cyclooctadienepalladium(II). This product undergoes several different further reactions with bases and nucleophiles. Sodium borohydride reduces the complex to cyclooctylmalonic ester. Sodium methoxide and dichloro-1,5-cyclooctadienepalladium(II) produce 4-cyclooctenylmalonic ester. This reaction probably involves methoxide attack at the metal, β elimination of metal hydride with formation of formaldehyde, and a subsequent shift of hydrogen from palladium to carbon (a reductive elimination). A mild base, sodium carbonate, converts the malonic ester adduct into the 3,5-cyclooctadienylmalonic ester by an elimination and double-bond shift. The strongly basic dimethyl sulfoxide anion reacts to form 9,9-dicarbomethoxybicyclo[6.1.0]nonene, apparently by way of the α anion of the malonic ester group which then undergoes a ring closure with displacement of palladium. Reaction of the cyclooctenyl-malonic ester–palladium derivative with a second mole of dimethyl malonate anion produces still a different reaction. The remaining double bond appears to be attacked giving a cyclooctadiylpalladium intermediate which decomposes by reductive elimination forming 2,6-bis(dimethyl malonyl)bicyclo-[3.3.0]octane (275) (see Scheme VI-7).

Open-chain nonconjugated dienes may complex and undergo reactions with nucleophiles, although very few examples have been studied. Dichloro-π-1,5-hexadieneplatinum(II), in one example, was reacted with S-1-phenyl-ethylamine. The adduct formed retains the hydrogen chloride, forming a dipolar complex, but only one pair of diasteromers is obtained since the amine attacks the side of the diene away from the metal (276).

F. ADDITION REACTIONS OF ORGANOTRANSITION METAL COMPOUNDS TO DIENES

Organotransition metal complexes generally add to conjugated dienes to place the organic substituent on a terminal carbon and form a π-allyl metal complex with the remaining three carbons of the diene system. In a typical example, "phenylpalladium chloride," prepared in the reaction mixture by the exchange reaction of phenylmercuric chloride with palladium(II) chloride or lithium tetrachloropalladate, reacts with butadiene at room temperature to form chloro-1-benzyl-π-allylpalladium(II) dimer in 48% yield (277).

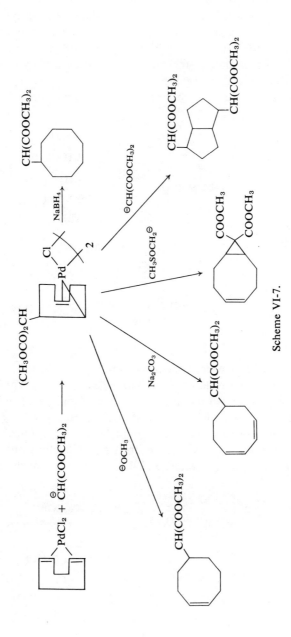

Scheme VI-7.

Pentacarbonylphenylmanganese(I) also reacts with butadiene. A vacant coordination position is apparently made available for the butadiene by a shift of the phenyl from manganese to one of the coordinated carbonyl groups and this is followed by additions of the benzoyl group. The initially formed adduct appears to rearrange since tetracarbonyl-π-1-benzoyl-3-methylallylmanganese(I) is the product isolated. On warming this product loses carbon monoxide and forms a 1-oxapentadienylmanganese complex (278).

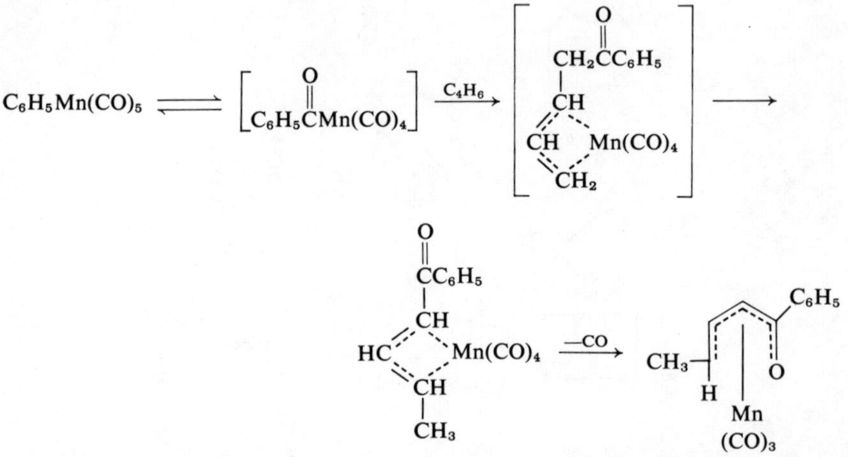

$$C_6H_5Mn(CO)_5 \rightleftharpoons \left[C_6H_5\overset{\overset{O}{\|}}{C}Mn(CO)_4 \right] \xrightarrow{C_4H_6} \left[\begin{array}{c} \overset{O}{\|} \\ CH_2CC_6H_5 \\ | \\ CH \\ \diagup\diagdown \\ CH \quad Mn(CO)_4 \\ \diagdown\diagup \\ CH_2 \end{array} \right] \longrightarrow$$

$$\begin{array}{c} \overset{O}{\|} \\ CC_6H_5 \\ | \\ CH \\ HC\diagup\diagdown Mn(CO)_4 \\ \diagdown\diagup \\ CH \\ | \\ CH_3 \end{array} \xrightarrow{-CO} \quad CH_3 \diagdown\diagup\diagdown_H \overset{C_6H_5}{\underset{\underset{(CO)_3}{Mn}}{\diagup\diagdown_O}}$$

The addition of tetracarbonylmethylcobalt(I) to butadiene is similar except that the initial adduct is isolable. Tetracarbonylacetylcobalt(I) reacts similarly but more slowly as carbon monoxide dissociation must occur initially (279). The adduct in this case undergoes an elimination reaction with various bases. A proton on the methylene between the π-allyl and acetyl groups is apparently lost first and then the anion eliminates tricarbonylcobaltate anion (probably solvated) forming 1-acetylbutadiene (101). The formation of acyldienes by

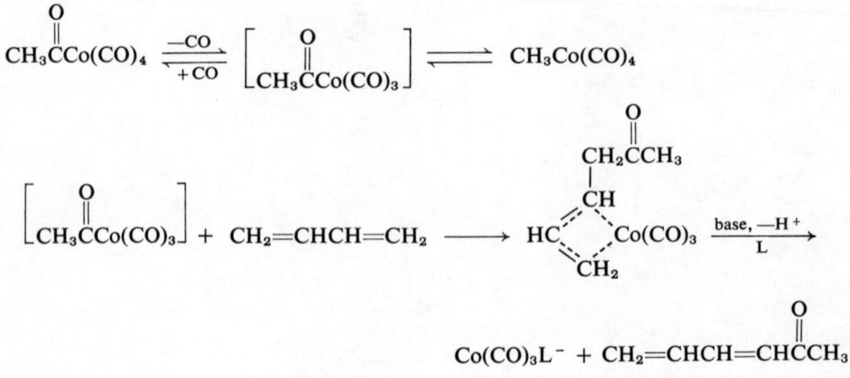

$$CH_3\overset{\overset{O}{\|}}{C}Co(CO)_4 \underset{+CO}{\overset{-CO}{\rightleftharpoons}} \left[CH_3\overset{\overset{O}{\|}}{C}Co(CO)_3 \right] \rightleftharpoons CH_3Co(CO)_4$$

$$\left[CH_3\overset{\overset{O}{\|}}{C}Co(CO)_3 \right] + CH_2=CHCH=CH_2 \longrightarrow HC\diagup\diagdown_{\begin{array}{c} CH_2CCH_3 \\ | \\ CH \\ \diagup\diagdown \\ Co(CO)_3 \\ \diagdown\diagup \\ CH_2 \end{array}} \xrightarrow[L]{base, -H^+}$$

$$Co(CO)_3L^- + CH_2=CHCH=CH\overset{\overset{O}{\|}}{C}CH_3$$

the above reaction is quite general and it can be carried out catalytically in tetracarbonylcobaltate anion. In the catalytic reaction an alkyl halide (or acyl halide), a conjugated diene, a hindered amine as the base, and carbon monoxide at atmospheric pressure are reacted in ether solution with a catalytic amount of the carbonylcobaltate anion. The halide reacts with the anion first, forming the alkyl-or acylcobalt derivative which then adds to the diene. The diene adduct next reacts with the amine forming the acyldiene and the tricarbonylcobaltate anion. The last compound finally reacts with the carbon monoxide present and reforms the catalyst (101).

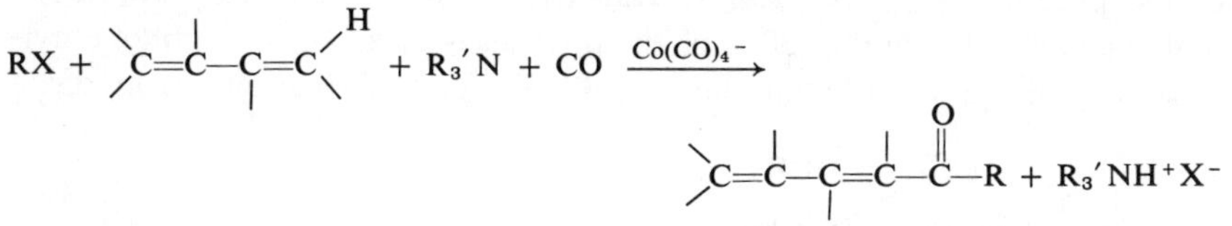

The addition to dienes can also be done internally with formation of cyclic π-allylic complexes. When tricarbonyltriphenylphosphine-*trans,trans*-2,4-pentadienoylcobalt(I) is warmed to 75° internal addition occurs and the π-methylcyclopentenonylcobalt complex is formed in 33% yield. The low yield probably occurs because isomerization of the 2-double bond from the trans to the cis isomer must occur during the reaction in order to close the ring and this is not favorable. The cyclopentenonyl product undergoes hydride elimination with triphenylmethyl tetrafluoroborate forming a cationic π-cyclopentadienone complex (277).

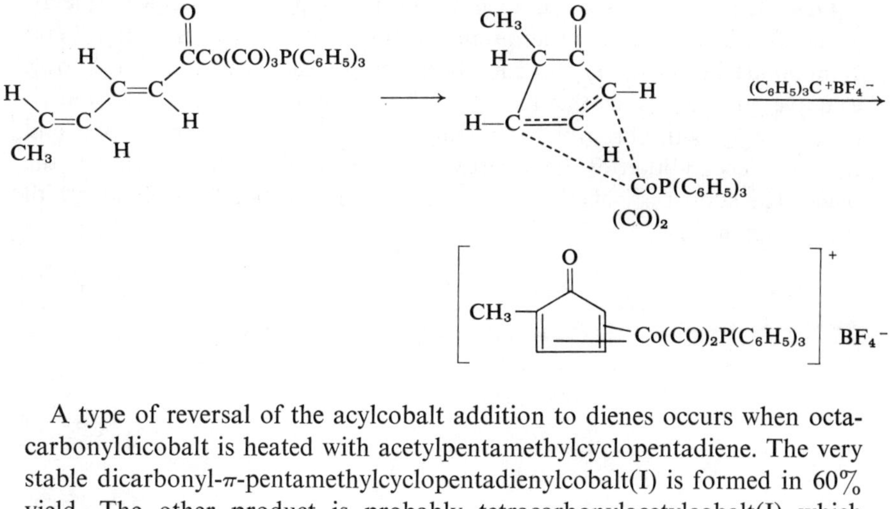

A type of reversal of the acylcobalt addition to dienes occurs when octa-carbonyldicobalt is heated with acetylpentamethylcyclopentadiene. The very stable dicarbonyl-π-pentamethylcyclopentadienylcobalt(I) is formed in 60% yield. The other product is probably tetracarbonylacetylcobalt(I) which decomposes under the reaction conditions (280). The mechanism of this reaction is not clear. Quite possibly, both cobalt atoms of the binuclear metal carbonyl are simultaneously involved. In the related reaction of

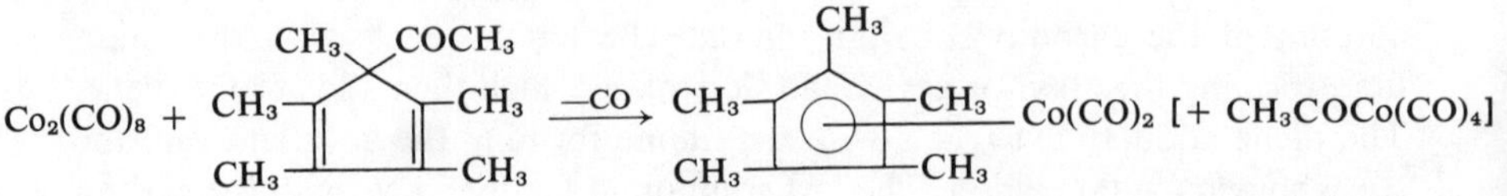

hexacarbonylmolybdenum(0) with the same cyclopentadiene derivative part of the product (43%) is tricarbonylmethyl-π-pentamethylcyclopentadienyl-molybdenum(II). Another 20% of the product is dicarbonyl-π-pentamethyl-cyclopentadienylmolybdenum dimer (280). The loss of other cyclopentadiene

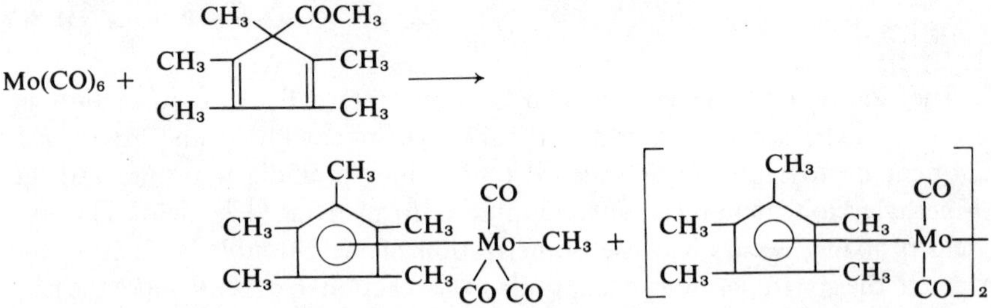

substituents such as the 1-chloroethyl (281) and the ethoxy group (282) has been observed when the appropriately substituted cyclic dienes are reacted with transition metal compounds (cf. Section A of this chapter).

Presumably nonconjugated dienes which can dicoordinate with metals generally will also undergo addition reactions with organotransition metal compounds, but so far very little has been done on such reactions. One known example of interest is the reaction of hexafluoroacetylacetonato-π-allyl-palladium(II) with norbornadiene. Only one of the double bonds reacts and there is a cis addition of the allylpalladium group from the exo side of the diene. The adduct is isolable because the allyl double bond coordinates with the palladium (283).

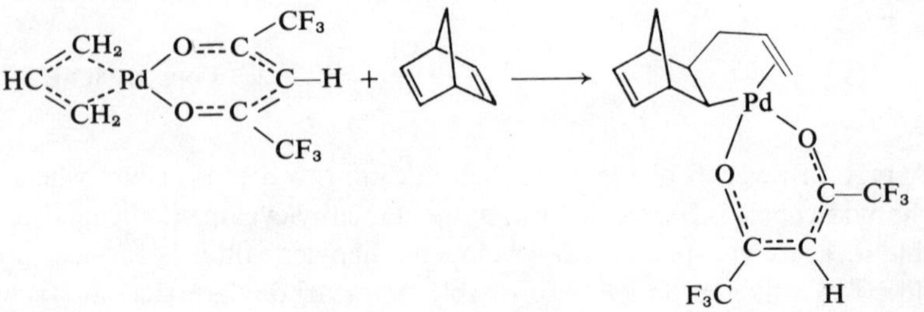

G. METAL-STABILIZED DIENYL AND RELATED CARBONIUM ION COMPLEXES

Metal-stabilized dienyl complexes have already been mentioned in the protonation of complexed polyenes and in hydride abstraction reactions (cf. Section D). These complexes may be prepared by several other methods as well. The simplest of these is by ionization of an allylic substituent from a diene π complex. A good example is the acid-catalyzed ionization of tricarbonyl-π-*trans,trans*-2,4-hexadien-1-oliron(0). The tricarbonyl-1-methylpentadienyliron(III) cation formed reacts reversibly with water forming the isomeric 1,3-hexadien-5-ol π complex (284). Isomerization of both the carbinol group from syn to anti and the hydroxyethyl group from syn to anti also must have occurred in the reactions.

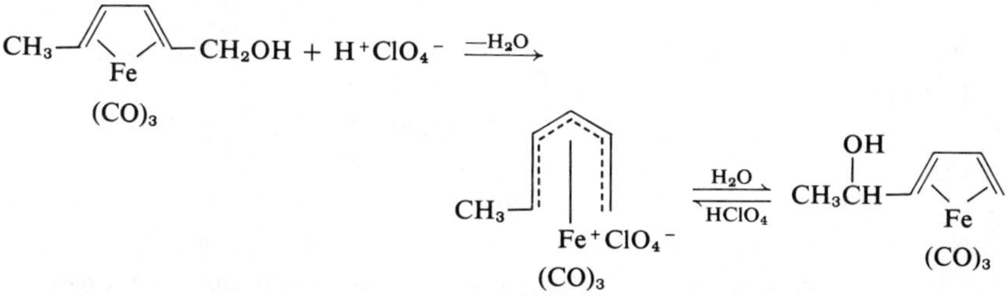

The stabilization of the pentadienyl system by the metal is observed whether the carbonium ion is on a carbon syn or anti to the π-complexed diene system. In the syn position stabilization is apparently by a resonance effect similar to that observed in benzyl carbonium ions, whereas with *anti*-carbonium ions a coordination of the ion to the metal occurs. The rate of ionization of potential pentadienyl π complexes may be very dependent on the isomeric structure of the complex. For example, the two enantiomeric forms of 4-methyl-2,4-heptadien-6-yl 3,5-dinitrobenzoate solvolyze in aqueous acetone at vastly different rates, yet both isomers yield mainly alcohol with the same structure as the starting alcohol.

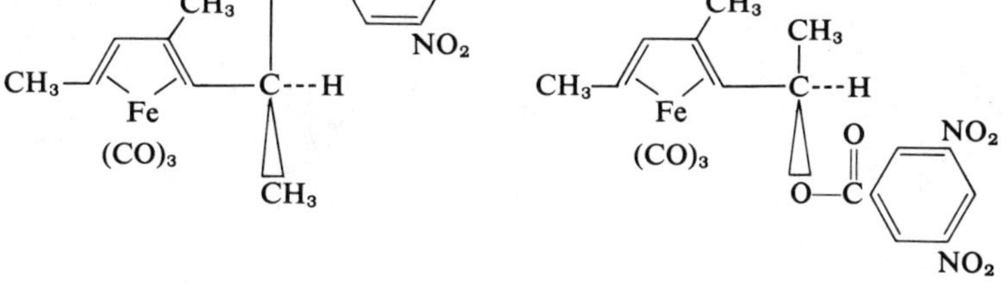

Relative rates of hydrolysis   2500   1

The isomer with the 3,5-dinitrobenzoate group most nearly trans to the tricarbonyliron group in its most stable conformation solvolyzes 2500 times faster than the other isomer (285).

The formation of a pentadienylcobalt system occurs when π-cyclopentadienyl-π-*exo*-5-[chloro(phenyl)methyl]-1,3-cyclopentadienecobalt(I) is warmed to 40° for 50 hr. The product is a cyclohexadienyl complex with the phenyl group at the endo position of the uncomplexed carbon (286).

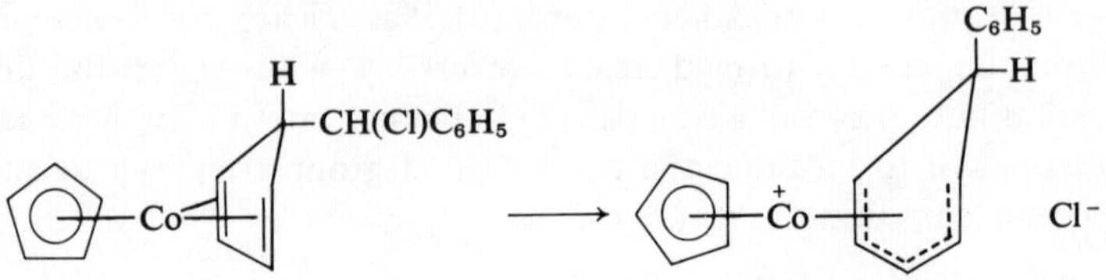

Pentadienyl complexes also may be formed by anionic attack on the appropriate compounds. An example of some synthetic interest is the reaction of bis(π-aromatic)iron(II) dications with organolithium reagents. The attack is exo, as usual, giving a monohexadienyl complex with one equivalent of organolithium complex and a bis derivative with two. Oxidation of the bis complexes with potassium permanganate removes the organic ligands from the iron and oxidizes them to the fully aromatic compounds. By means of this reaction, quite hindered benzene derivatives have been made in yields which may be higher than obtained by alternative, conventional methods (287). The starting complexes are readily available from the reaction between ferrous chloride, aluminum chloride, and the aromatic compound (see Scheme VI-8).

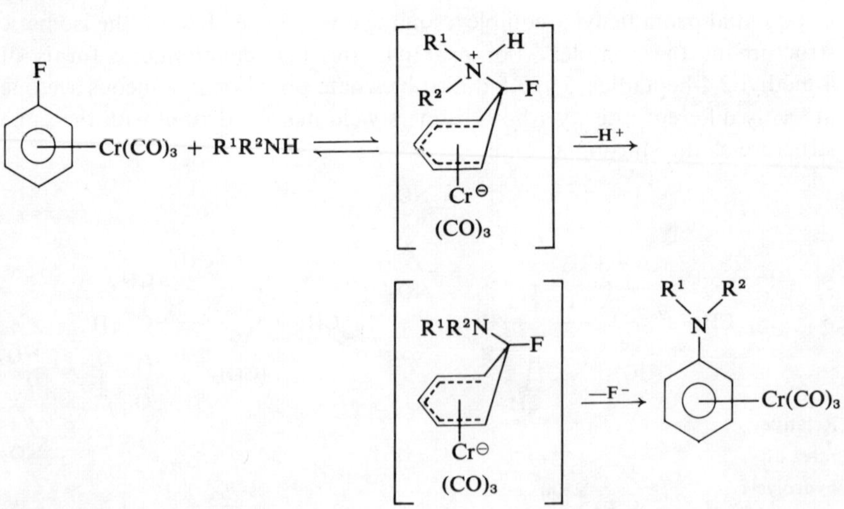

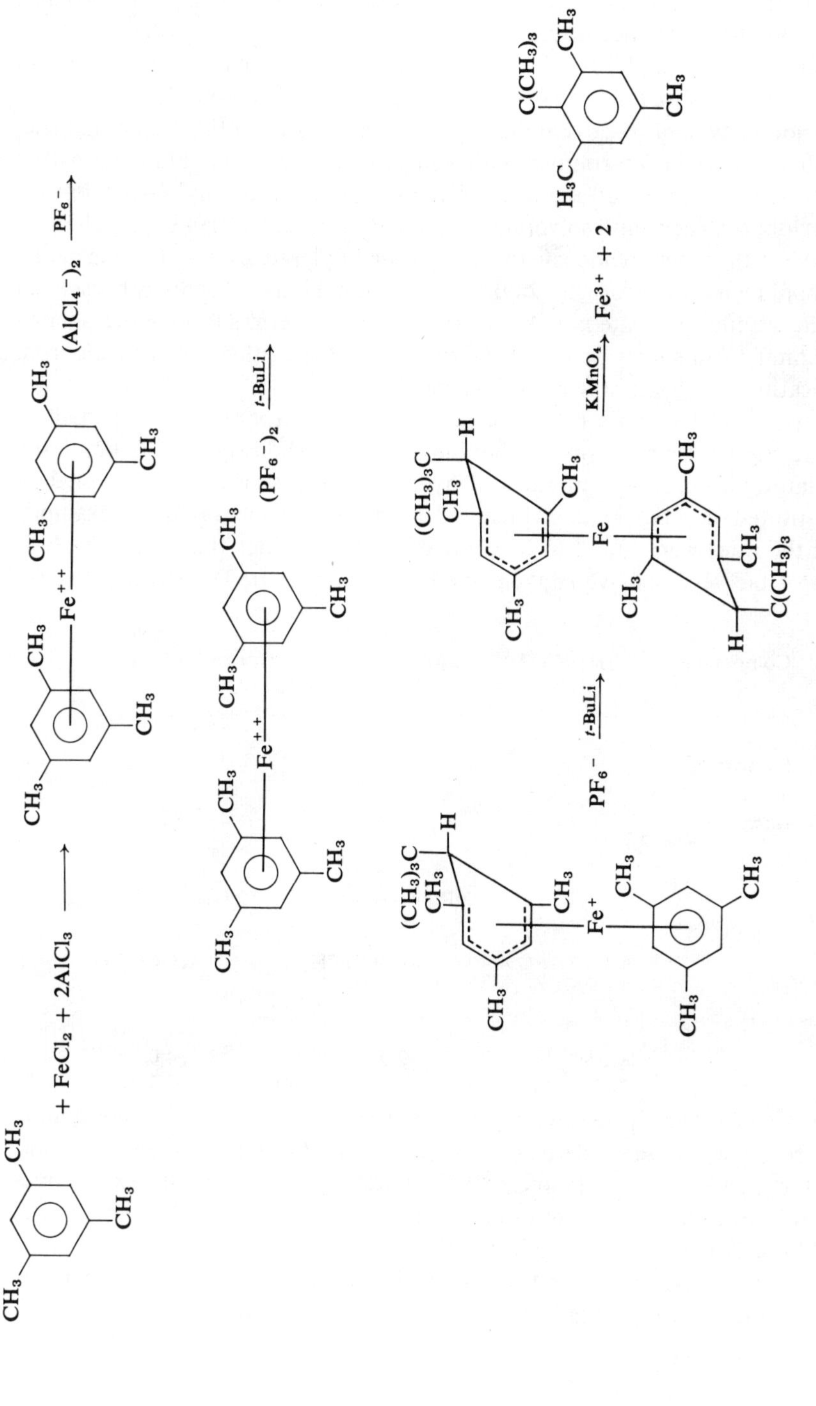

Scheme VI-8.

Chromium-stabilized cyclohexadienyl intermediates are likely involved in the second-order halide displacement reactions of tricarbonyl-π-halobenzene-chromium(0) complexes. The complexed halides undergo displacement reactions far more easily than the uncomplexed halides do (288).

Another type of metal-stabilized pentadienyl system is the cyclobutadiene-methyl carbonium ion complex with iron. (π-Cyclobutadiene complexes will be discussed in more detail in the following section.) Tricarbonyl-π-chloromethylcyclobutadieneiron(0) solvolyzes about 10^8 times faster than benzyl chloride showing the great tendency for the iron-complexed cyclobutadienemethyl carbonium ion to be formed (289). The carbonium ion may actually be isolated as the tetrafluoroborate salt. An X-ray structural determination of the isolated carbonium ion shows no direct interaction of the methyl ion with the iron, indicating stabilization is by resonance.

Several related types of complexes have been prepared which also show the strong tendency to form carbonium ions on carbons next to π-complexed (unsaturated) groups. The rates of hydrolysis of a series of acetoxymethyl-substituted dicyclopentadienyl metal derivatives were measured, for example, and the rates were found to be comparable to or faster than the hydrolysis rate of the very reactive triphenylmethyl acetate (290). The stereochemistry

Compound	$(C_6H_5)_3COAc$	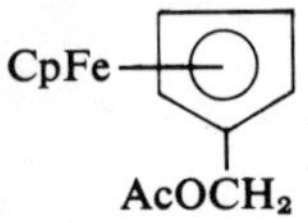 CpFe—⬡ AcOCH$_2$	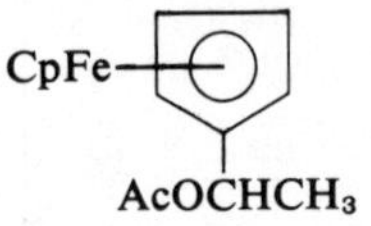 CpFe—⬡ AcOCHCH$_3$
Relative hydrolysis rates	1.0	0.63 	6.7 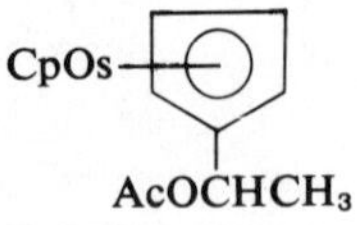

	CpRu—⬡ AcOCHCH$_3$	CpOs—⬡ AcOCHCH$_3$
	9.0	34.0

at the leaving group relative to the metal is very important in determining the reactivity of these derivatives as it is with the related dienyl systems. In the acetoxycyclopentanoferrocene series, for example, the exo isomer hydrolyzes 2240 times more rapidly than the endo compound. Both isomers yield the exo alcohols as products (291).

A smaller activation has been noted in the reactions of tricarbonyl-π-1-acetoxymethyl(cyclopentadienyl)manganese(I) derivatives (292).

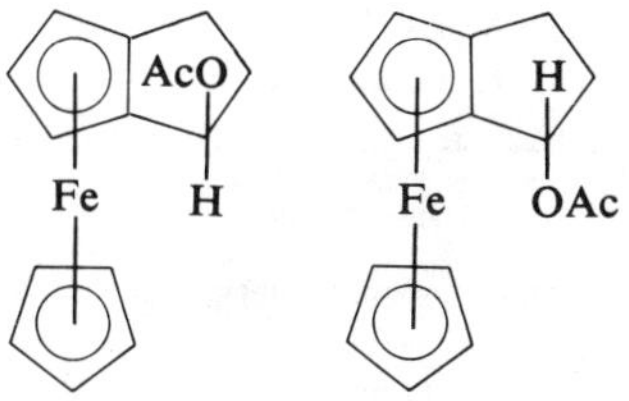

Relative hydrolysis rates   2240   1.0

H. AROMATIC π COMPLEXES AND THEIR REACTIONS

Two kinds of complexes will be considered in this section: (1) complexes of compounds that are already aromatic and (2) complexes of compounds which become aromatic when they are complexed with metals. Many reactions of both types of compounds have been given previously in other connections. The π complexes of various aromatic compounds with the Group VI metal carbonyls have been by far the most studied as they are readily available by heating the hexacarbonylmetal(0) complexes with the aromatic. The π complexes undergo some of the same reactions as the free aromatic but at different rates and with different isomer distributions in substitution reactions. The Friedel–Crafts reaction with these complexes has been studied by several groups. Some interesting differences in isomer distribution appear between the complexed and uncomplexed aromatics. Table VI-1 gives the isomeric composition of products of the aluminum chloride-catalyzed acetylation of toluene and t-butylbenzene free and in a chromium carbonyl complex (293). The selectivity in the acylation of toluene is much reduced on complexation, while the position of substitution is altered when t-butylbenzene is complexed. The large tertiary butyl group apparently fixes the aromatic ring in an eclipsed configuration with the t-butyl group between two carbonyls of the metal complex. In this configuration the eclipsed ortho and para positions would have difficulty in losing hydrogen from the acyl adduct intermediate because there would be carbonyl groups in the way. Substitution, therefore, prefers the meta position where there is less hindrance to departure of the proton.

A curious ring enlargement occurs when π-benzene-π-cyclopentadienylchromium(I) is acetylated or benzoylated; a methyl- or phenyl-substituted π-cycloheptatrienyl-π-cyclopentadienylchromium(III) cation is formed in low yield (294).

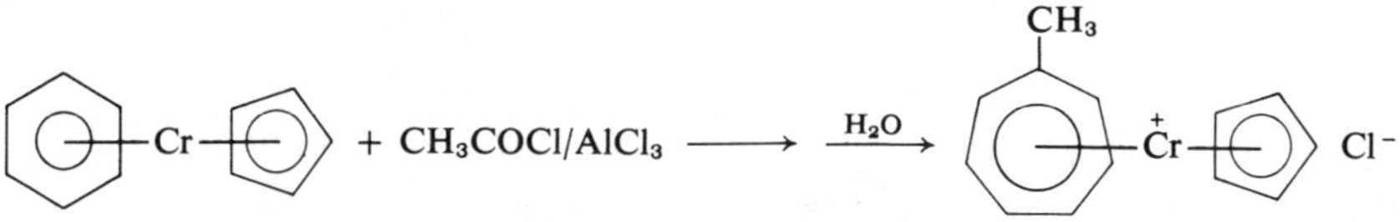

TABLE VI-1
Acetylation Products[a]

Compound acetylated	Reaction products (%)		
	Ortho	Meta	Para
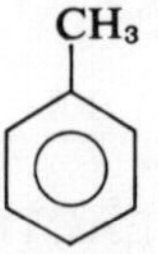 CH₃ benzene ring	1.2	2.0	96.8
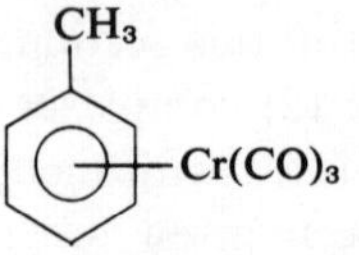 CH₃ benzene ring–Cr(CO)₃	43.0	17.0	40.0
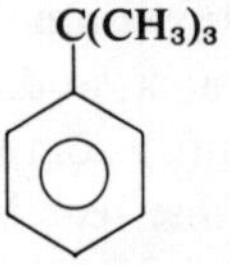 C(CH₃)₃ benzene ring	0.0	4.3	95.7
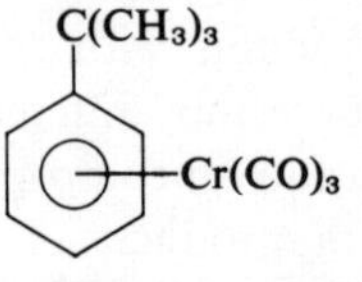 C(CH₃)₃ benzene ring–Cr(CO)₃	0.0	87.0	13.0

[a] Data taken from Jackson and Jennings (293).

A catalytic alkylation reaction has been found to occur when reactive halides and aromatics are heated with catalytic amounts of tricarbonyl-π-aromaticmolybdenum(0) complexes. For example, toluene and *t*-butyl chloride react to give an 82% yield of *t*-butyltoluene (isomeric composition not reported) (295). This reaction is probably a normal Friedel–Crafts reaction, however, and does not involve an alkylation of the π-complexed aromatic.

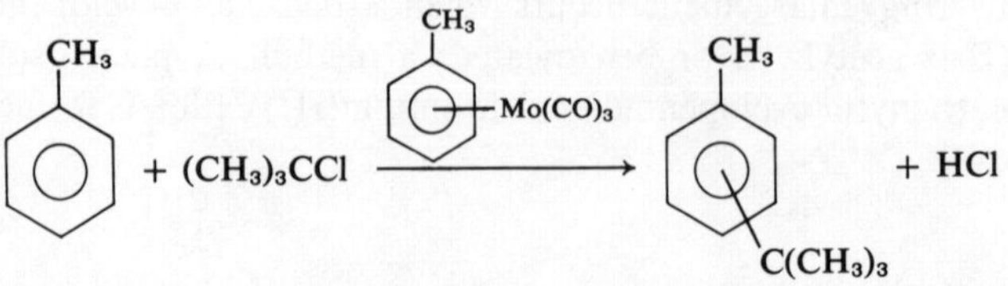

Metalation of dibenzenechromium may be done with amylsodium, but the reaction is not selective; only a few percent of the monosodium salt is formed along with polysodium derivatives. The sodium derivatives react with carbonyl compounds and carbon dioxide in the usual manner (296).

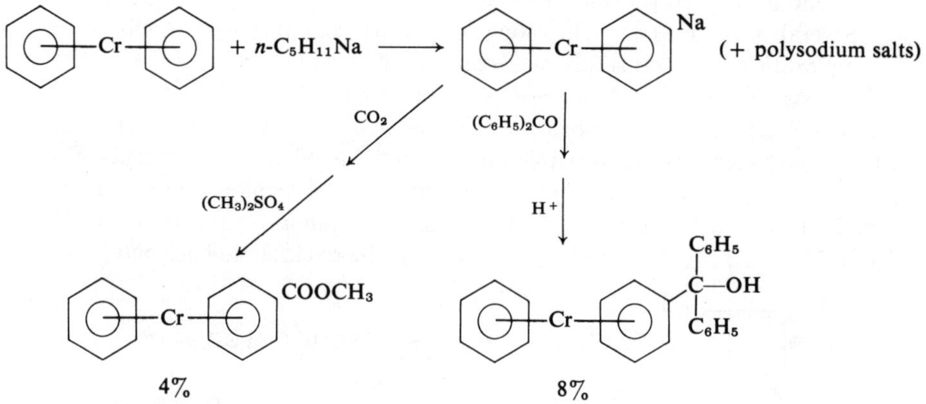

The tricarbonylchromium group complexed with aromatics can be used to control reactions sterically on positions adjacent to aromatic groups. Metal hydride reduction of the carbonyl group in optically active 3-methyl-1-indanone complexed with chromium carbonyl occurs very specifically from the exo side giving endo alcohols. Two isomeric chromium complexes are formed in the preparation as the chromium carbonyl group can add to either side of the aromatic ring. These diastereomers must first be separated and then optically pure products are obtained on reduction. The pure alcohols

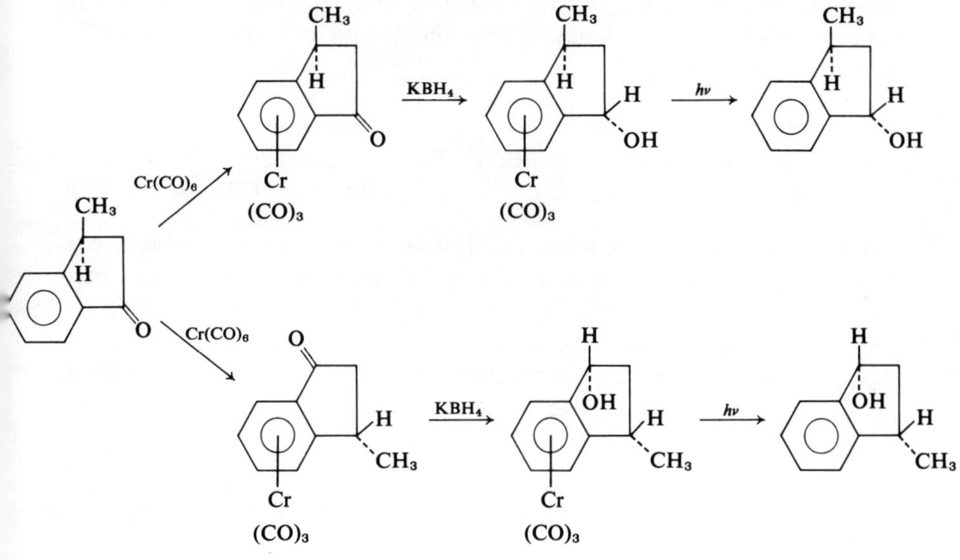

can be freed from the chromium group by irradiation (297). The reduction of
the uncomplexed 3-methyl-1-indanone with lithium aluminum hydride gives
only one of the two possible isomers (297).

Several unsaturated cyclic compounds are significantly stabilized on
complexation with transition metals and could be called aromatic, at least on
the basis that they now undergo substitution reactions. Perhaps the best
known examples of this effect are with π-cyclobutadiene metal complexes.
Four main methods are available for the synthesis of these compounds. The
most commonly used method is the reaction of 3,4-dichlorocyclobutene or
its derivatives with a metal carbonyl or carbonyl anion. Nonocarbonyldiiron
and 3,4-dichlorocyclobutene give tricarbonyl-π-cyclobutadieneiron(0) and

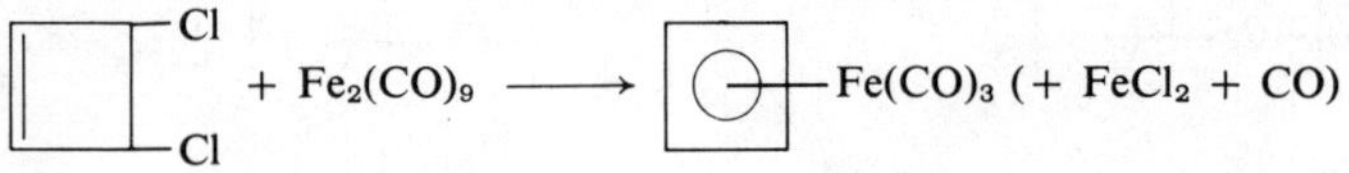

presumably ferrous chloride and carbon monoxide (298). The product is a
stable, yellow crystalline solid (m.p. 26°). Sodium tetracarbonylcobaltate(—I)
reacts similarly, but the product has a cobalt–cobalt bond in order to form
a diamagnetic complex (299).

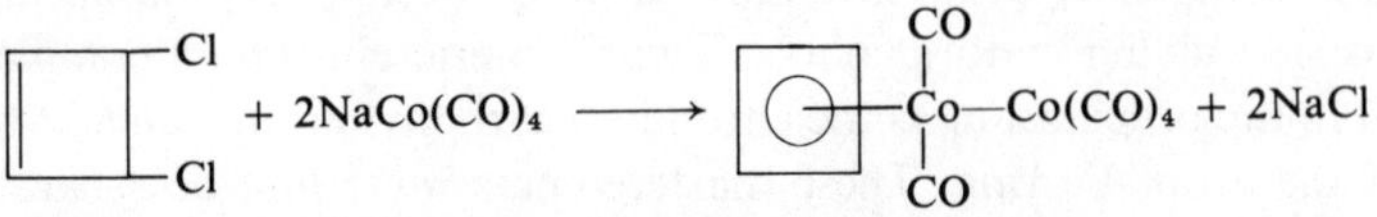

A second method of preparation is a ligand exchange reaction of a
cyclobutadienyl metal complex with a metal carbonyl derivative. The exchange
of diiodo-π-tetramethylcyclobutadienenickel(II) with octacarbonyldicobalt(0)
is an example (300).

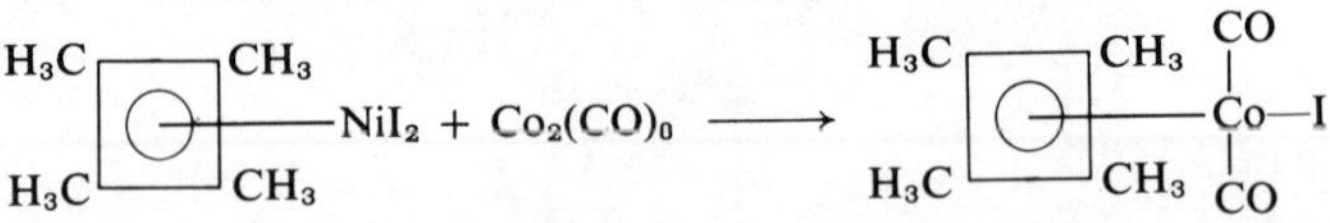

A third method involves photochemical generation of the cyclobutadiene
derivative in the presence of a reactive metal carbonyl. For example, dicar-
bonyl-π-cyclopentadienylcobalt(I) when irradiated with the β-lactone of
2-hydroxy-3-cyclobutenecarboxylic acid yields the "mixed sandwich" complex, π-cyclobutadiene-π-cyclopentadienylcobalt(I) (301). The latter complex
can be acetylated and metalated at the cyclobutadiene ring, although yields
are low (301).

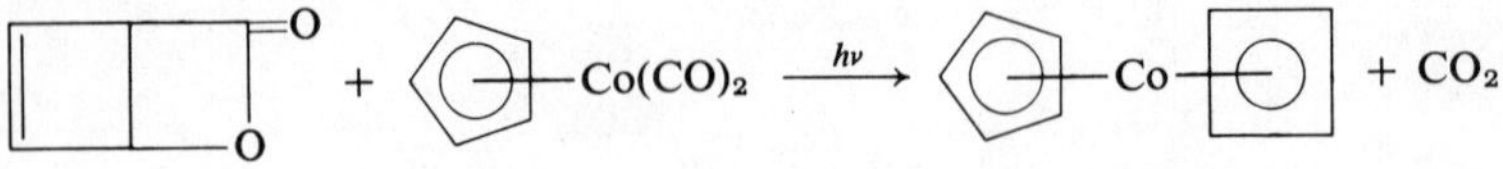

The fourth method of cyclobutadiene metal complex synthesis is a cyclic dimerization of acetylenes on the metal. A typical example is the reaction of trimethylsilylphenylacetylene with π-1,5-cyclooctadiene-π-cyclopentadienyl-cobalt(I) resulting in the formation of the two possible isomeric cyclobutadiene complexes. The trimethylsilyl groups are readily removed with hydrogen chloride yielding the two isomeric diphenylcyclobutadiene complexes (302).

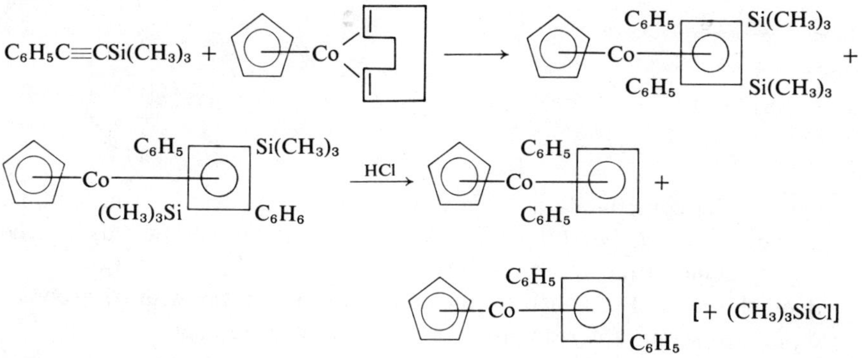

Some very useful applications of the tricarbonyl-π-cyclobutadieneiron(0) complexes in organic syntheses have been found. Two basic kinds of reactions have been studied: (1) reactions catalyzed by irradiation where a carbonyl group dissociates from the complex and another reactant enters and reacts with the cyclobutadiene ligand and (2) oxidation of the complex with formation of "free cyclobutadiene" which may then be reacted with other reagents present in the solution.

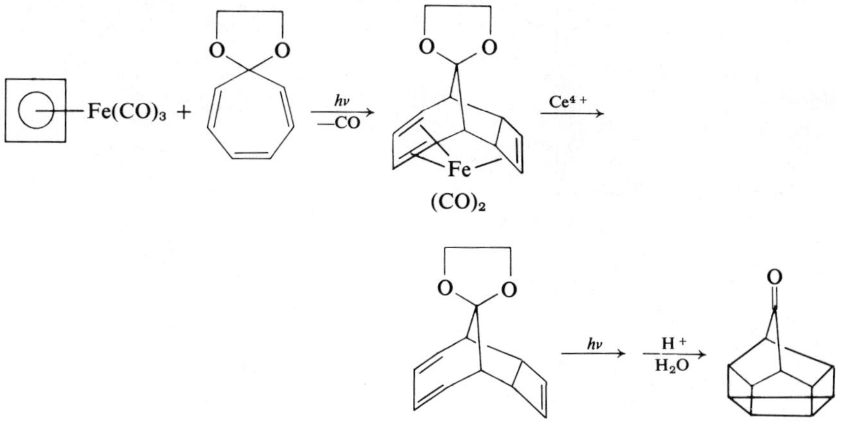

An example of the first kind is the Diels–Alder-type reaction which occurs between tricarbonylcyclobutadieneiron(0) and cycloheptatrienone ethylene ketal on irradiation. Subsequent oxidation of the complex liberates the new triene ligand. Further irradiation of the triene and acid hydrolysis of the ketal group gives a complex saturated polycyclic ketone (303).

The second kind of reaction is illustrated by the reaction of "free cyclobutadiene" with quinone. The tricyclodecadienedione obtained cyclizes on irradiation to another saturated polycyclic compound (304). The synthesis

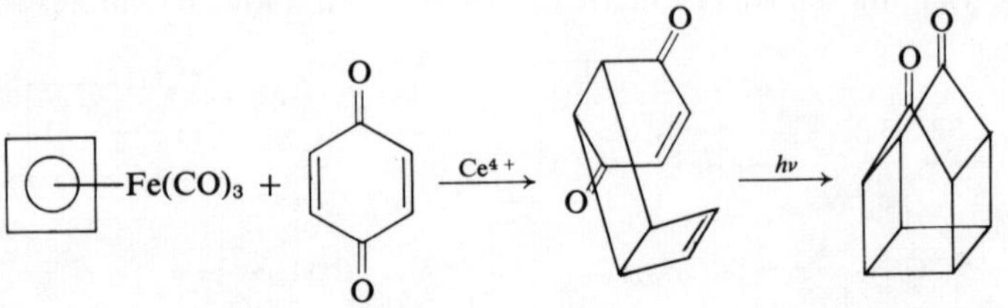

of polycyclic compounds is often a very tedious process by conventional organic chemistry, but becomes relatively easy if the compounds can be made by either of the above methods.

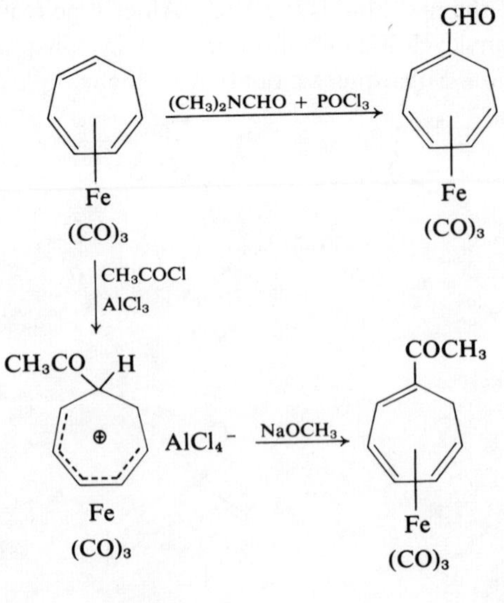

Complexation of cycloheptatriene and cyclooctatetraene with tricarbonyliron groups significantly stabilizes the ligands. In these complexes only two double bonds in each ligand are coordinated to the iron at any one time, and yet some electrophilic substitution reactions can be carried out without excessive decomposition. The cycloheptatriene complex can be formylated with dimethylformamide and phosphorus oxychloride and acetylated with acetyl chloride–aluminum chloride. In the acetylation an intermediate acetylcycloheptadienyliron cation can be isolated, which on treatment with base gives the neutral acetylcycloheptatriene complex (305). Similarly, the cyclooctatetraeneiron π complex can be formylated with dimethylformamide and phosphorus oxychloride in 60% yield (306).

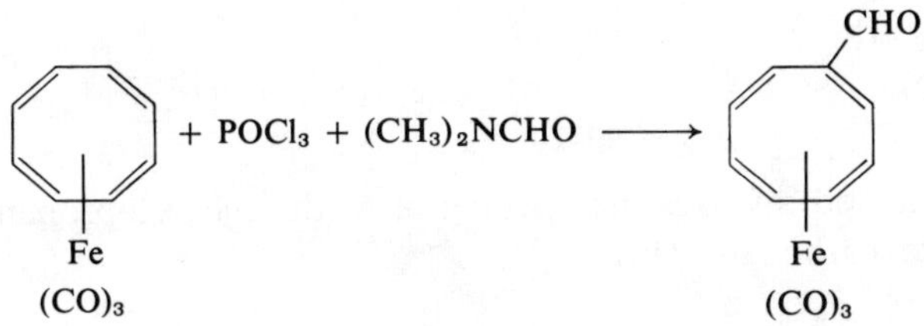

Cyclooctatetraene also forms stable π complexes with cerium and uranium. Probably other rare earth and f-transition elements will do the same. Di-π-cyclooctatetraeneuranium(0) is a particularly interesting compound since it is relatively stable and appears to be a homolog of the di-π-cyclopentadienyl complexes of the d-transition metals. The uranium complexes are prepared from the tetrachloride and cyclooctatetraene dianions (307).

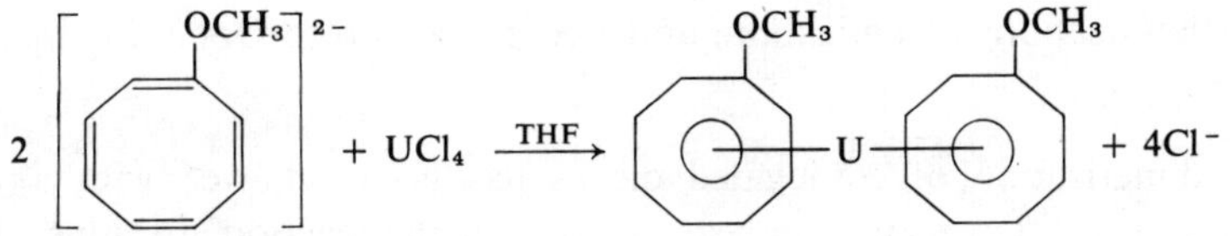

I. TRIMETHYLENEMETHANE COMPLEXES

Trimethylenemethane is a neutral four-electron-donating ligand or a six-electron-donating dianion which forms stable complexes with several transition metals, although little work has been done with these. The complexes have been produced in low yields in three ways. One method of preparation is the reaction of 2-chloromethyl-3-chloropropene with metal carbonyl dianions (308). Oxidation of the iron complex, prepared in this manner, liberates the ligand. Free trimethylenemethane apparently can be "trapped" by reaction with tetracyanoethylene but only in low yield (308). The second

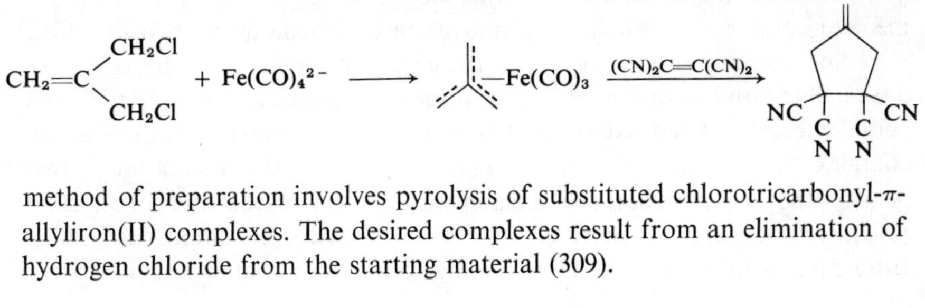

method of preparation involves pyrolysis of substituted chlorotricarbonyl-π-allyliron(II) complexes. The desired complexes result from an elimination of hydrogen chloride from the starting material (309).

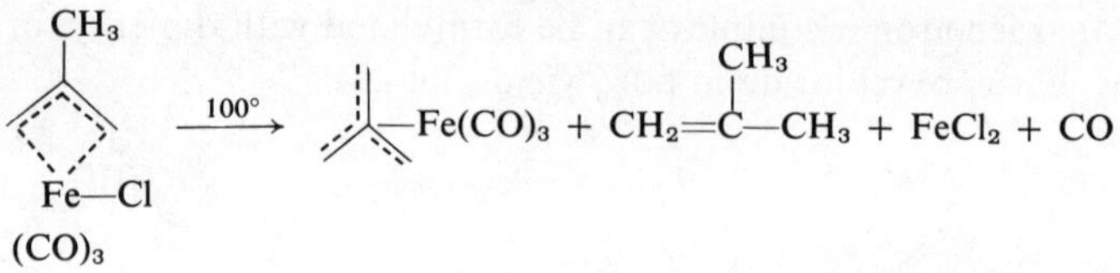

$$\text{20\%}$$

A third method involves the opening of methylenecyclopropane rings by reactive metal carbonyls (310).

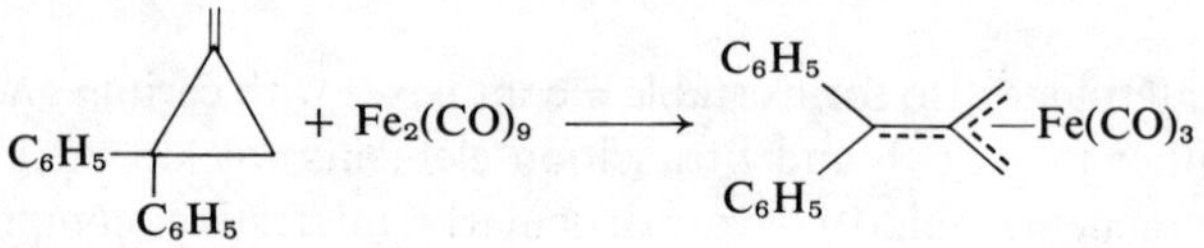

J. DIENE DIMERIZATION, TRIMERIZATION, AND RELATED REACTIONS

1. Dimerization, Trimerization, and Polymerization to Open-Chain Products

The dimerization of conjugated dienes has been achieved with a wide variety of transition metal catalysts. Typical of the reactions forming open-chain products is the butadiene dimerization to a mixture of 3-methyl-1,4,6-heptatriene and 1,3,6-octatriene with a triethylaluminum–cobalt chloride catalyst (311). The probable mechanism of formation of the major product

$$\underset{}{CH_2{=}CHCH{=}CH_2} \xrightarrow[C_6H_6]{Et_3Al—CoCl_2} \overset{CH_3}{\underset{}{CH_2{=}CHCHCH{=}CHCH{=}CH_2}} +$$

$$CH_2{=}CHCH{=}CHCH_2CH{=}CHCH_3$$

can be formulated on the basis of the structure of an intermediate cobalt complex, isolated from a similar reaction mixture, and its reactions. A red crystalline complex was obtained by reducing a solution of cobalt(II) chloride and butadiene in ethanol at $-30°$ with sodium borohydride (312). The intermediate contained a cobalt atom and three butadiene units, two of which were linked together and bonded to the cobalt through a π-allyl system and a π-olefinic bond at opposite ends of a methylheptadienyl chain. At $60°$ the complex catalyzed the butadiene dimerization. An interesting feature of the complex found in an X-ray crystallographic study was that one of the hydrogens on the fourth carbon of the chain was only 3.1 Å from the cobalt atom (312). This distance is short enough that the hydrogen could easily shift from carbon to cobalt and subsequently to the coordinated butadiene unit to produce a complex of the product triene. Ligand replacement with more butadiene would then give the major triene product (see Scheme VI-9). The

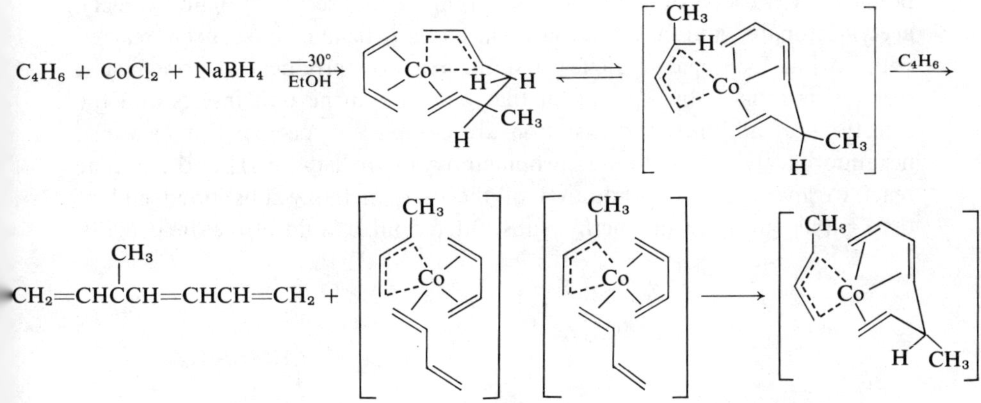

Scheme VI-9.

mechanism of formation of the intermediate is less clear, but probably a hydridobisbutadienecobalt(I) complex is formed in the initial reduction. This complex could then form π-butadiene-π-1-methylallylcobalt(I). Reaction of this compound with another butadiene would give a product which could undergo a shift of the 1-methylallyl group to one of the coordinated butadiene ligands. The addition could occur at either end of the allyl group and apparently does since two isomeric products are formed. The major product is formed by reaction at the methyl-substituted end of the allyl group and the adduct then undergoes the hydrogen shift and ligand displacement as described above with reformation of the π-methylallyl-π-butadiene complex to complete the catalytic cycle.

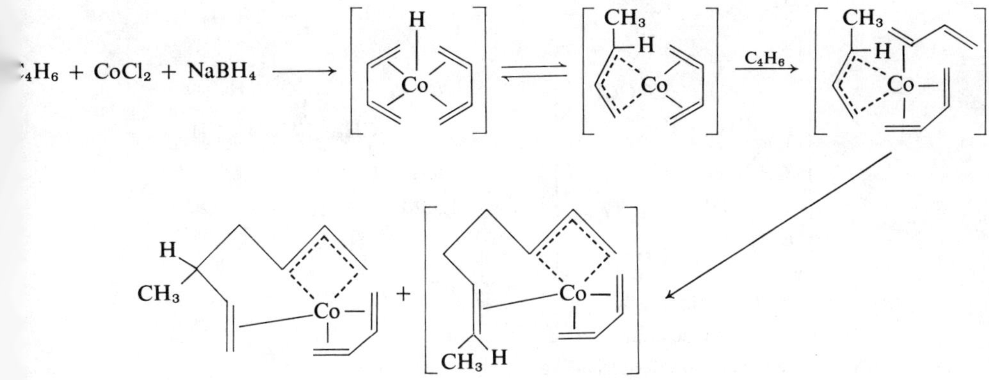

The addition of the π-allylic group to the conjugated diene appears to involve a more complex mechanism than a simple 1:2 shift of an allylic carbon from the metal to the diene, at least with certain palladium compounds. A variety of π-allylic palladium complexes with halo (dimer), acetylacetonate, and hexafluoroacetylacetonate ligands have been reacted with various conjugated dienes and the reaction products identified. The reaction is remarkably specific in that only one diene unit inserts and the directions of addition and insertion are completely specific. For example, hexafluoroacetylacetonato-π-1-carbomethoxyallylpalladium(II) and isoprene react to give exclusively addition of the carbomethoxy-substituted end of the π-allylic group to the methyl-substituted double bond of the diene (313).

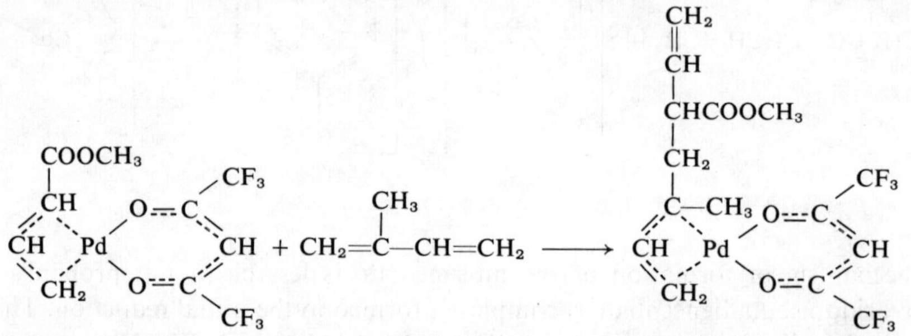

An analysis of the relative rates of reaction of several different dienes, determination of product structures, and NMR studies of intermediates, apparently σ-bonded allylic palladium species, have led to the proposal of the

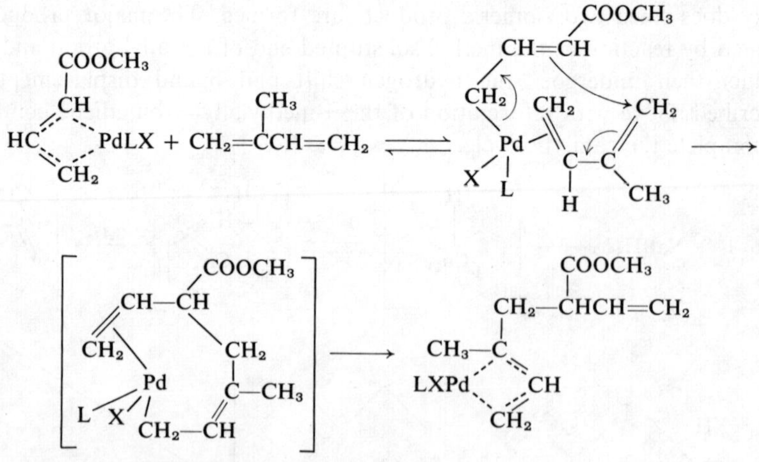

L = solvent or bridging X
X = halide or carboxylate
LX = an acetylacetonate derivative

following mechanism for the addition of the π-allylic palladium group to conjugated dienes (313). π-Allyl formation could also occur directly in the second step rather than by way of a σ-allyl intermediate. The important feature of this reaction is the electrocyclic addition step in which uncomplexed double bonds react. This mechanism may well be involved in other diene dimerization, trimerization, oligomerization, and polymerization reactions, but if so, it is not as specific with some other metals.

A binuclear butadiene dimer complex has been obtained by the reaction of *trans*-bromobistriisopropylphosphine(methyl)nickel(II) with butadiene (314). The fate of the methyl groups has not been determined, nor has a mechanism for the reaction been proposed.

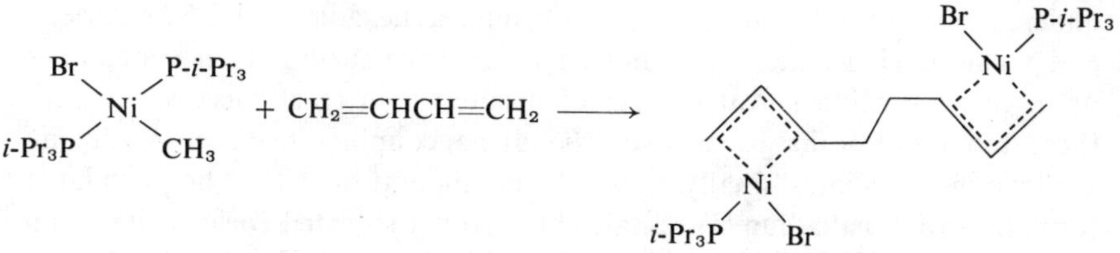

Reaction of butadiene with a catalytic amount of π-allylpalladium acetate produces a large amount of a butadiene trimer, 1,3,6,10-dodecatetraene, and some linear dimer (315). The dimer, 1,3,7-octatriene, may possibly be formed by way of a bis-π-allylic-type intermediate analogous to nickel complexes which have been isolated (see below). Palladium(0) formed by a reductive elimination or any other reaction can coordinate with two butadiene units. The coordinated ligands can then combine to give a bis-π-allylic palladium(II) intermediate. A hydride transfer from the carbon atom in the 4-position to an allylic position would yield a complex of the linear triene. A final replacement of the coordinated triene by more butadiene would complete the catalytic cycle.

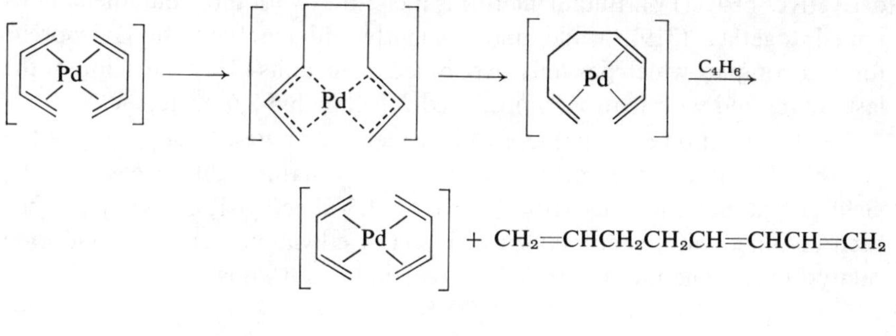

Another kind of intermediate appears to be involved in the π-allylpalladium acetate dimer trimerization reaction. A possible intermediate containing two palladium atoms, two acetate groups, and a linear twelve-carbon organic ligand has been isolated from the reaction mixture (315). The organic ligand is bonded to different palladium atoms at each end of the chain by means of π-allylic groups and an isolated double bond is present in the middle of the chain. Treatment of the complex with carbon monoxide causes reductive elimination to occur forming 3,10-diacetoxy-1,6,11-dodecatriene. The formation of the dodecatetraene in the catalytic reaction can be imagined as occurring by way of a hydrogen shift from the fourth carbon to the nearest palladium, then to the other palladium, and finally to one of the π-allylic carbons joined to the second palladium atom. The initial step in the catalytic reaction is probably an addition of the π-allylic ligand of the catalyst to butadiene with formation of a 1-butenyl-π-allylpalladium(II) acetate dimer as is known to occur in other similar reactions (316). The last compound could undergo an internal hydrogen shift forming a hexadiene, 1,3,6-hexatriene and palladium(I) acetate, an unknown species. The palladium(I) acetate dimer or more probably a solvated form of it could then react successively with three butadiene units by way of the diene coupling reaction producing π-allylic intermediates, finally giving the compound isolated. The postulated formation of a palladium(I) acetate dimer (or a solvated form of it) as an intermediate is very speculative and this mechanism can only be called a guess until more information is obtained. It is, of course, also possible that the isolated dimeric complex is not involved at all in the catalytic trimerization reaction, but is merely formed in a subsequent reaction (see Scheme VI-10).

Biscyclooctatetraeneiron(0) also catalyzes the butadiene trimerization to the same product. The mechanism has not been investigated (317).

The butadiene dimers are conjugated dienes and can themselves be dimerized to sixteen-carbon pentaenes. This has been done with trans-1,3,7-octatriene using bis-π-allylpalladium(II) as catalyst and there was obtained a 70% yield of 1,5,7,10,15-hexadecapentaene (318). The cis-1,3,7-octatriene does not dimerize under the same conditions.

$$\xrightarrow{(\pi\text{-}C_3H_5)_2Pd}$$

70%

Ruthenium trichloride reacts with conjugated dienes to give bis-π-allylic derivatives of Ru(IV). Butadiene forms a complex with three butadiene units joined together (319), while under slightly different conditions isoprene forms a complex which has only two joined diene units (320). Reaction of the last compound with hydrogen produced 2,7-dimethyl-2,6-octadiene.

The dimerization of isoprene is of considerable interest because one of the possible dimers has the carbon skeleton of the natural diterpenes. A 50% yield of a dimer with the natural skeleton 2,6-dimethyl-1,3,6-octatriene has been obtained with a tetraallylzirconium(IV)–ethylaluminum sesquichloride catalyst (321). The mechanism of the reaction is not known.

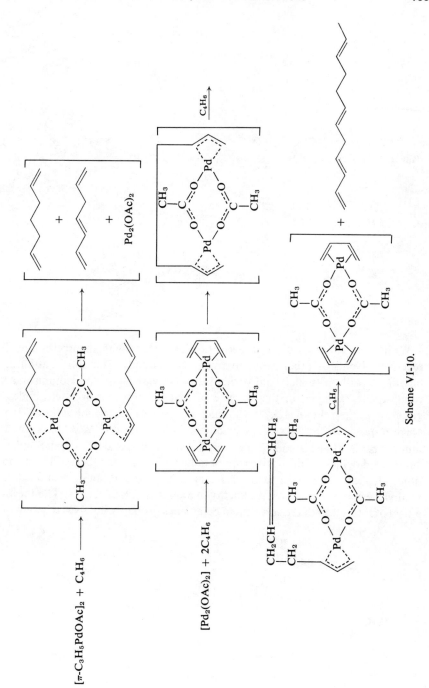

Scheme VI-10.

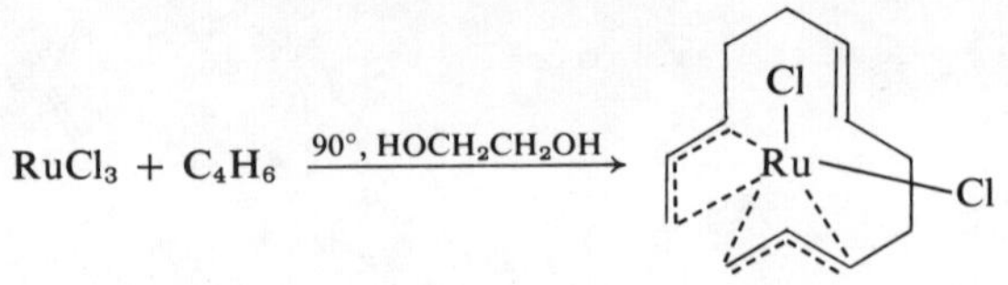

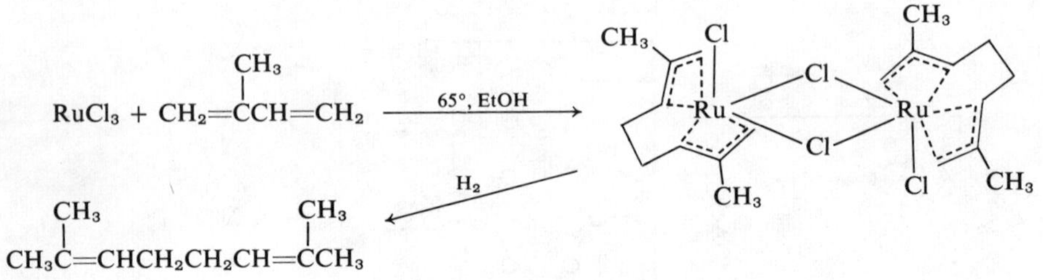

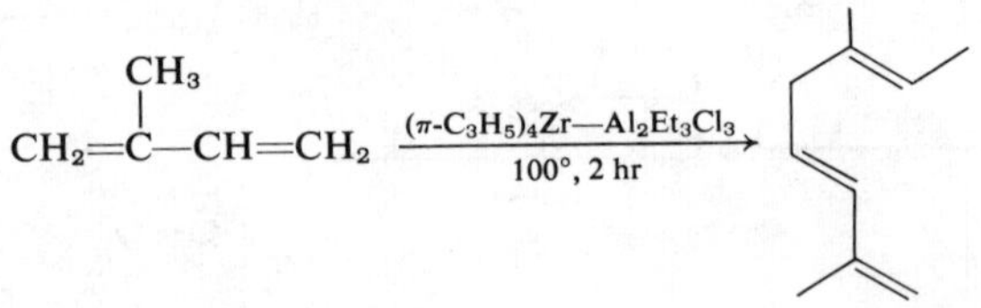

$$RuCl_3 + C_4H_6 \xrightarrow{90°, HOCH_2CH_2OH}$$

$$RuCl_3 + CH_2=\overset{\overset{\displaystyle CH_3}{|}}{C}CH=CH_2 \xrightarrow{65°, EtOH}$$

$$\xleftarrow{H_2}$$

$$\overset{\overset{\displaystyle CH_3}{|}}{CH_3C}=CHCH_2CH_2CH=\overset{\overset{\displaystyle CH_3}{|}}{CCH_3}$$

$$CH_2=\overset{\overset{\displaystyle CH_3}{|}}{C}-CH=CH_2 \xrightarrow[100°, 2\ hr]{(\pi\text{-}C_3H_5)_4Zr-Al_2Et_3Cl_3}$$

With some catalysts diene dimerization can occur with incorporation of reactive solvents or other active hydrogen compounds. The dimerization of butadiene with π-(maleic anhydride)bistriphenylphosphinepalladium(0) as catalyst, for example, in the absence of reactive solvents or nucleophiles, gives an 86% yield of 1,3,7-octatriene, whereas in methanol solution the reaction produces 1-methoxy-2,7-octadiene in 85% yield (322, 323). A small amount of the allylic isomer, 3-methoxy-1,7-octadiene, is also formed. Similar, substituted octadiene derivatives are obtained with this or other catalysts and other alcohols, carboxylic acids, primary and secondary amines, and active methylene compounds such as dimethyl malonate (322, 324, 325). The mechanism of formation of the substituted octadienes probably involves reaction of

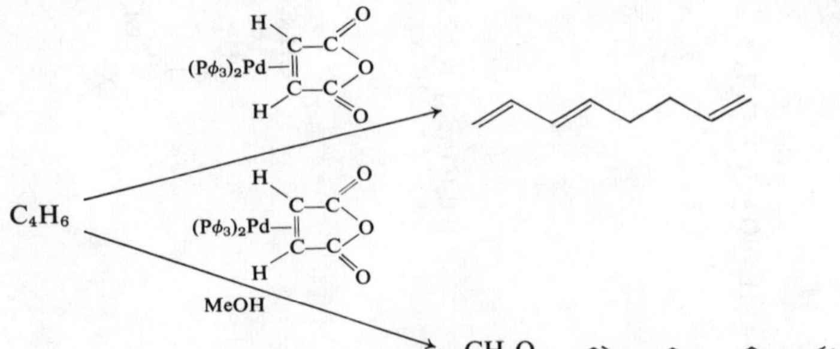

$$CH_3O\diagdown\diagup\diagdown\diagdown\diagup\diagdown\diagup$$

an intermediate bis-π-octadienedilypalladium(II) species with the active hydrogen compound. Intermediate four-valent palladium oxidative addition products may be involved which undergo two reductive eliminations, or possibly a nucleophilic displacement on a π-allylic carbon may take place with reagents which are nucleophilic, followed by reductive elimination (Scheme VI-11).

Butadiene and other conjugated dienes with few substituents can often be polymerized by the proper transition metal catalysts (or other catalysts) to very long-chain polyenes which, in some instances, are useful rubbery materials. Four isomeric polymers can be obtained from butadiene. If linkage in the polymer is through the 1- and 4-carbons of the diene, then the remaining double bond may be either cis or trans, whereas a polymerization in a 1:2 manner leads to a polymer with vinyl side chains. The latter polymer may also exist in two forms, one in which the asymmetric vinyl-bearing carbons have predominantly the same R or S configuration in a given chain (isotactic) or segment of a chain, and another form where the R and S molecules tend to alternate along each chain (sydiotactic) (326). Both forms are optically inactive since the total product in either reaction contains the same number of R and S carbons. Whether *cis-* or *trans-*1,4-polymers are formed presumably depends on whether the π-allylic metal complex intermediates prefer to have an anti or syn arrangement of allylic (polymer chain) substituents, respectively. A reasonable explanation why some catalysts give isotactic and others syndiotactic polymers has not yet been offered, but it likely is the net result of a competition between steric effects of ligands around the catalyst molecule and the tendency to form the energetically favored configuration of the polymer chain. Typical diene polymerization catalysts are combinations of titanium, chromium, or vanadium compounds with aluminum alkyls. Most of these catalysts are heterogeneous.

2. Cyclic Dimerization and Trimerization of Conjugated Dienes

Mechanistically the linear and cyclic dimerizations and trimerizations are closely related. The important difference arises in the final step. If the eight or twelve carbon atom ligands attached to the metal in the intermediates react by a hydrogen shift, open-chain products result, whereas if the allylic groups in the ligand are able to react with each other more rapidly than the hydrogen shift occurs, cyclic products are formed. Nickel(0) catalysts, extensively studied at the Max Planck Institute in Mülheim, Germany, have proven to be the most versatile and useful cyclization catalysts. Whether cyclic dimers or trimers are formed depends on whether the nickel catalyst has at least one firmly attached ligand (an organophosphine).

The cyclic dimerization of butadiene is readily achieved with catalysts prepared by reducing nickel(II) acetylacetonate with aluminum alkyls in the

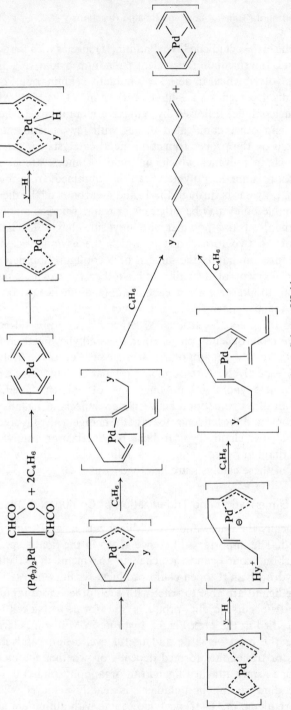

Scheme VI-11.

presence of an organophosphine or by adding certain nickel(0) compounds to organophosphines. A probable intermediate nickel complex which is catalytically active can be isolated and its structure suggests how the reaction occurs. The intermediate has been isolated from the reaction of tricyclohexyl-phosphine-π-cyclododecatrienenickel(0) with butadiene (90, 327) or from the reaction of bis(ditricyclohexylphosphine)nickel(0) dinitrogen with butadiene (90). The complex has an eight-carbon chain with allylic attachments to nickel at each end of the chain. One allylic attachment is σ-bonded and the other is π-bonded to the nickel. The fourth coordination position is occupied by a tricyclohexylphosphine group. Heating the isolated intermediate with

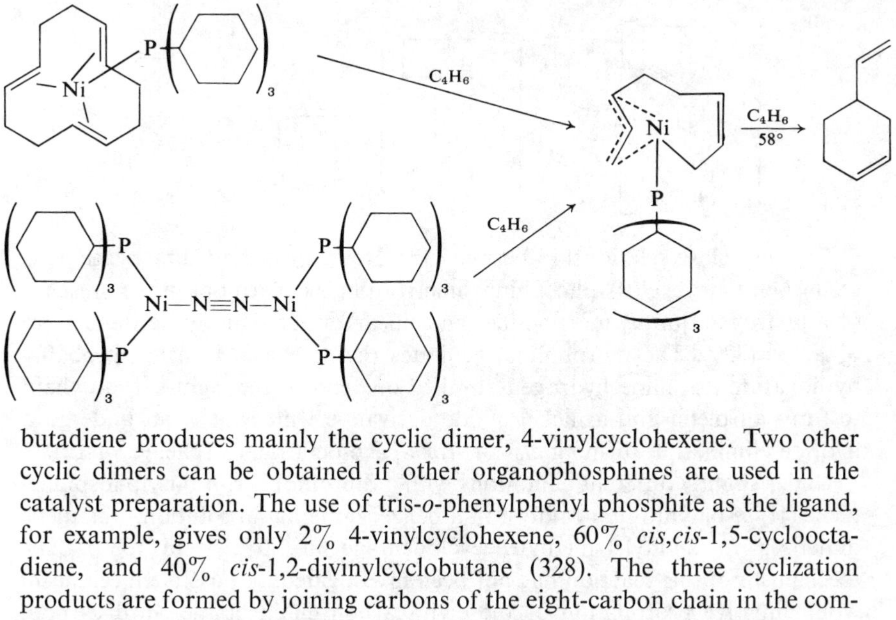

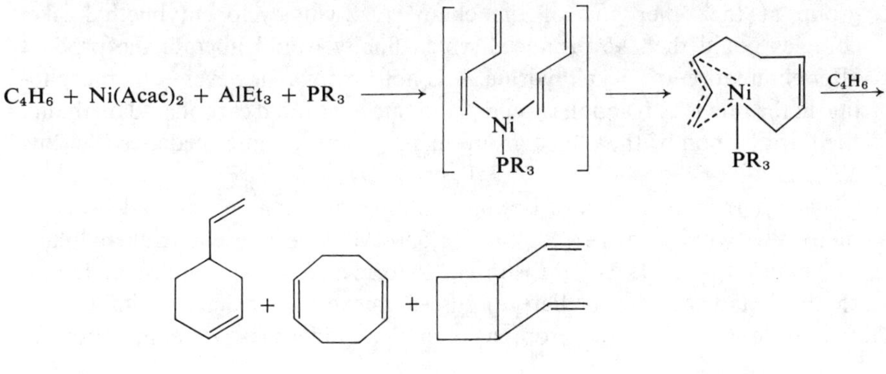

butadiene produces mainly the cyclic dimer, 4-vinylcyclohexene. Two other cyclic dimers can be obtained if other organophosphines are used in the catalyst preparation. The use of tris-o-phenylphenyl phosphite as the ligand, for example, gives only 2% 4-vinylcyclohexene, 60% cis,cis-1,5-cyclooctadiene, and 40% cis-1,2-divinylcyclobutane (328). The three cyclization products are formed by joining carbons of the eight-carbon chain in the complex at either the 1- or 3-carbon of the allylic groups. Steric effects seem to be the dominant factor in determining how the ring closes. The reacting form of the intermediate may be a bis-π-allylic (syn or anti), a mono-π-allylic-mono-σ species (syn or anti allylic substituents and with either a 1- or 3-σ attachment) or a di-σ derivative (with either a 1- or 3-attachment and cis and trans isomers if it is attached to the 1-carbon).

Although it is not clear which forms yield which products, guesses can be made. The bis-primary-σ form would probably be favored by the largest phosphine ligands and this form would probably close to the cyclobutane product. The σ–π-allyl forms may give vinylcyclohexene and the bis-π-allyl form the cyclooctadiene.

The very close relationship between the cyclic and linear dimerizations is seen when the nickel(0)–phosphine dimerization is carried out in the presence of a hydrogen donor, morpholine, and linear rather than cyclic dimers are obtained (329). The morpholine facilitates the hydrogen transfer, probably by donating its amine hydrogen atom to one end of the eight-carbon chain to form an olefin and a nickel amide derivative which then can undergo a hydride elimination forming *cis*- or *trans*-1,3,6-octatriene (Scheme VI-12).

Under slightly different conditions with a dibromobistri-*n*-butylphosphine-nickel(II)–*n*-butyllithium catalyst in a benzene–methanol solution, butadiene dimerizes to 2-vinyl-1-methylenecyclopentane in 90% yield (330). The mechanism of this reaction has not been investigated. A possible mechanism would involve protonation of the eight-carbon chain at the third carbon, with formation of a terminal olefin which could then react with the allylnickel group at the other end of the chain. A 2-vinylcyclopentylmethylnickel complex could then be produced which finally would liberate the product after elimination of nickel hydride. It is not obvious, however, why morpholine in the previous reaction should protonate a terminal carbon and methanol the third carbon of the allylic group in similar nickel intermediates (Scheme VI-13).

The cyclic dimerization of isoprene to dipentene can be achieved stoichiometrically with a nickel complex. 1,5,9-Cyclododecatriene(triphenylphosphine)nickel(0) reacts with the isoprene to form an isolable complex analogous to the butadiene intermediate discussed above. On reaction with carbon monoxide at $-30°$, the isoprene complex undergoes an elimination of

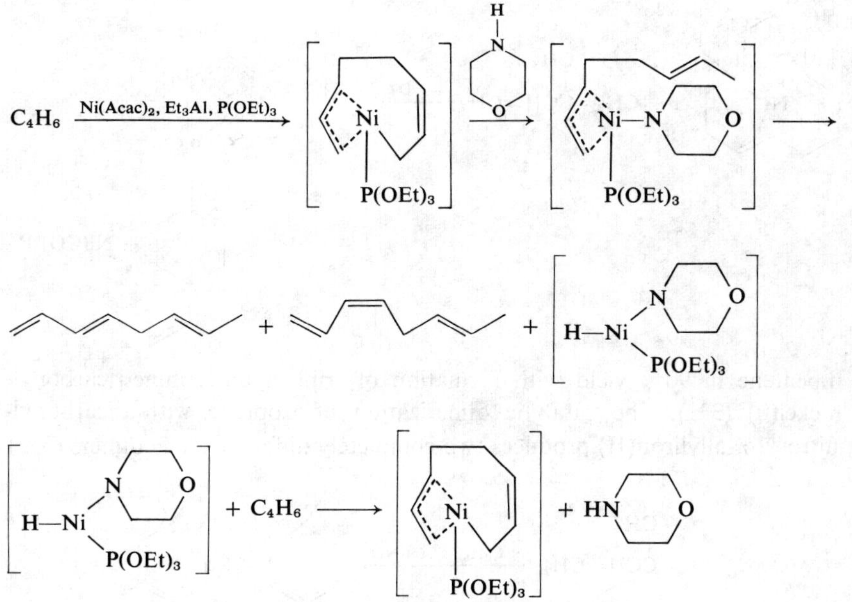

Scheme VI-12.

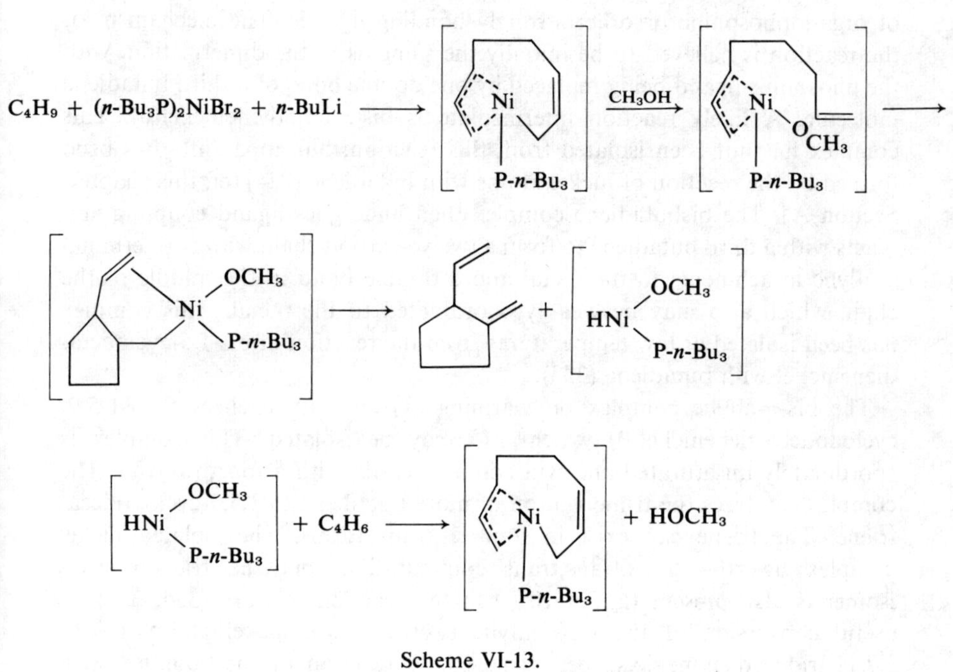

Scheme VI-13.

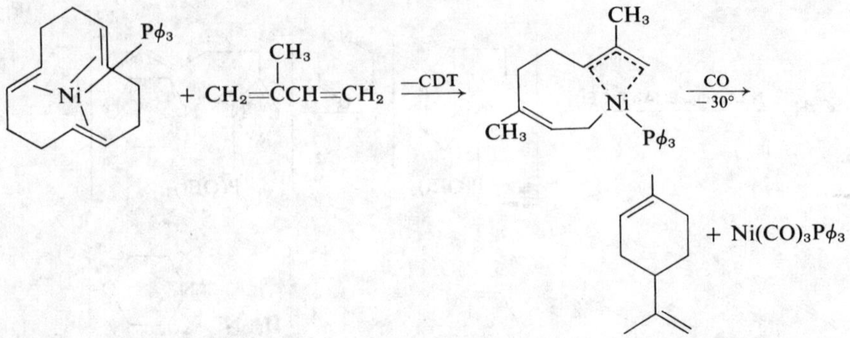

dipentene in 90% yield with formation of triphenylphosphinetricarbonyl-nickel(0) (331). The catalytic dimerization of isoprene with dicarbonyl-nitrosyl-π-allyliron(II) produces two completely different cyclic dimers (332).

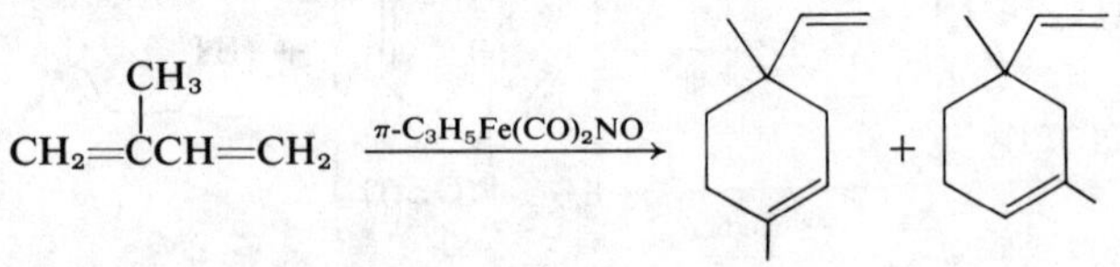

Butadiene is cyclically trimerized by reduced nickel catalysts in the absence of organophosphine or other strongly bonding ligands. The mechanism of the reaction is believed to be initially the same as in the dimerization, with the phosphine ligand being replaced by one double bond of a third butadiene molecule. A likely reaction intermediate is bis-π-butadienenickel(0). This complex has not been isolated from this reaction, but apparently has been formed by the reaction of nickel atoms with butadiene (224) (cf. this chapter, Section A). The bisbutadiene complex then undergoes ligand coupling and reacts with a third butadiene to form a twelve-carbon chain with two terminal π-allylic attachments to the metal and a double bond in the middle of the chain which also may be weakly coordinated to the metal. This complex has been isolated at low temperatures from the reaction of bis-1,5-cycloocta-dienenickel with butadiene (333).

The bis-π-allylic complex on warming above $-40°$ cyclizes to π-1,5,9-cyclododecatrienenickel(0) which also may be isolated. This complex is coordinately unsaturated and will combine easily with a fourth ligand. The complex catalyzes the trimerization of more butadiene to 1,5,9-cyclododeca-triene. The triene can exist in four cis,trans forms. The isolated nickel complex has the trans,trans,trans configuration, but the trans,trans,cis isomer is also present (about 10%) in the reaction mixture (334, 335). A useful conversion of the bis-π-allylic twelve-carbon nickel complex into 3,7,11-tridecadienone has been achieved by reaction of the complex with *t*-butyl isocyanide (333). The ends of the carbon chain add to the carbon of the isocyanide group forming a thirteen-carbon imide which on hydrolysis gives the thirteen-carbon cyclic ketone in 70% yield. About 3% of the ring closure reaction gives the 2-vinyl-5,9-undecadiene imine. The amount of the eleven-membered ring closure that occurs depends on the isocyanide used. Carbon monoxide in place of an isocyanide produces only the eleven-membered ring product (333) (Scheme VI-14).

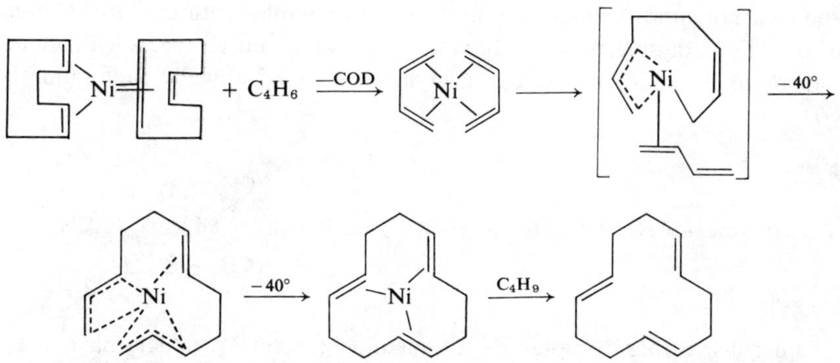

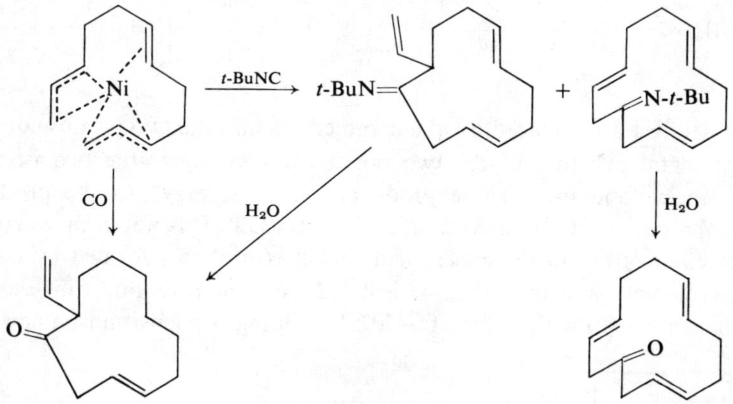

Scheme VI-14.

Larger ring polyenes, with 16, 20, and 24 carbons, have been obtained from butadiene with a combination of bis-π-allylnickel(II) and chloro-π-allylnickel dimer as catalysts (336).

3. Dimerization, Trimerization, and Oligomerization of Allene

Allene undergoes 1:2-addition reactions with organotransition metal compounds. In the few reactions studied the organic group adds to the middle carbon, and the metal forms a π-allylic complex with the three allenic carbons. For example, acetylacetonato-π-allylpalladium(II) reacts with allene to form as the major product acetylacetonato-π-2-allylallylpalladium(II) (337).

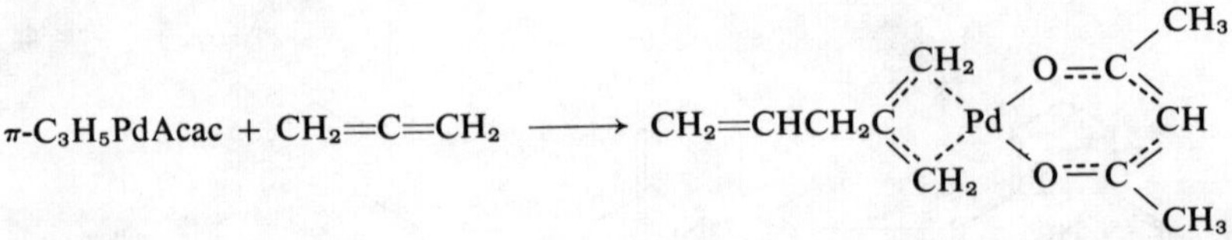

π-C3H5PdAcac + CH2=C=CH2
CH2=CHCH2C
CH2
CH2
Pd
O=C
O=C
CH3
CH
CH3

Numerous, rather complex organometallic reaction products have been isolated from many allene reactions. Perhaps the most interesting reactions involve the formation of cyclic oligomers. When allene is passed over *o*-bisdiphenylphosphinobenzenedicarbonylnickel at 200°, a mixture of 10% 1,2-dimethylenecyclobutane, 36% 1,3-dimethylenecyclobutane, and 20% 1,2,4-trimethylenecyclohexane is obtained (338). Possibly the carbonyl groups

CH2=C=CH2
ϕ2
P
Ni
CO
CO
P
ϕ2
200°
+
+

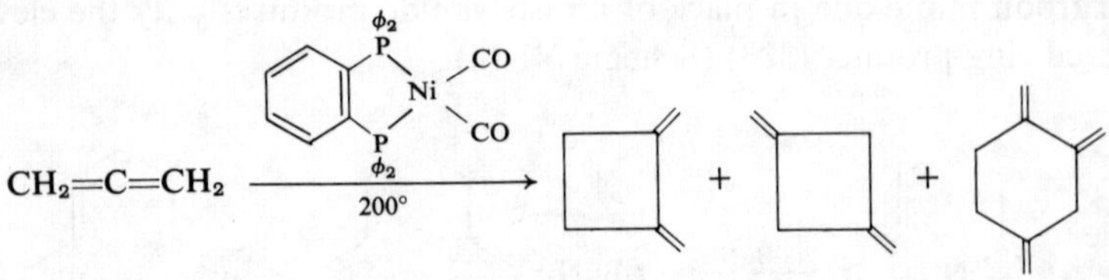

in the catalyst are replaced by allene molecules and the two groups combine on the metal in either of the two possible ways to give the two isomeric methylenecyclobutanes. These products, however, may also be produced under the same conditions without a catalyst (339). It is not clear where the trimer comes from in the experiment, but it is probably formed by way of an intermediate similar to the one isolated from the reaction of bis-π-cyclooctadienenickel(0) with allene at −70°, by adding triphenylphosphine (340).

Ni
+ 3CH2=C=CH2
− 70°
Pϕ3
− 30°
CH2
H2C
C
C
CH2
Ni
Pϕ3
CS2
CH2
CH2
C
CH2

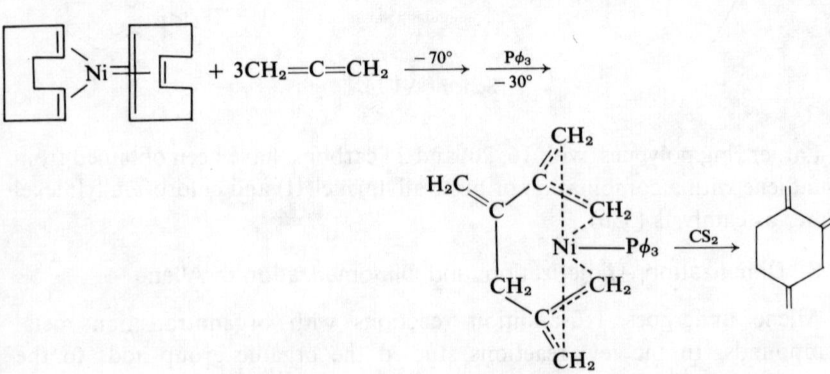

Treatment of the complex with carbon disulfide removes the metal. The cyclic trimerization then would involve the coupling of the two terminal allylic groups on the nickel. The trimer can be obtained catalytically using bis(tri-*o*-phenylphenyl phosphite)nickel(0) as the catalyst (341). Heating the triphenylphosphine allene trimer–nickel complex with more allene at 50° produces an isolable complex with four connected allene groups. With carbon disulfide this complex loses the metal and yields the cyclic tetrameric allene, 1,2,4,7-tetramethylenecyclooctane (342). The reaction of bis-1,5-

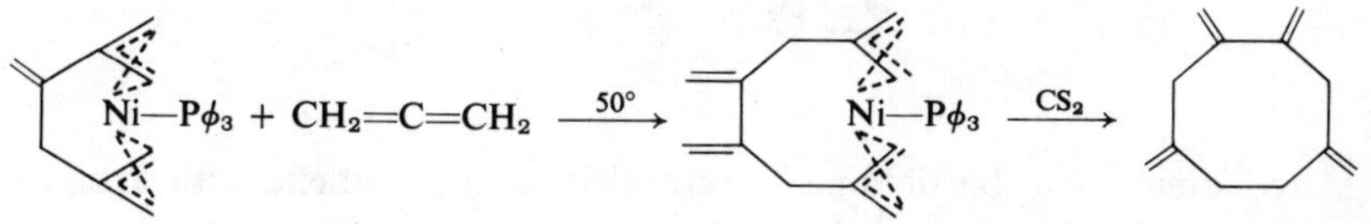

cyclooctadienenickel(0) with allene at 40° gives a 55% yield of a cyclic allene pentamer (343). Presumably the pentamer is produced, mainly, because it forms a fairly stable coordination complex with the catalyst. The mechanism of the pentamer synthesis probably involves formation of the tetramer complex shown above with allene in place of triphenylphosphine and, then, addition of the chain to the fifth allene unit with subsequent coupling of the allylic groups. The proposed product–nickel complex has not been isolated, however.

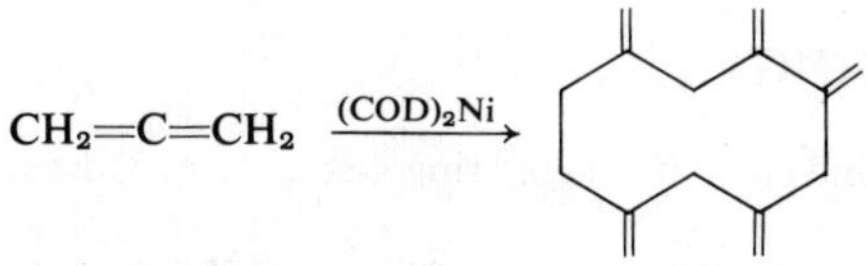

A different isomeric cyclic pentamer complex is formed from allene when chloro(bisethylene)rhodium(I) dimer is used as the catalyst (344, 345). The complex is isolable in this case and is quite stable. The ligand is displaced from the complex by bisdiphenylphosphinoethane, however. Chlorotristri-

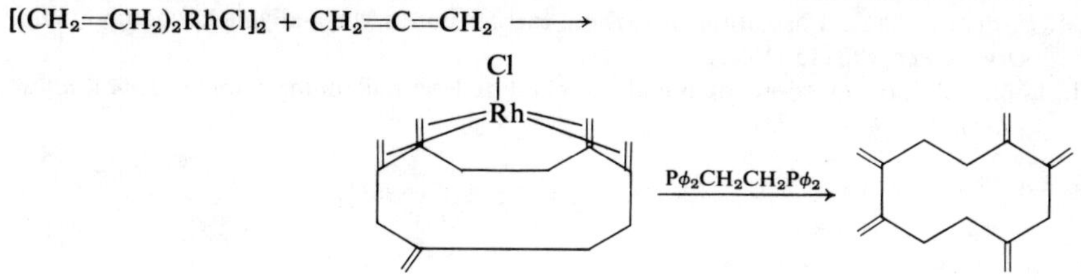

phenylphosphinerhodium(I) catalyzes the formation of another tetramer from allene, a spirane compound (344), whereas chlorobiscarbonylrhodium(I)

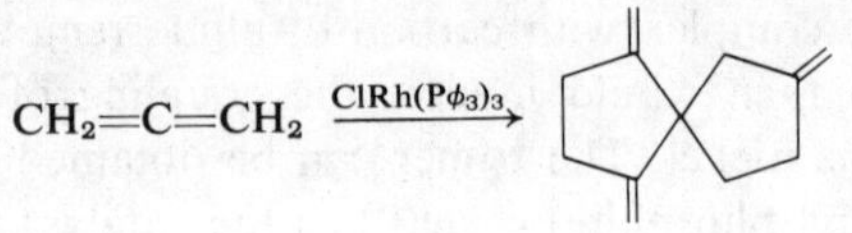

$$CH_2\!=\!C\!=\!CH_2 \xrightarrow{ClRh(P\phi_3)_3}$$

dimer and two moles of triphenylphosphine convert allene into a mixture of products from which an unusual tricyclic hexamer has been isolated (346).

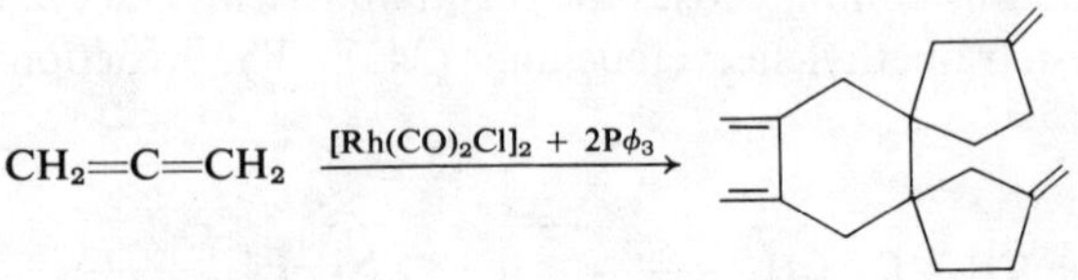

$$CH_2\!=\!C\!=\!CH_2 \xrightarrow{[Rh(CO)_2Cl]_2 + 2P\phi_3}$$

Mixed dimers can be obtained from allene and butadiene with a bistriphenylphosphine-π-(maleic anhydride)palladium(0) catalyst, but yields are not high (347).

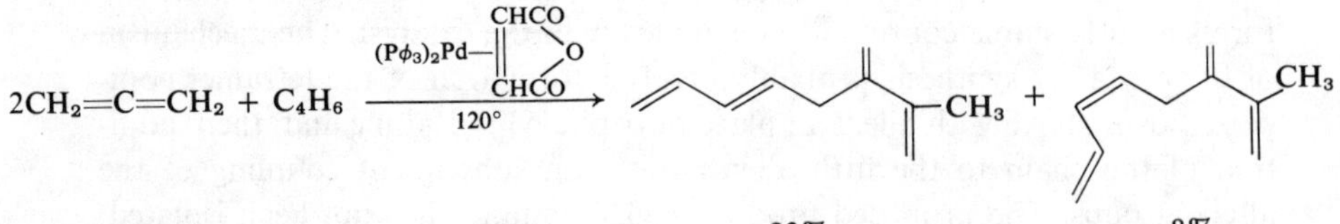

$$2CH_2\!=\!C\!=\!CH_2 + C_4H_6 \xrightarrow[120°]{(P\phi_3)_2Pd-\overset{CHCO}{\underset{CHCO}{\|}}\!\!>\!\!O}$$

23%   8%

FURTHER READING

M. A. Bennett, Seven and eight membered ring π-complexes. *Advan. Organometal. Chem.* **4**, 353 (1966).

E. O. Fischer, Diene *pi* complexes. *Angew. Chem., Int. Ed. Eng.* **2**, 80 (1963).

P. Heimbach, P. W. Jolly, and G. Wilke, π-Allylnickel intermediates in organic syntheses. *Advan. Organometal. Chem.* **8**, 29 (1970).

W. Keim, π-Allyl systems in catalysts. "Transition Metals in Homogeneous Catalysts," Chapter 3. Dekker, New York, 1971.

P. M. Maitlis and K. W. Eberius, Cyclobutadiene-metal complexes. *Nonbenzenoid Aromat.* **2**, 360 (1971).

M. F. Semmelhack, Formation of carbon–carbon bonds via π-allylnickel compounds. *Org. React.* **19**, 115 (1972).

J. Tsuji, Addition reactions of butadiene catalyzed by palladium complexes. *Accounts Chem. Res.* **6**, 8 (1973).

Chapter VII

Transition Metal Reactions with Acetylenes

A. π-ACETYLENIC COMPLEXES

Relatively few π-acetylenic complexes of transition metals are known. This is partly because some are quite unstable and generally more reactive than the corresponding olefin complexes and partly because few attempts have been made to prepare them. The bistriorganophosphine-π-acetyleneplatinum complexes are about the most stable and best known of the group. These complexes can be prepared by a simple replacement of a phosphine group from trisorganophosphineplatinum(0) derivatives with the acetylene. Only two electrons of the triple bond appear to be used in bonding resulting, therefore, in products which are coordinately unsaturated (348). Very stable

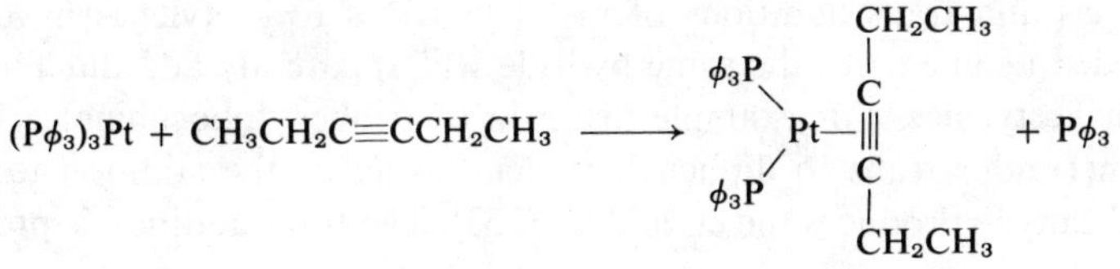

$$(P\phi_3)_3Pt + CH_3CH_2C{\equiv}CCH_2CH_3 \longrightarrow \begin{array}{c} CH_2CH_3 \\ | \\ \phi_3P \diagdown \; C \\ \quad\; Pt{-}||| \\ \phi_3P \diagup \; C \\ | \\ CH_2CH_3 \end{array} + P\phi_3$$

complexes with highly strained, cyclic acetylenes can be formed if the acetylene is produced in the presence of the platinum(0) complex (349).

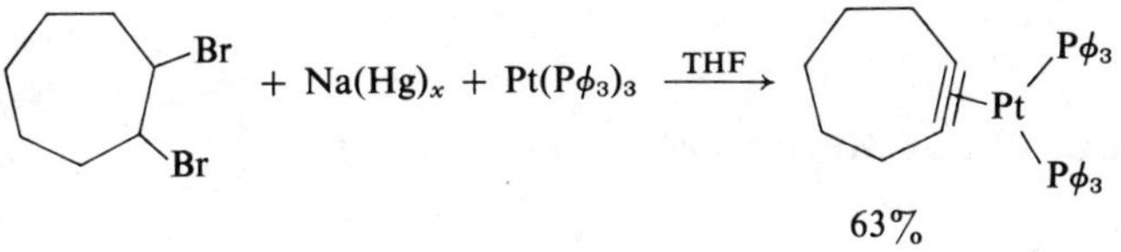

$$63\%$$

167

Two series of complexes with a different kind of "π-acetylenic" group are known. In these complexes all four acetylenic π electrons are used in bonding to two metal atoms which are also bonded to each other. Octacarbonyldicobalt(0) reacts with acetylenes producing one series (350) and carbonyl-π-cyclopentadienylnickel dimer and acetylenes give the other (351).

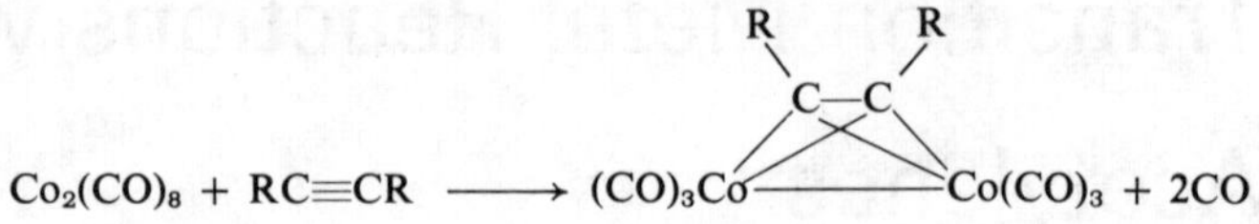

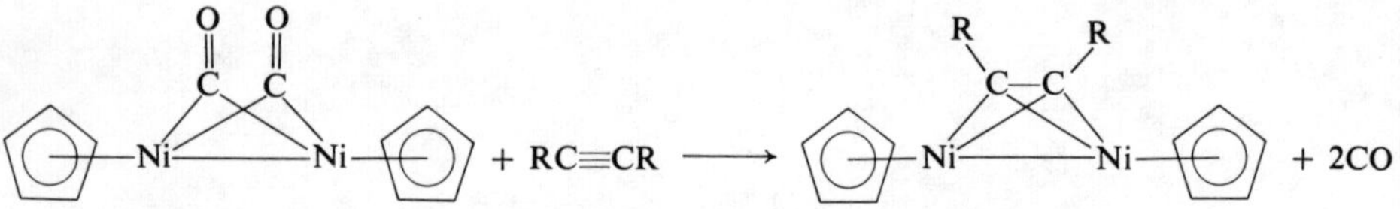

$$Co_2(CO)_8 + RC{\equiv}CR \longrightarrow (CO)_3Co\overset{\overset{\displaystyle R \quad R}{\underset{\displaystyle C{-}C}{\diagdown \diagup}}}{}Co(CO)_3 + 2CO$$

A third type of acetylenic complex has been isolated from reactions of a nickel–iron binuclear compound with acetylenes. These complexes appear to have the acetylenic carbons attached to the metal atoms as though they were coordinated dicarbenes (352).

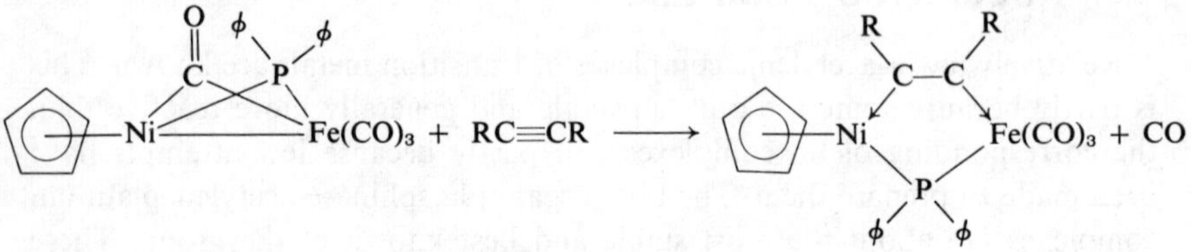

B. ADDITION REACTIONS OF ACETYLENES

1. Metal Hydride Additions

Both cis and trans additions of metal hydrides to acetylenes have been observed. In some cases the same hydride will apparently add differently to different acetylenes. For example, hydrido(tristriphenylphosphine)carbonylrhodium(I) adds trans to diphenylacetylene, whereas the addition to hexafluoro-2-butyne produces the cis adduct (353). The trans addition is probably

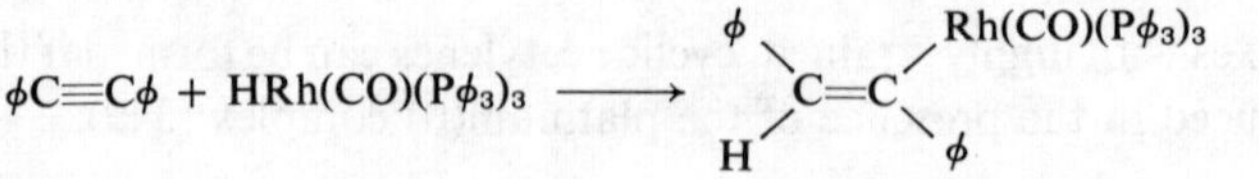

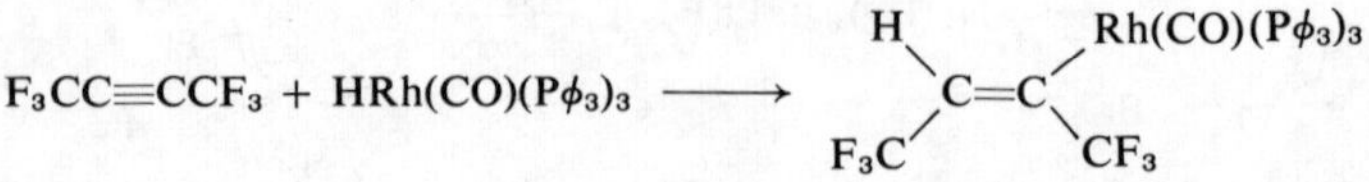

$$\phi C{\equiv}C\phi + HRh(CO)(P\phi_3)_3 \longrightarrow \overset{\phi}{\underset{H}{\diagdown}}C{=}C\overset{Rh(CO)(P\phi_3)_3}{\underset{\phi}{\diagup}}$$

$$F_3CC{\equiv}CCF_3 + HRh(CO)(P\phi_3)_3 \longrightarrow \overset{H}{\underset{F_3C}{\diagdown}}C{=}C\overset{Rh(CO)(P\phi_3)_3}{\underset{CF_3}{\diagup}}$$

ionic with protonation of the acetylene occurring first. The cis addition probably occurs by a covalent mechanism and it is preferred with hexafluoro-butyne because it is a much less basic acetylene than diphenylacetylene.

Reaction of bistriphenylphosphine-π-hexafluoro-2-butyneplatinum(0) with trifluoroacetic acid likewise produces the cis adduct probably by way of the intermediate metal hydride (354). Reaction of the diphenylacetyleneplatinum

$$
\begin{array}{c}
\phi_3P \diagdown \quad \overset{\displaystyle CF_3}{\underset{\displaystyle C}{|}} \\
\phi_3P \diagup Pt{-}||| \\
\quad \underset{\displaystyle CF_3}{\overset{\displaystyle C}{|}}
\end{array}
+ F_3CCOOH \longrightarrow
\left[
\begin{array}{c}
\phi_3P \diagdown \quad \diagup H \;\; {}^-OCOCF_3 \\
\qquad Pt^+ \\
\phi_3P \diagup \quad \diagdown\!\!= CCF_3 \\
\qquad F_3CC
\end{array}
\right] \longrightarrow
$$

$$
\begin{array}{c}
\phi_3P \diagdown \quad \diagup OCOCF_3 \\
\qquad Pt \qquad\qquad H \\
\phi_3P \diagup \quad \diagdown C{=}C \diagup \\
\qquad\quad CF_3 \qquad CF_3
\end{array}
$$

complex with trifluoroacetic acid produces only *trans*-stilbene, however, by way of an unstable vinylplatinum intermediate (354). The intermediate appears to be the cis adduct which must subsequently undergo isomerization with trifluoroacetic acid (150). A plausible isomerization mechanism involves protonation of the vinyl group beta to the metal, forming a carbonium ion adjacent to the metal which then reverts to the more stable *trans*-vinyl complex (355). These carbonium ions are known to be unusually stable because of probable back bonding from the metal to form cationic carbene complexes (cf. Chapter X, Section C,1).

$$
\begin{array}{c}
\phi_3P \diagdown \quad \overset{\displaystyle \phi}{\underset{\displaystyle C}{|}} \\
\phi_3P \diagup Pt{-}||| \\
\quad \underset{\displaystyle \phi}{\overset{\displaystyle C}{|}}
\end{array}
+ CF_3COOH \longrightarrow
\begin{array}{c}
\phi_3P \diagdown \quad \diagup OCOCF_3 \\
\qquad Pt \qquad\qquad H \\
\phi_3P \diagup \quad \diagdown C{=}C \diagup \\
\qquad\quad \phi \qquad \phi
\end{array}
\underset{\displaystyle \rightleftharpoons}{\overset{\displaystyle H^+}{\;}}
$$

$$
\left[
\begin{array}{c}
\phi_3P \diagdown \quad \diagup OCOCF_3 \\
\qquad Pt \\
\phi_3P \diagup \quad \overset{+}{\diagdown} C{-}CH_2\phi \\
\qquad\qquad \phi
\end{array}
\right]
\longleftrightarrow
\left[
\begin{array}{c}
\phi_3P \diagdown \quad \diagup OCOCF_3 \\
\qquad \overset{+}{Pt} \\
\phi_3P \diagup \quad \diagdown C{-}CH_2\phi \\
\qquad\qquad \phi
\end{array}
\right]
\underset{\displaystyle \rightleftharpoons}{\overset{\displaystyle -H^+}{\;}}
$$

$$
\left[
\begin{array}{c}
\phi_3P \diagdown \quad \diagup OCOCF_3 \\
\qquad Pt \\
\phi_3P \diagup \quad \diagdown C{=}C \diagup\!\!\phi \\
\qquad\quad \phi \qquad H
\end{array}
\right]
\xrightarrow{\; CF_3COOH \;}
\begin{array}{c}
H \qquad\qquad \phi \\
\diagdown C{=}C \diagup \\
\phi \diagup \qquad \diagdown H
\end{array}
+ (P\phi_3)_2Pt(OCOCF_3)_2
$$

A similar reduction of diphenylacetylene by hydridodichloro(trisdimethyl sulfoxide)iridium(III) ion in the presence of hydrogen chloride produces the cis adduct and on further reaction only *cis*-stilbene (152) (cf. Chapter IV, Section E). The stability of the possible carbonium ion intermediate in the isomerization probably varies significantly from complex to complex and the degree of isomerization observed may very well be quite dependent on the complex being reacted and the reaction conditions.

Hydridopentacarbonylmanganese yields trans adducts even with hexafluorobutyne (356) suggesting an ionic or possibly radical addition is much preferred by this hydride.

$$
HMn(CO)_5 + CF_3C{\equiv}CCF_3 \longrightarrow
\begin{array}{c} F_3C \\ \diagdown \\ H \diagup \end{array} C{=}C \begin{array}{c} Mn(CO)_5 \\ \diagup \\ \diagdown CF_3 \end{array}
\quad + \quad
\begin{array}{c} H \\ \diagdown \\ F_3C \diagup \end{array} C{=}C \begin{array}{c} CF_3 \\ \diagup \\ \diagdown H \end{array}
$$

Obviously, much more needs to be learned about hydride additions before the structures of their reaction products can be predicted with certainty.

2. Metal–Carbon, Metal–Oxygen, and Metal–Chlorine Additions

A few examples of the addition of alkyl or acyl transition metal complexes to acetylenes are known. Tristriphenylphosphinemethylrhodium(I) adds to diphenylacetylene to give the cis adduct. Cleavage of the metal–carbon bond in the product with acid, however, gives a mixture of *cis*- and *trans*-olefins. The isomerization has been explained as occurring by way of a protonated vinyl metal intermediate which loses configuration (355). Protonation of the carbon atom alpha to the metal was proposed, but β protonation as suggested in the preceding section would also explain the isomerization.

Triphenylchromium in tetrahydrofuran (THF) reacts with acetylene to form 1,6-diphenyl-1,3,5-hexatriene in 18% yield (stereochemistry not reported) and polymeric products (357). A triple insertion appears to be involved, but why little or no mono- or diacetylenic insertion product is

$$
(C_6H_5)_3Cr(THF)_3 + HC{\equiv}CH \longrightarrow C_6H_5{-}\hspace{-0.2em}\diagup\hspace{-0.5em}\diagdown\hspace{-0.5em}\diagup\hspace{-0.5em}\diagdown\hspace{-0.5em}\diagup\hspace{-0.5em}\diagdown{-}C_6H_5
$$

produced has not been explained. Perhaps all three THF molecules are replaced by acetylene and they link up one by one, but the stereochemistry must be correct or mono- and diinsertion products would be formed also.

If coordinated carbonyl groups are present in the metal alkyl complex, acyl addition may be observed. The reaction of pentacarbonylmethyl-manganese(I) with phenylacetylene, for example, produces a *cis*-acetyl-manganese adduct and a complex with a cyclic pentadienyl ligand containing two acetylenic units and the acetyl group (358). Pentacarbonylacetylman-

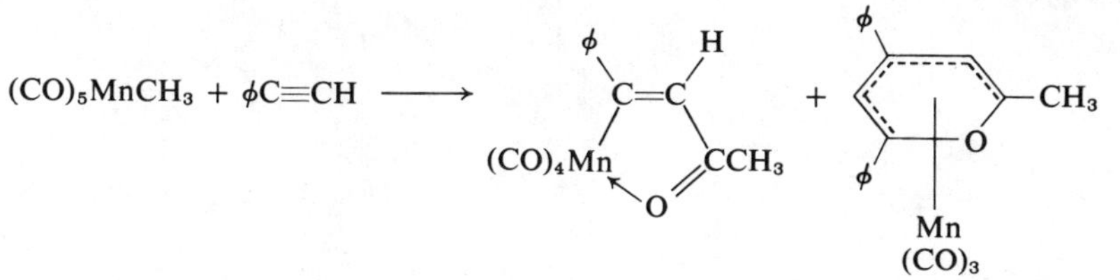

ganese reacts similarly. The same one-to-one adduct can be obtained by the (trans) addition of hydridopentacarbonylmanganese(I) to 4-phenyl-3-butyn-2-one (358).

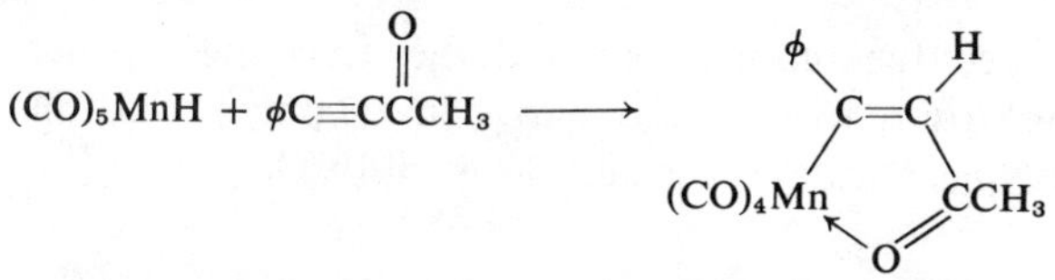

Tetracarbonylacetylcobalt(I) apparently adds cis to disubstituted acetylenes also, but the initial adduct is not isolable. The adduct must rapidly undergo a carbon monoxide insertion and a cyclization (94b). The cyclization seems to involve an addition of the acylcobalt group to the ketonic carbonyl four carbons away, with cobalt going onto carbon and the acyl group onto oxygen. The cyclization produces an allylic group which then forms the stable π-allylic cobalt structure (Scheme VII-1). This reaction has been used as the basis for a catalytic synthesis of 2,4-pentadienolactones. The catalytic reaction requires the use of an acylcobalt derivative with a proton-activating group on the α-carbon of the acyl group. This substituent allows a base to abstract a proton from the acetylene adduct which results in the elimination of the tricarbonylcobaltate anion with formation of the free lactone. The reaction is made catalytic by carrying it out under carbon monoxide to regenerate the tetracarbonylcobaltate anion, in the presence of a base, usually a hindered amine, and with the acetylene and the halide necessary to form the π-allylic intermediate (94b). The formation of the lactone ring in the acetylene reaction

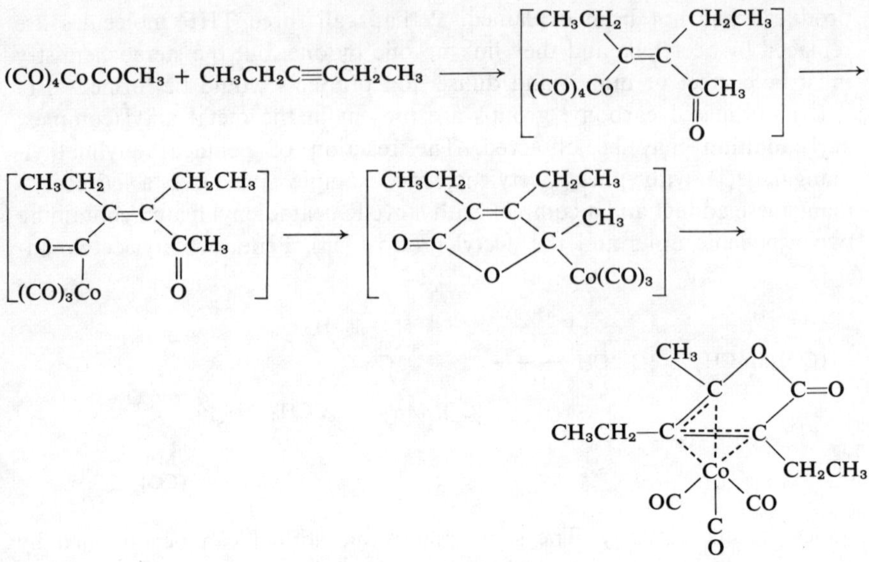

$(CO)_4CoCOCH_3 + CH_3CH_2C{\equiv}CCH_2CH_3 \longrightarrow$

$$\left[\begin{array}{c} \underset{}{CH_3CH_2} \qquad CH_2CH_3 \\ C{=}C \\ (CO)_4Co \qquad\quad CCH_3 \\ \underset{O}{\|} \end{array} \right] \longrightarrow$$

$$\left[\begin{array}{c} CH_3CH_2 \qquad CH_2CH_3 \\ C{=}C \\ O{=}C \qquad\quad CCH_3 \\ | \qquad\qquad \| \\ (CO)_3Co \qquad\quad O \end{array} \right] \longrightarrow \left[\begin{array}{c} CH_3CH_2 \qquad CH_2CH_3 \\ C{=}C \\ O{=}C \qquad CH_3 \\ \quad\searrow \quad C \\ O \qquad Co(CO)_3 \end{array} \right] \longrightarrow$$

(structure: CH₃, O, C=O, CH₃CH₂—C, C, CH₂CH₃, Co, OC, CO, C, O)

Scheme VII-1.

is more favored the more substituted the acetylene is since substituents facilitate ring closure over the alternative linear polymerization of the acetylene or copolymerization of the acetylene with CO.

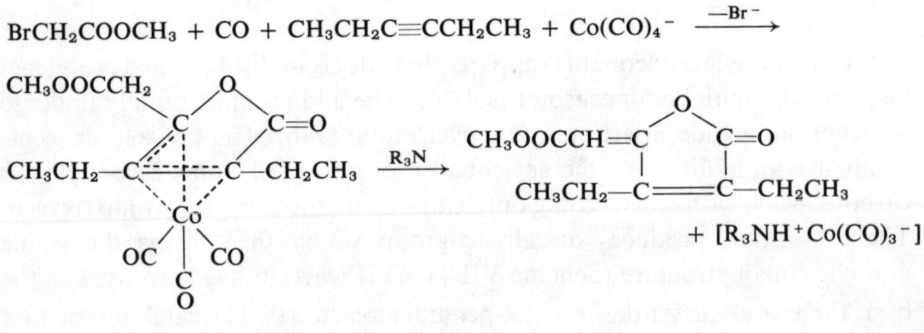

$$BrCH_2COOCH_3 + CO + CH_3CH_2C{\equiv}CCH_2CH_3 + Co(CO)_4^- \xrightarrow{-Br^-}$$

(structure: CH₃OOCCH₂, O, C, C=O, CH₃CH₂—C, C—CH₂CH₃, Co, OC, CO, C, O) $\xrightarrow{R_3N}$

$$\begin{array}{c} O \\ CH_3OOCCH{=}C \qquad C{=}O \\ | \qquad\qquad | \\ CH_3CH_2{-}C{=\!=\!=}C{-}CH_2CH_3 \end{array}$$

$$+ \; [R_3NH^+Co(CO)_3^-]$$

$$[Co(CO)_3^-] + CO \longrightarrow Co(CO)_4^-$$

An example of the trans addition of a halide to an acetylene is the reaction of palladium(II) chloride with 2-butynyldimethylamine. The adduct is stabilized by coordination of the amine group with the metal (359).

Trans-methoxyplatination is observed when cationic complexes such as *trans*-bisdimethylphenylphosphine-π-(acetylenedicarboxylic acid)methylplatinum(II) hexafluorophosphate are produced in methanol (360). The methyl-

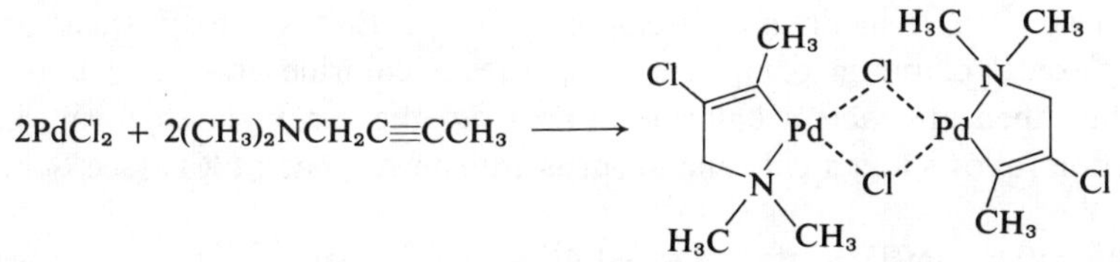

$$2PdCl_2 + 2(CH_3)_2NCH_2C\equiv CCH_3 \longrightarrow$$

platinum group apparently does not add to the acetylenic group because it is trans to it and the complex would have to isomerize first.

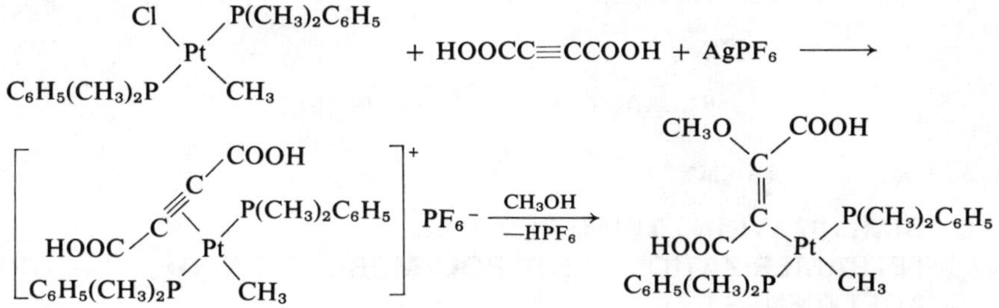

$$+ HOOCC\equiv CCOOH + AgPF_6 \longrightarrow$$

The above reaction without the silver hexafluorophosphate also forms a π complex. Interestingly, insertion will occur in this five-coordinate complex by way of a radical-type intermediate in chloroform solution but not in several other solvents. Apparently, a chlorine atom must be removed from the complex before insertion can take place. The reaction is complicated by the simultaneous formation of a dimethylplatinum(IV) complex and the dichloro-bisphosphineplatinum complex, which reaction occurs in the absence of the

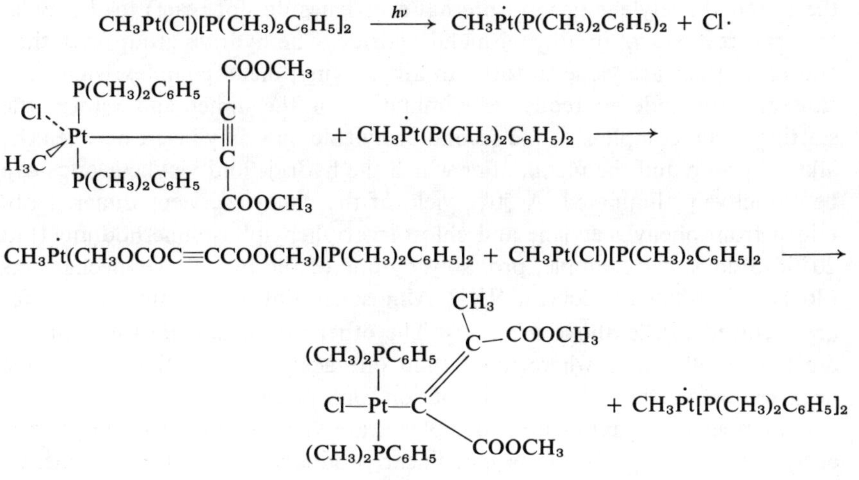

$$CH_3Pt(Cl)[P(CH_3)_2C_6H_5]_2 \xrightarrow{h\nu} CH_3\overset{\bullet}{P}t(P(CH_3)_2C_6H_5)_2 + Cl\cdot$$

$$+ CH_3\overset{\bullet}{P}t(P(CH_3)_2C_6H_5)_2 \longrightarrow$$

$$CH_3\overset{\bullet}{P}t(CH_3OCOC\equiv CCOOCH_3)[P(CH_3)_2C_6H_5]_2 + CH_3Pt(Cl)[P(CH_3)_2C_6H_5]_2 \longrightarrow$$

$$+ CH_3\overset{\bullet}{P}t[P(CH_3)_2C_6H_5]_2$$

Scheme VII-2.

acetylene also. Kinetic measurements indicate that some of the "un-π-complexed" platinum compound first loses a chlorine atom and that this product then abstracts a chlorine atom from the π-complexed form. Then insertion follows and a chlorine atom is transferred back (361) (see Scheme VII-2).

The ruthenium(III)-catalyzed hydration of acetylenes to form ketones is presumably an example of the addition of the elements of a metal hydroxide to an acetylene. Since the intermediate has not been isolated, the mechanism of addition is uncertain (362).

$$RC\equiv CH + H_2O \xrightarrow{\text{Ru(III)}} RC\overset{\displaystyle O}{\overset{\displaystyle \|}{C}}CH_3$$

C. DIMERIZATION, TRIMERIZATION, TETRAMERIZATION, AND POLYMERIZATION OF ACETYLENES

The linear dimerization of acetylenes by an addition mechanism is rarely observed, probably because there is no particularly favorable intermediate complex formed which could stabilize the two acetylenes to one metal atom ligand involved and further polymerization usually occurs. The β-hydride elimination is much less favorable with vinyl metal derivatives than it is with the alkyls since acetylenes are more strained than olefins. When linear acetylene dimerizations or oligomerizations are observed, it is likely that they usually proceed by a different mechanism, by an oxidative addition of the terminal acetylene group (internal acetylenes do not react) to the metal to form, reversibly, an alkynyl metal hydride. The hydride group may then add to another acetylene to form an alkynylvinyl metal complex which can subsequently undergo reductive elimination of the dimer and reform the starting metal complex, or the second acetylenic unit may insert between the alkynyl group and the metal, after which the hydride and vinylic groups can be reductively eliminated. A 50% yield of the phenylacetylene dimer is obtained from phenylacetylene and chlorotristriphenylphosphinerhodium(I) at 20° in 3 days, for example, probably by one of the two above mechanisms (363) and as shown in Scheme VII-3. Almost certainly acetylenic π complexes are involved in the above reactions. The other products from this reaction are higher oligomers where more than one acetylene insertion must have occurred before the reductive elimination took place.

A number of organometallic complexes are known which will polymerize acetylene to low molecular weight linear products. Simply heating phenylacetylene with dichlorobistriphenylphosphineplatinum(II), for example, will

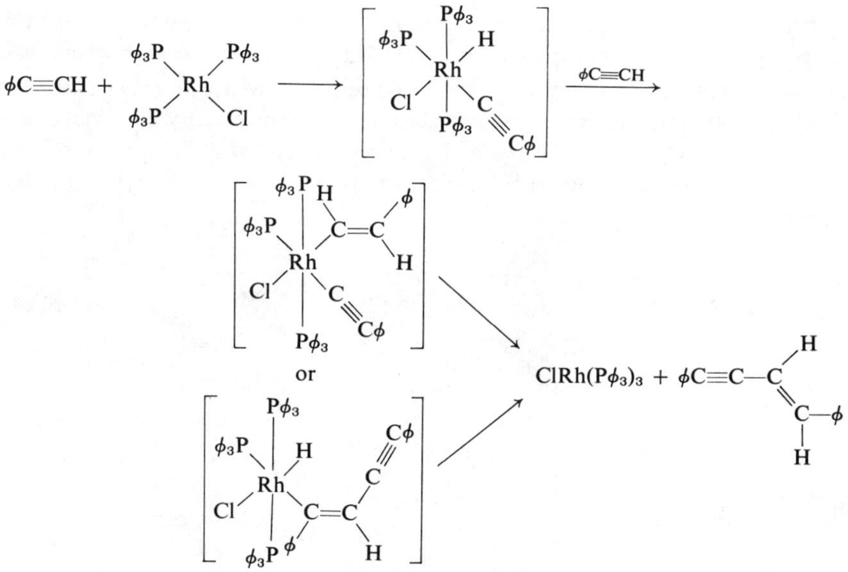

$$\phi C\equiv CH + \quad \begin{array}{c}\phi_3P \quad P\phi_3 \\ \diagdown Rh \diagup \\ \phi_3P \quad Cl\end{array} \quad \longrightarrow \quad \left[\begin{array}{c}P\phi_3 \\ \phi_3P \mid H \\ \diagdown Rh \diagup \\ Cl \mid C \\ P\phi_3 \quad \parallel \\ C\phi\end{array}\right] \quad \xrightarrow{\phi C\equiv CH}$$

$$\left[\begin{array}{c}\phi_3P \; H \quad \phi \\ \phi_3P \diagdown \mid \diagup C=C \\ Rh \quad H \\ Cl \diagup \mid C \\ \quad \parallel \\ P\phi_3 \quad C\phi\end{array}\right]$$

or

$$\left[\begin{array}{c}P\phi_3 \quad C\phi \\ \phi_3P \mid H \quad \parallel \\ \diagdown Rh \diagup \quad C \\ Cl \diagup \mid C=C \\ \phi_3P \; \phi \quad H\end{array}\right]$$

$$\longrightarrow ClRh(P\phi_3)_3 + \phi C\equiv C-C\begin{array}{c}H \\ \diagdown \\ C-\phi \\ \diagup \\ H\end{array}$$

Scheme VII-3.

give polymers in the molecular weight range of 600–1600 (364). Presumably, an addition of a platinum chloride group to the acetylene initiates the reaction although an alkynyl metal hydride complex from oxidative addition could also add and begin the polymerization.

Cyclic products are often obtained from acetylenes and transition metal compounds rather than linear ones. Perhaps, the simplest example is the reaction of nickelocene with dimethyl acetylenedicarboxylate in which a Diels–Alder addition occurs. The reaction takes place to form a product having the double bond from the reacting acetylene coordinated to the metal suggesting that the Diels–Alder reaction occurs within the coordination sphere of the metal (365). Presumably, this reaction can occur with nickelocene

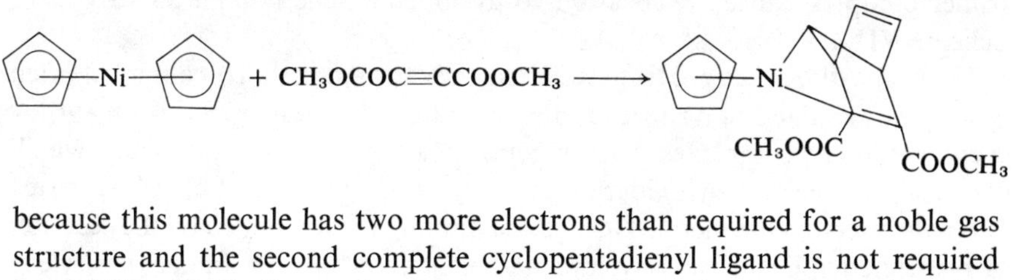

$$\bigcirc\!\!-Ni-\!\!\bigcirc + CH_3OCOC\equiv CCOOCH_3 \longrightarrow \bigcirc\!\!-Ni \underset{CH_3OOC \quad COOCH_3}{\diagup}$$

because this molecule has two more electrons than required for a noble gas structure and the second complete cyclopentadienyl ligand is not required to stabilize it. Cyclic products may also be formed from linear insertion products as seems to occur in the reaction of palladium(II) chloride with diphenylacetylene. If carried out in ethanol solution the reaction forms an ethoxycyclobutenyl π complex (366, 367). The mechanism probably involves a trans alkoxypalladation of one acetylenic unit followed by a cis addition to another acetylenic molecule and cyclization by a conrotatory electrocyclization specifically to give the *endo*-alkoxy product (367). Treatment of the product with hydrochloric acid produces dichloro-π-tetraphenylcyclobutadienepalladium(II).

If the reaction of the acetylene and palladium chloride is carried out in an aprotic solvent, the cyclobutadiene complex is formed directly along with hexaphenylbenzene (368). The cyclic trimerization does not go by way of the cyclobutadiene intermediate, since it does not react further with the acetylene (371), but by an open-chain triacetylene intermediate as determined by the isolation of this intermediate from the reaction of 2-butyne with dichlorobisbenzonitrilepalladium(II) and an investigation of its decomposition reaction (369). An initial cis addition of the palladium chloride group is proposed. This step is followed by two successive additions to more 2-butyne. The open-chain triene product is very labile and readily decomposes to hexamethylbenzene and palladium chloride. The addition of triphenylphosphine to the trimer complex caused cyclization to cyclopentadiene complexes (370) (see Scheme VII-4).

The cyclic dimerization, trimerization, and tetramerization occur most often by way of metalacarbon rings, starting with ligand coupling to form metallocyclopentadiene complexes. Such reactions can only occur with metals which can easily increase their oxidation state by two, since ligand coupling forms two new bonds to the metal. The ligand coupling is exemplified by the reaction of bistriphenylphosphine-π-cyclopentadienylcobalt(I) with diphenylacetylene (92, 372, 373). Initially, a triphenylphosphine is replaced by an acetylene forming an isolable complex. The π complex then reacts with a second acetylene

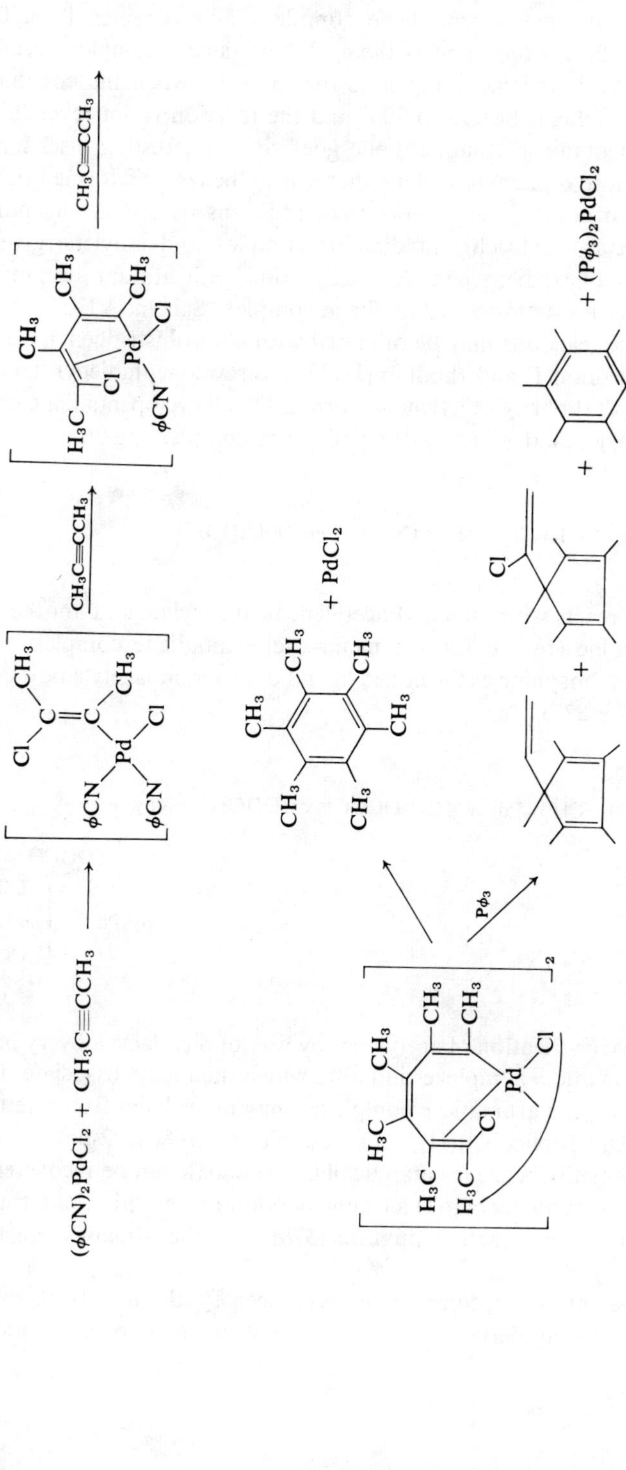
(φCN)2PdCl2 + CH3C≡CCH3
CH3C≡CCH3
CH3C≡CCH3
+ PdCl2
Pφ3
+ (Pφ3)2PdCl2

Scheme VII-4.

forming the cobaltacyclopentadiene complex. This complex is neither in equilibrium with nor the same as the cyclobutadiene π complex because the very stable cyclobutadiene complex is formed only when the cobaltacyclopentadiene complex is heated to 200° and the reaction is not reversible. The cobaltacyclopentadiene complex undergoes some interesting insertion reactions. With more diphenylacetylene, hexaphenylbenzene is formed, but only to the extent of an 8% yield under the conditions tried. Carbon monoxide forms a π-tetraphenylcyclopentadienone complex and ethylene gives a π-tetraphenylcyclohexadiene complex. Reduction with lithium aluminum hydride produces a π-tetraphenylbutadiene complex (Scheme VII-5).

Very similar reactions may be obtained with chlorobistriphenylphosphine-π-dinitrogeniridium(I) and rhodium(I) (374). Biscyclopentadienyltitanium(II) also reacts with diphenylacetylene to form a metalacyclopentadiene complex (375). The very reactive zerovalent palladium complex, bisdibenzalacetone-

$$(\pi\text{-}C_5H_5)_2Ti + 2\phi C{\equiv}C\phi \longrightarrow (\pi\text{-}C_5H_5)_2Ti \overset{\phi \diagup \phi}{\underset{\phi \diagdown \phi}{\Big|}}$$

dipalladium, reacts with dimethyl acetylenedicarboxylate and another ligand such as pyridine (py) to form a metal cyclopentadiene complex, whereas with triphenylphosphine as the added ligand only a monoacetylenic π complex is formed (376, 377).

$$(\phi CH{=}CHCCH{=}CH\phi)_2Pd_2 + 4CH_3OCOC{\equiv}CCOOCH_3 + 4py \longrightarrow$$

$$2py_2Pd \begin{array}{c} COOCH_3 \\ \diagup\!\!=\!\!-COOCH_3 \\ \diagdown\!\!=\!\!-COOCH_3 \\ COOCH_3 \end{array}$$

The cyclic trimerization of acetylenes by way of metalacycles very probably produces aromatic π complexes initially, which then may dissociate. In other cases, however, the aromatic π complexes may be isolated from the reaction mixtures. "Dimesitylcobalt(II)," for example, reacts with 2-butyne at $-10°$ to form hexamethylbenzene catalytically. The cobalt can be recovered in the form of bishexamethylcobalt(I) tetraphenylborate by addition of tetraphenylborate anion to the reaction mixture (378). "Diphenylmanganese(II)" behaves similarly (378).

Polynuclear aromatic compounds have been synthesized from two types of aromatic acetylene derivatives and chlorotristriphenylphosphinerhodium(I).

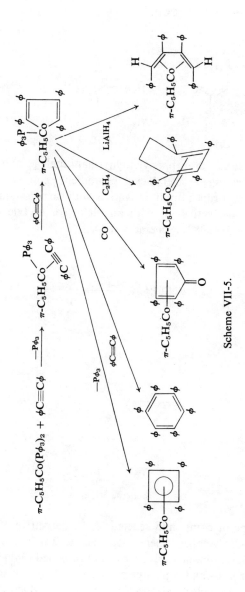

Scheme VII-5.

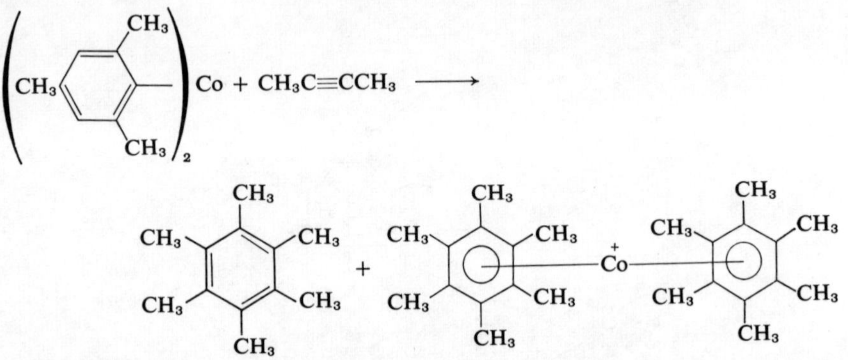

One reaction involves 1,8-di(phenylethynyl)naphthalene and the rhodium complex, first forming a metalacyclopentadiene complex which when reacted with oxygen forms a furan derivative, with carbon monoxide forms a cyclopentadienone derivative, or with an olefin (or acetylene) produces a benzene derivative (379, 380) (Scheme VII-6).

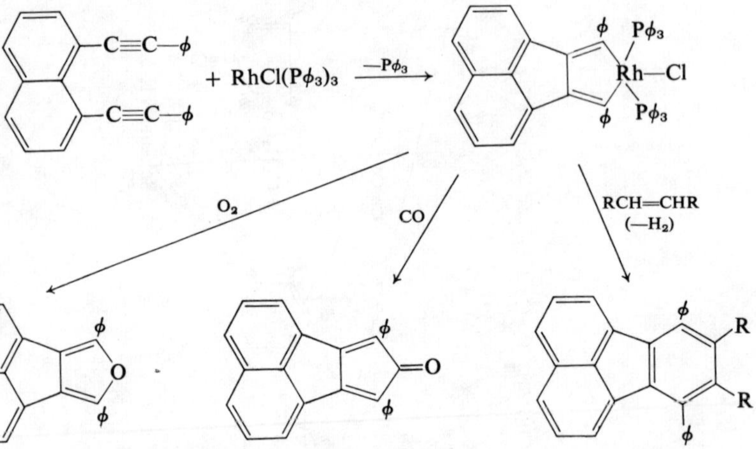

Scheme VII-6.

The second type of compound reacted was a derivative of *o*-dipropiolyl-benzene. The rhodiacyclopentadiene intermediate was reacted with another acetylene and anthraquinone derivatives were obtained (381).

Pentacarbonyliron(0) and diphenylacetylene under the proper conditions will form mainly a dinuclear ferracyclopentadiene in which the second iron atom is bonded to the first iron atom and π-bonded to the two ferracyclopentadiene double bonds (382). This complex reacts with certain dihalides to form other unusual heterocyclic compounds (382).

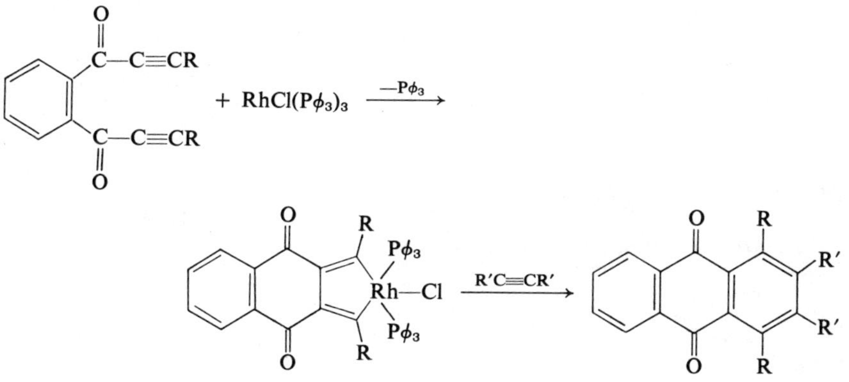

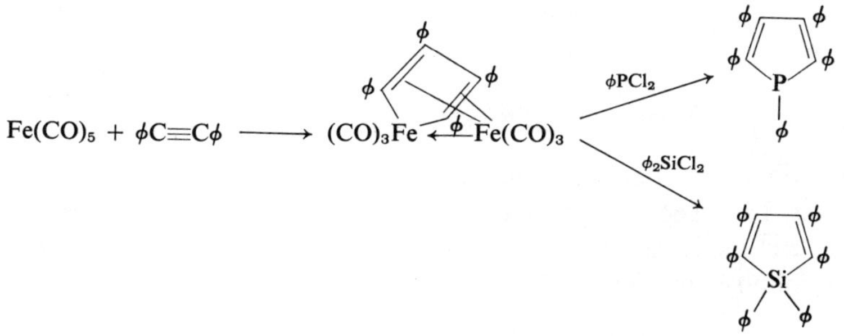

The cyclic trimerization of acetylenes by octacarbonyldicobalt probably goes by way of the known monoacetylenehexacarbonylcobalt complexes and a complex analogous to the dinuclear iron complex, but this intermediate has not been detected. A three acetylene to two cobalt atom complex has been isolated from the reaction, however. An X-ray structural determination shows the complex prepared from *t*-butylacetylene to have the three acetylenic units joined in a chain with each end attached to a different cobalt atom by a σ bond and the three terminal carbons at each end attached by π-allylic bonding to the other cobalt atom (the one that that end of the carbon chain is not σ-bonded to). Heating the complex liberated the ligand in the form of the benzene derivative (108, 109, 383). The preparation of 1,2,4-tri-*t*-butyl-benzene (Scheme VII-7) was the first reported synthesis of this strained molecule. 1,2,4,5-Tetra-*t*-butylbenzene was also prepared in a similar way by reacting di-*t*-butylacetylenehexacarbonyldicobalt with mono-*t*-butylacetylene.

A wide variety of mono- and disubstituted acetylenes have been trimerized to benzene derivatives with various transition metal catalysts (384). With unsymmetrical acetylenes two isomeric trimers are possible and often both are formed. Cyclic tetramers of acetylenes may also be produced, particularly with nickel catalysts and with acetylene itself or esters of propiolic acid. The

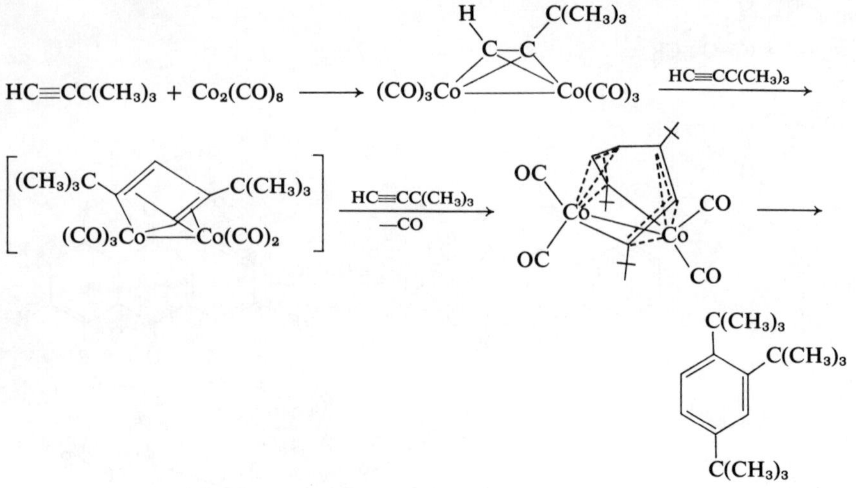

Scheme VII-7.

formation of cyclooctatetraene from acetylene with nickel catalysts was discovered by Reppe many years ago (385). Methyl propiolate forms two isomeric cyclic tetramers and two isomeric trimers on heating with tetrakis-trichlorophosphinenickel(0) (386).

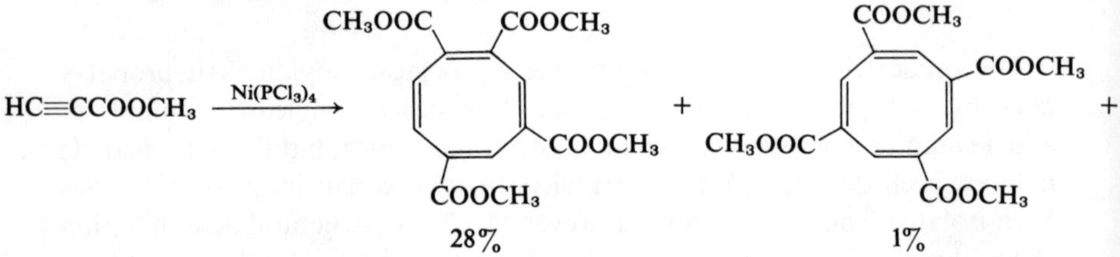

D. ACETYLENIC CYCLIZATIONS WITH INCORPORATION OF ALKYL AND ARYL LIGANDS

Several examples of the reaction of alkyl metal complexes with acetylenes are known in which the alkyl group becomes incorporated into a ring with the acetylenes, usually two acetylenes. Heating dicarbonyl-π-cyclopentadienylmethyliron(II) with diphenylacetylene, for example, is reported to produce 1,2,3,4-tetraphenylferrocene in 10% yield (387, 388). The cyclic diacetylene, 1,8-cyclotetradecadiyne, and pentacarbonyliron(0) also give a

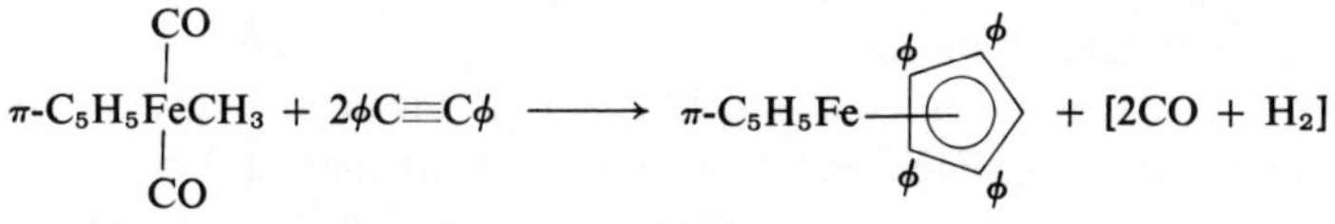

$$\pi\text{-}C_5H_5\overset{\overset{CO}{|}}{\underset{\underset{CO}{|}}{Fe}}CH_3 + 2\phi C\equiv C\phi \longrightarrow \pi\text{-}C_5H_5Fe\text{—}\left[\underset{\phi\ \ \phi}{\overset{\phi\ \ \phi}{\bigcirc}}\right] + [2CO + H_2]$$

cyclopentadienyliron derivative (389). Dicarbonylcyclopentadienylcobalt(I)

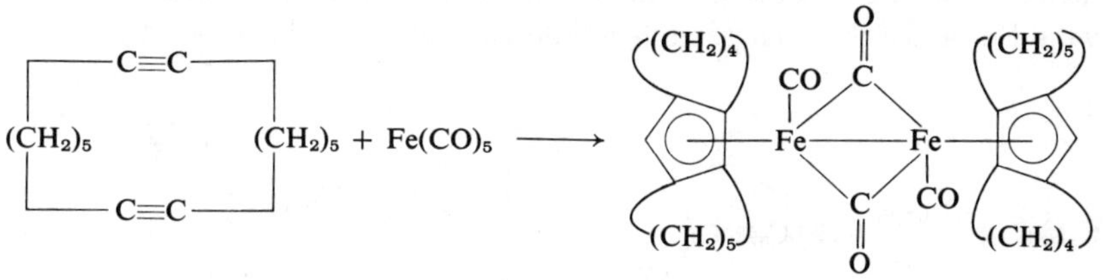

$$\begin{array}{c} \text{—C}\equiv\text{C—} \\ \text{(CH}_2)_5 \qquad \text{(CH}_2)_5 \\ \text{—C}\equiv\text{C—} \end{array} + Fe(CO)_5 \longrightarrow$$

on the other hand, with the same acetylene forms a π-cyclobutadiene complex

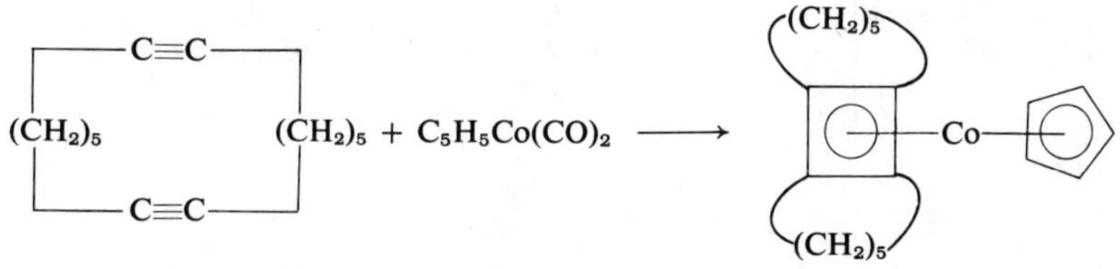

$$\begin{array}{c} \text{—C}\equiv\text{C—} \\ \text{(CH}_2)_5 \qquad \text{(CH}_2)_5 \\ \text{—C}\equiv\text{C—} \end{array} + C_5H_5Co(CO)_2 \longrightarrow$$

(390). "Trimethylchromium" in tetrahydrofuran reacts with diphenylacetylene to form tetraphenylcyclopentadiene and hexaphenylbenzene (391, 392). "Triethylchromium" in the same reaction forms 1,2,3,4-tetraphenylbenzene and hexaphenylbenzene. Tricarbonyl-π-cyclopentadienylmethylmolybdenum(II) also forms tetraphenylcyclopentadiene on heating with diphenylacetylene, but the ethyl derivative gives some 1-methyl-2,3,4,5-tetraphenylcyclopentadiene as well as the 1,2,3,4-tetraphenylbenzene π complex with a π-cyclopentadienylmolybdenum group (387, 388).

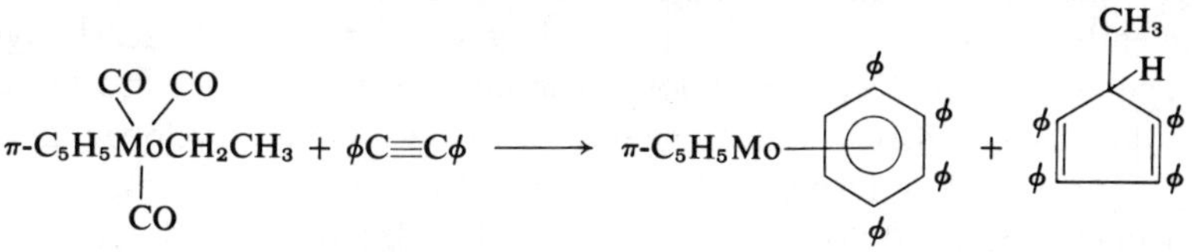

$$\pi\text{-}C_5H_5\overset{CO\ \ CO}{\underset{\underset{CO}{|}}{\overset{\diagdown\ \diagup}{Mo}}}CH_2CH_3 + \phi C\equiv C\phi \longrightarrow \pi\text{-}C_5H_5Mo\text{—} + $$

The formation of the six-membered ring products from the ethyl derivatives probably occurs by way of ethylene (from a β-hydride elimination) adding to a metalacyclopentadiene intermediate followed by dehydrogenation. The formation of the five-membered ring products is not as easily explained. Two mechanisms seem reasonable depending on the metal. The α elimination of methane from "trimethylchromium" forming a carbene (methylene) chromium complex is a possible reaction (393). Reaction of this complex with two acetylene molecules could give the tetraphenylcyclopentadiene observed. This reaction seems less likely with the iron and molybdenum alkyls, although radical abstraction of α-hydrogens may still be a favorable reaction. More probably two acetylenes are inserted into the metal alkyl group initially. Next, the *cis*-methyl may oxidatively add to the metal and then ring closure may occur. A final elimination of a second hydridic hydrogen would produce the cyclopentadienyl derivatives (see Scheme VII-8).

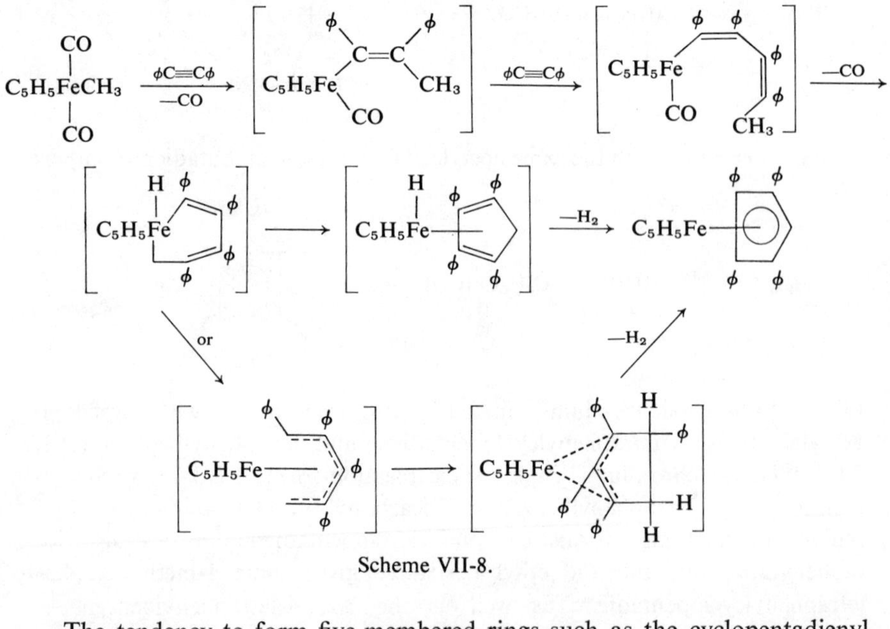

Scheme VII-8.

The tendency to form five-membered rings such as the cyclopentadienyl ring, which can strongly complex with the metal, is encountered fairly frequently. Another example is found in the reaction of decacarbonyldimanganese(0) with acetylene at 150°. The main product (40%) is a bicyclo-[3.3.0]octane complex in which one five-membered ring is a cyclopentadienyl ring π-complexed to the metal (394). Possibly the reaction proceeds via a cyclooctatetraene metal complex as an intermediate. Interestingly, decacarbonyldirhenium(0) reacts with 1,5-cyclooctadiene at 250° to give a reduced form of a similar complex in 5% yield (236) (see Chapter VI, Section A). A related

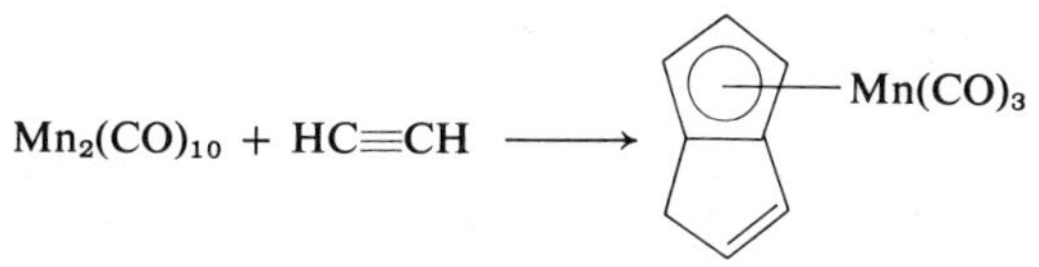

$$Mn_2(CO)_{10} + HC\equiv CH \longrightarrow$$

ring contraction occurs in the reaction between 1,5,9-cyclododecatriene and *cis*-bis(trimethylgermyl)tetracarbonylruthenium(II) (237) (see Chapter VI, Section A). In another example diphenylacetylene was reacted with a combination of aluminum bromide, vanadium tetrachloride, and zinc metal and a 41% yield of a triphenylazulene was obtained (395).

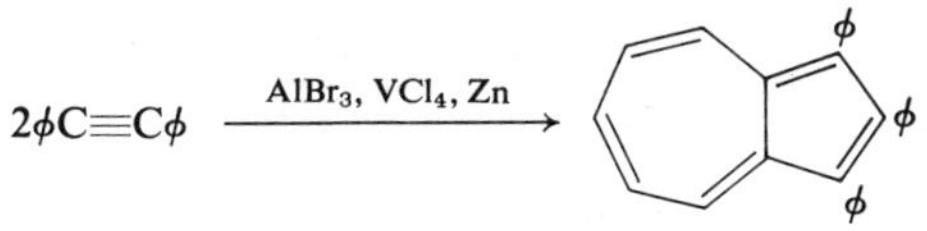

$$2\phi C\equiv C\phi \xrightarrow{\text{AlBr}_3,\ \text{VCl}_4,\ \text{Zn}}$$

Reactions related to the alkyl metal–acetylene reactions also seem to occur with aryl metal derivatives. The reaction of triphenylchromium in tetrahydrofuran with 2-butyne produces 1,2,3,4-tetramethylnaphthalene and hexamethylbenzene (393). An intermediate benzynechromium π complex was initially proposed for this reaction (396), but more recent evidence appears to rule that out (397). A new mechanism was proposed in which the first step is a cis 1:2-addition of phenylchromium to the 2-butyne. The styrylchromium derivative formed then is proposed to undergo an internal metalation at the *o*-phenyl carbon presumably forming an unstable chromium(V) hydride which then reductively elimiminates benzene. The resulting benzochromacyclopentadiene complex finally adds to another 2-butyne and phenylchromium(I) is reductively eliminated (397). The elimination of phenylchromium(I) seems unusual and perhaps the last step is more complicated than it appears.

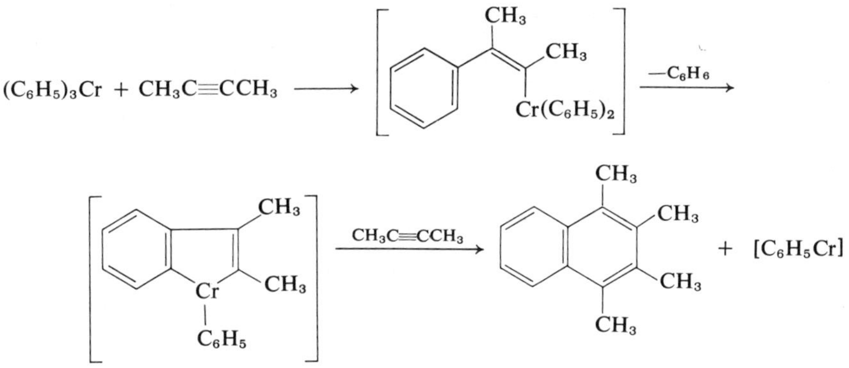

$$(C_6H_5)_3Cr + CH_3C\equiv CCH_3 \longrightarrow$$ $$\xrightarrow{-C_6H_6}$$

$$\xrightarrow{CH_3C\equiv CCH_3}$$ $$+ \ [C_6H_5Cr]$$

An example which does seem to be explicable in terms of a benzyne intermediate is the reaction of biscyclopentadienyldiphenyltitanium(IV) with diphenylacetylene. The product is a metalacycle corresponding to one of the proposed intermediates in the above chromium reaction (398). The intermediate benzyne complex apparently also may react with carbon dioxide giving a chelated *o*-titanobenzoate complex (399).

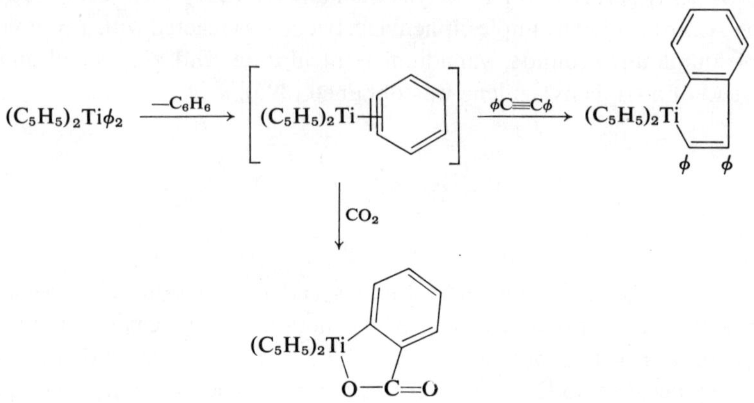

FURTHER READING

M. A. Bennett, Acetylene complexes. *Chem. Rev.* **62**, 611 (1962).

C. W. Bird, Oligomerization of acetylenes. "Transition Metal Intermediates in Organic Syntheses," Chapter 1. Academic Press, New York, 1967.

F. L. Bowden and A. P. B. Lever, Transition metal-acetylene chemistry. *Organometal. Chem. Rev.* **3**, 227 (1968).

F. R. Hartley, Olefin and acetylene complexes of platinum and palladium. *Chem. Rev.* **69**, 799 (1969).

W. Reppe, N. von Kutepow, and A. Magin, Cyclization of acetylenes. *Angew. Chem., Int. Ed. Eng.* **7**, 727 (1969).

Chapter VIII

Mixed Olefin, Diene, and Acetylene Combination Reactions

The combination of two or more unsaturated molecules with themselves has been considered in the preceding chapters. The coupling of two or more different unsaturated compounds of the same type is possible also, but relatively few examples have been studied because complex mixtures of the various possible products usually result. The combination of unsaturated molecules of different types, i.e., olefins, dienes, or acetylenes, with each other, however, is often quite specific and a number of useful reactions of this kind are known.

A. OLEFIN–DIENE REACTIONS

Two basic mechanisms of reaction are generally observed with olefin–diene reactions. Ligand coupling may occur as illustrated by the reaction of tricarbonyl-π-butadieneiron(0) with tetrafluoroethylene catalyzed by ultraviolet irradiation (91). This reaction results in a formal increase in oxidation state of the metal by two and, therefore, will occur only with metals that have two stable oxidation states two electrons apart. Fluorinated organic ligands,

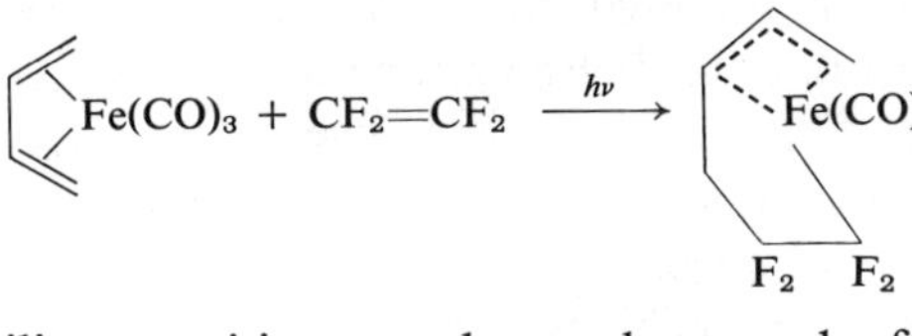

in general, stabilize transition metal complexes and often reactive intermediates can be isolated when fluorinated groups are present, whereas they cannot be with the unfluorinated compounds. The ligand coupling may ultimately give cyclic products if the initial adduct reductively eliminates both hydrocarbon groups. This kind of reaction occurs with the nonconjugated diene, norbornadiene, and activated olefins such as acrylonitrile with bisacrylonitrilenickel (0) as catalyst (400). Possibly a π-homoallylic nickel(II) intermediate is involved.

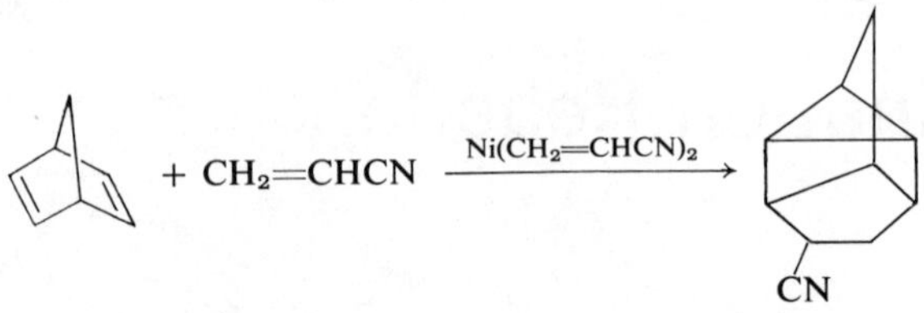

The second basic olefin–diene reaction is an addition of a π-allyl group, formed in an initial step from the diene, to the olefin. There is no change in the oxidation state of the metal in this reaction. The reaction with two allyl groups is illustrated by the coupling of bis-π-methallylnickel(II) with tetrafluoroethylene (401).

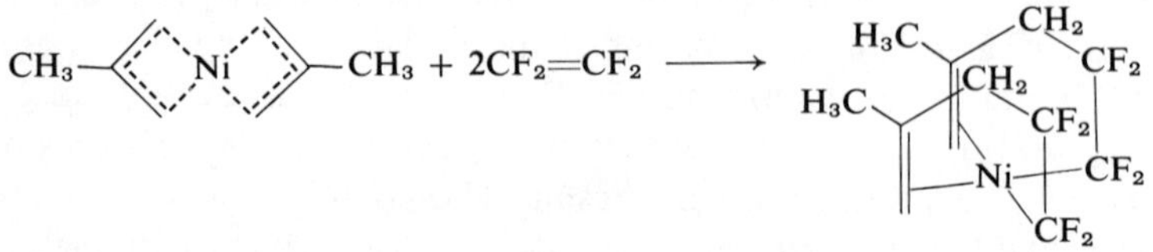

The monoallylic–olefin addition is an important step of the rhodium trichloride-catalyzed reaction of ethylene with butadiene in ethanol solution which forms a cis-trans mixture of 1,4-hexadienes in 90% yield (402). 1,4-

$$CH_2{=}CH_2 + C_4H_6 \xrightarrow[\text{EtOH}]{\text{RhCl}_3} CH_2{=}CHCH_2CH{=}CHCH_3$$

Hexadiene is of commercial interest as a monomer for copolymerization with ethylene and propylene. The disubstituted double bond in the 1,4-hexadiene remains after the polymerization and provides a cite for sulfur vulcanization of the otherwise saturated and unreactive polymer. When the proper ratios of the three monomers are used, the vulcanized polymers are rubbery and can be used as substitutes for vulcanized natural rubber.

The mechanism of the ethylene–butadiene reaction has been studied in detail. The initial reaction is a reduction of the rhodium trichloride hydrate by ethylene to the known chlorobis-π-ethylenerhodium(I) dimer. The last compound then reacts reversibly with hydrogen chloride to form a dichloroethylrhodium(III) species. The hydrogen chloride is available from the reduction step or preferably is added in excess to the reaction mixture. The reaction intermediates are assumed to be saturated coordinately with ligands present in the reaction mixture—chloride ions, olefins, dienes, or solvent molecules. As these ligands are not specifically known and may not matter anyway, they are shown only as L's in the equations below. The σ-ethyl-rhodium complex next coordinates with butadiene and a reversible hydrogen shift from the ethyl group to the diene occurs forming a π-ethylene-π-crotyl complex. This complex then undergoes the addition step; the crotyl group adds to the ethylene. A 4-hexenylrhodium(III) complex is produced. Coordination of another diene molecule and a second hydrogen transfer, this time from the hexenyl ligand to the diene, gives a π-1,4-hexadiene-π-crotylrhodium(III) complex. A replacement of the 1,4-hexadiene with ethylene and another ligand, L, completes the catalytic cycle (402) (Scheme VIII-1).

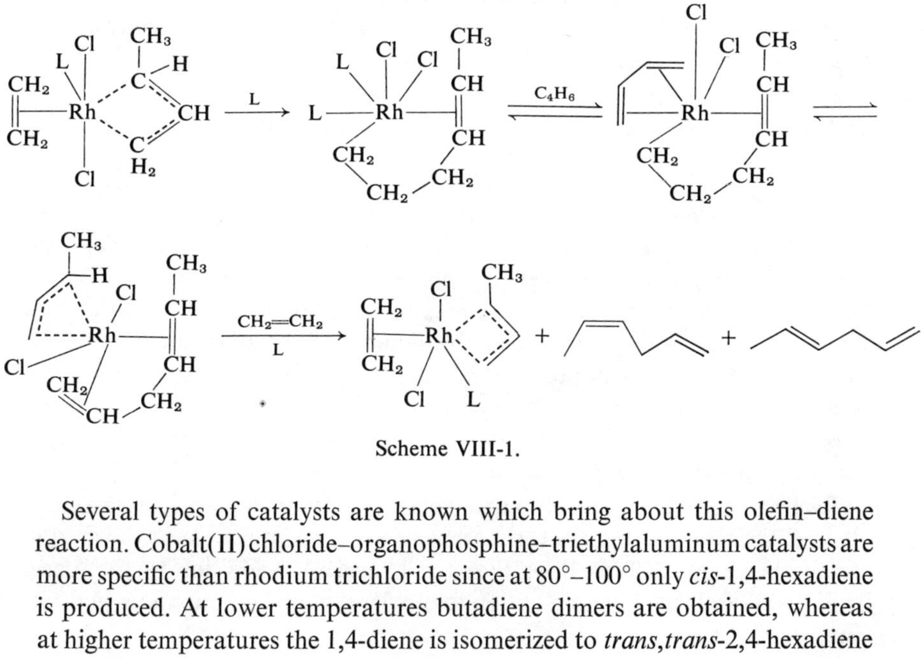

Scheme VIII-1.

Several types of catalysts are known which bring about this olefin–diene reaction. Cobalt(II) chloride–organophosphine–triethylaluminum catalysts are more specific than rhodium trichloride since at 80°–100° only *cis*-1,4-hexadiene is produced. At lower temperatures butadiene dimers are obtained, whereas at higher temperatures the 1,4-diene is isomerized to *trans,trans*-2,4-hexadiene (403, 404).

The cobalt catalysts have also been used to combine substituted olefins and dienes. Isoprene and ethylene give a mixture, apparently because both possible isomeric π-allylic complexes are formed from the isoprene (405). Propylene and butadiene produce a mixture of the *trans*-1,4- and the *trans*-1,3-dienes (405). The last compound probably arises from an isomerization of the first product. Methyl acrylate and butadiene with a similar cobalt

$$C_4H_6 + CH_3CH{=}CH_2 \xrightarrow{\text{CoCl}_2\text{—PCl}_3\text{—AlEt}_3}$$

$$\underset{25\%}{CH_2{=}\underset{\underset{CH_3}{|}}{C}{-}CH_2CH{=}CHCH_3} + \underset{2\%}{CH_2{=}\underset{\underset{CH_3}{|}}{C}{-}CH{=}CHCH_2CH_3}$$

catalyst also produce the expected 1,4-diene (406). The more positive carbon of the double bond in the olefin appears to be attacked preferentially by the least substituted end of the π-allylic group.

$$C_4H_6 + CH_2{=}CHCOOCH_3 \xrightarrow{\text{Co(Acac)}_3\text{—AlEt}_3} \underset{30\%}{CH_3CH{=}CHCH_2CH{=}CHCOOCH_3}$$

A third mechanism of diene–olefin coupling appears to occur in at least one reaction, the reaction of norbornadiene, the olefin in this example, with butadiene, using a cobaltic acetylacetonate–triethylaluminum catalyst (407). The 2-butadienylnorbornene obtained is probably formed by an initial hydridocobalt(I) addition to one double bond of the norbornadiene and then an addition of this cobalt alkyl to the butadiene. A π-butadienylmethylallyl-cobalt complex is probably an intermediate. The last compound then may coordinate with norbornadiene and undergo hydrogen transfer and ligand replacement with solvent or reactants (Scheme VIII-2).

Cyclic combinations of olefins and dienes also occur. An *o*-phenanthroline-dialkyltitanium catalyst converts ethylene and butadiene into vinylcyclo-butane in about 20% yield. A titanium hydride 1:4-addition to the diene is probably followed by an addition to ethylene, a cyclization, and a hydride elimination (408).

$$C_4H_6 + [HTiL_3] \rightleftharpoons [CH_3CH{=}CHCH_2TiL_3] \xrightarrow{\text{C}_2\text{H}_4}$$

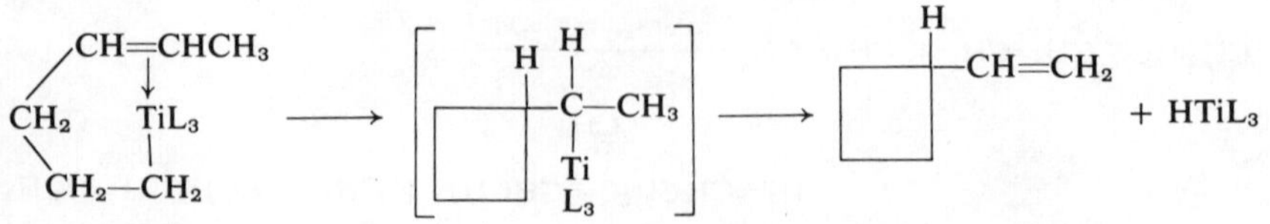

L = a ligand

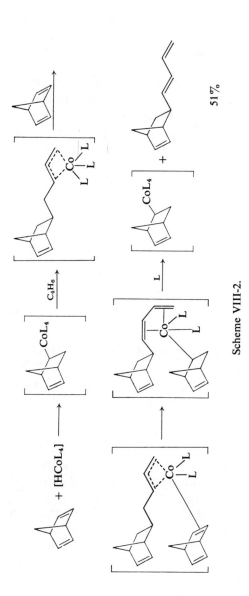

Scheme VIII-2.

Nickel(0) catalysts are useful reagents for reacting two diene units with one olefinic group to form cyclodecadiene derivatives. Tetrakistriphenyl-phosphinenickel(0) or bis-1,5-cyclooctadienenickel(0) have been used to convert ethylene and butadiene at about 30 atm and 25° in 80% yield into *cis,trans*-1,5-cyclodecadiene (409). The mechanism of the reaction appears to be related closely to the one proposed for the 1,5,9-cyclododecatriene synthesis (Chapter VI, Section J), but with the last part of the ring formed from an ethylene unit instead of another butadiene (409). *trans*-1,4,9-Deca-triene is also produced in the above reaction; the relative amount increases rapidly with increasing temperature. The hydrogen transfer reaction in the intermediate two-diene–one-ethylene adduct is apparently favored by higher temperatures.

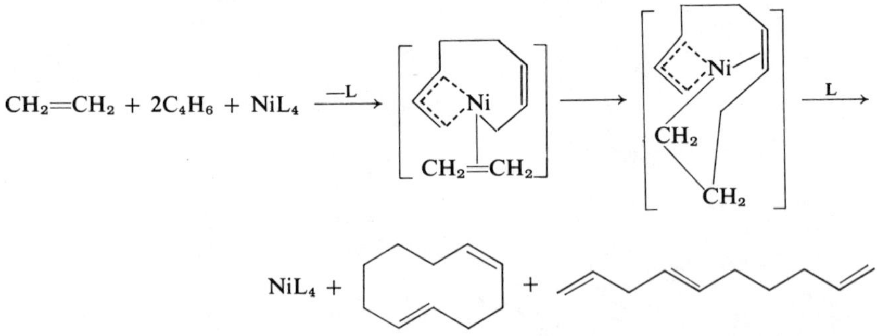

A related nickel catalyst prepared from nickel(II) chloride, bisdiphenyl-phosphinomethane, and triethylaluminum converts ethylene and butadiene at 110° into *cis*-1,4-hexadiene if the phosphine to nickel ratio is greater than one, but to *trans*-1,4,9-decatriene if the ratio is less than one (410, 411). The hexadiene is probably formed by the π-allylic–ethylene coupling reaction as shown in the rhodium(III) chloride reaction above. The 1,4,9-decatriene appears to be coming from the same type of intermediate as proposed in the preceding example. Elevated temperatures apparently cause Ni(0) catalysts to prefer hydride mechanisms rather than the reductive eliminations favored at lower temperatures.

A mixed cyclic coupling of dienes with allene has also been carried out with nickel(0) catalysts. Both ten- and twelve-membered ring products were formed. The ten-carbon product was composed of two diene units and one allene, while the twelve-carbon product contained two dienes and two allenes. The intermediates were presumably cyclooctadienediylnickel(II) species π-complexed with an allene molecule. An addition of the π-allylic group to the allene occurs followed either by an addition to another allene and cyclization or by cyclization before the second allene reacts. Since the allene units are unsymmetrical, two directions of addition are possible and both are observed (412) (Scheme VIII-3).

$2C_4H_6 + CH_2=C=CH_2 + Ni(O) \longrightarrow$

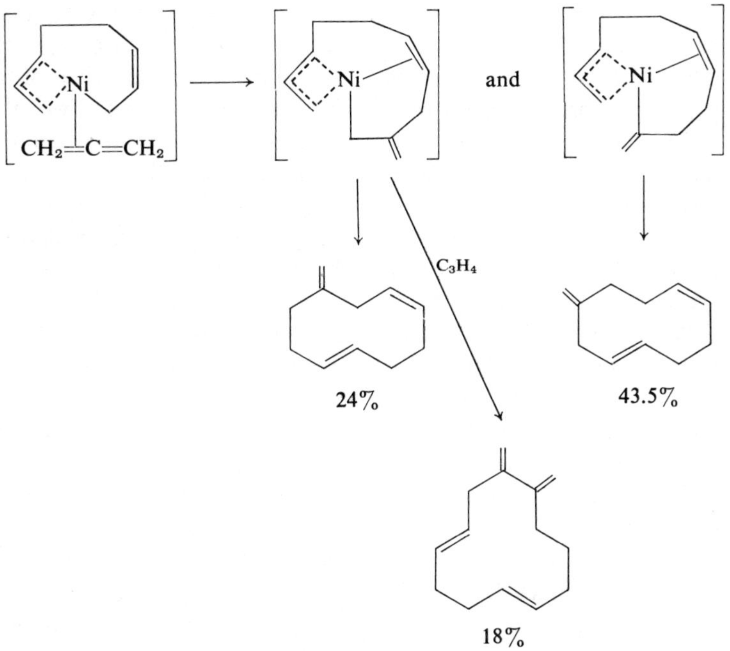

Scheme VIII-3.

Cyclic couplings with two different dienes and ethylene have been carried out with nickel(0) catalysts and, as might be expected, the reaction is not very selective as to which diene it uses and complex mixtures of isomers are obtained (413). The problems of separating and identifying the products are difficult, but if these can be solved, the reaction provides a relatively simple route to many cyclodecadiene derivatives that would be very difficult to obtain by more conventional methods. Some of the products obtained from ethylene, butadiene, and isoprene are shown below.

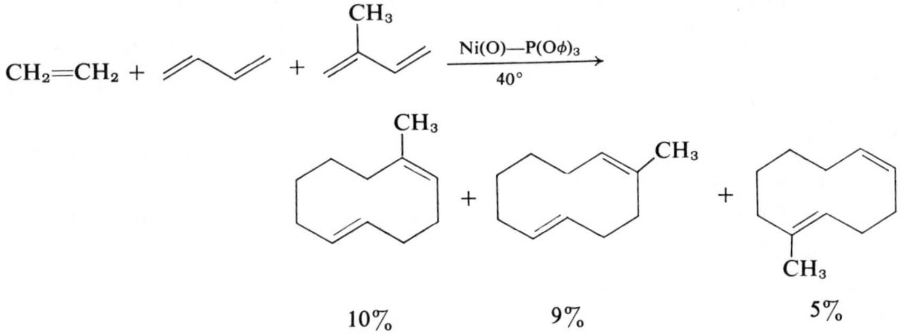

B. ACETYLENE–OLEFIN REACTIONS

Specific couplings of acetylenes with olefins are rare. Only two examples have been published. It has been reported that diphenylacetylene and ethylene or various other olefins will couple in the presence of dichlorobis-benzonitrilepalladium(II) to give butadiene derivatives. Two olefinic units couple with one acetylene. Prolonged heating converts the dienes into indene derivatives apparently by a simple acid-catalyzed (HCl) reaction (414). The mechanism of the coupling has not been investigated, but reasonable guesses can be made (415). A first step could be the addition of palladium chloride to the acetylene, cis or trans, followed by addition of the vinylpalladium species formed to ethylene. An elimination and readdition of a hydrido-palladium group in the reverse direction would give a π-allylic palladium intermediate. An addition of the π-allylic group to another ethylene molecule followed by hydride elimination and readdition in the reverse manner would allow a final elimination of dichloropalladium with formation of the conjugated diene product observed (Scheme VIII-4).

A second example of an acetylene–olefin reaction is observed when diphenylacetylene reacts with bisacrylonitrilenickel(0) and 2,3,4,5-tetraphenylbenzonitrile is produced in about 50% yield along with hexaphenylbenzene (416). A metalacyclopentadiene intermediate is probably involved which may react either with acrylonitrile to give the benzonitrile product after loss of H_2 (or perhaps propionitrile) or with more diphenylacetylene to form hexaphenylbenzene.

$$\phi C{\equiv}C\phi + Ni(CH_2{=}CHCN)_2 \longrightarrow$$

C. ACETYLENE–DIENE REACTIONS

The addition of π-allylic groups to acetylenes is an important step in several of the known diene–acetylene reactions. Simple examples of this reaction are not known, but there is an example of a more complicated one involving two allylic compounds and two acetylene molecules in which an allylic halide coupling is the final step. This reaction produces very reactive 1,4,6,9-decatetraene derivatives (417). The coupling step probably involves a diorganonickel intermediate or a four-centered transition state rather than a nickel(IV) oxidative addition product (see Scheme VIII-5).

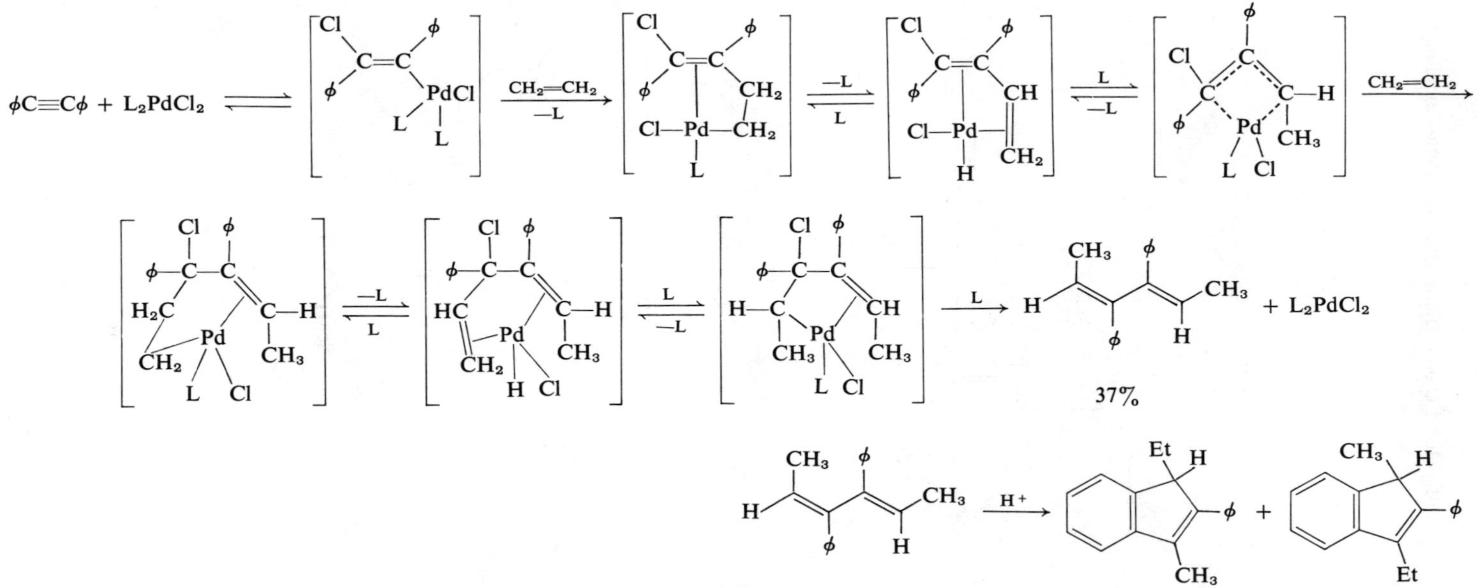

Scheme VIII-4.

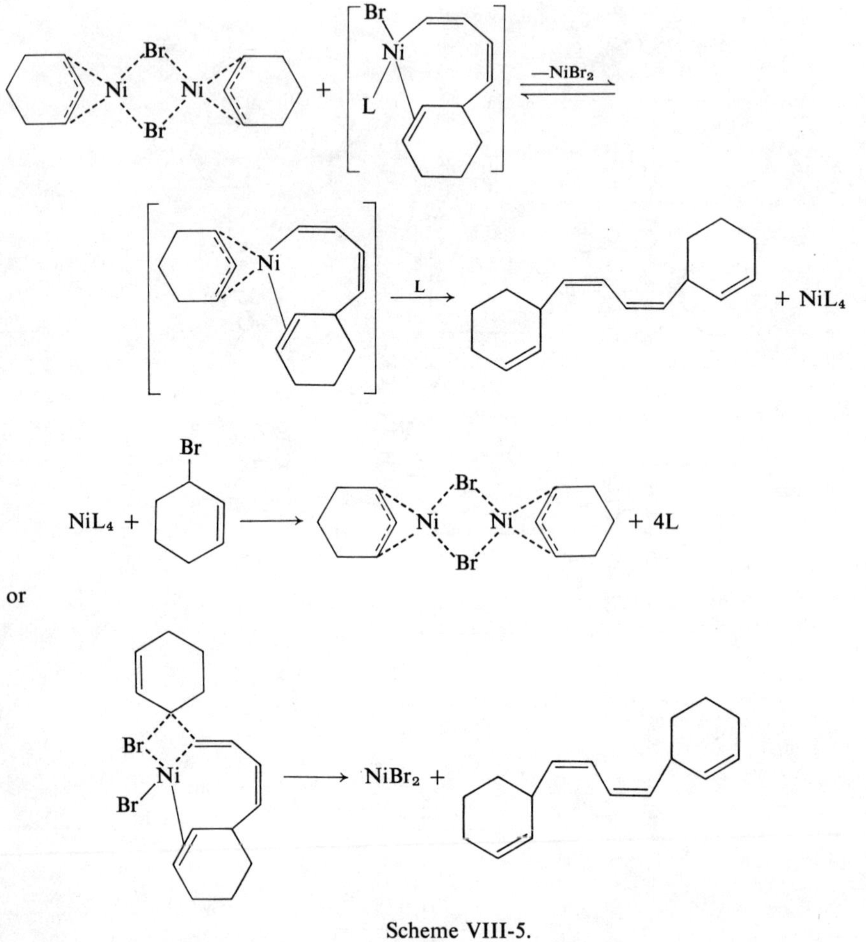

Scheme VIII-5.

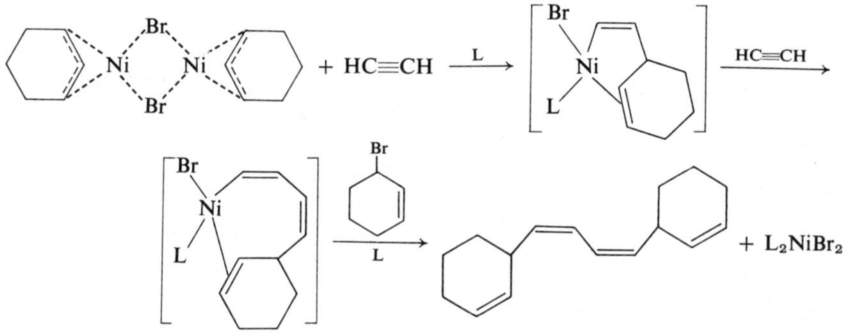

25%

Cyclic compounds are the usual products of diene–acetylene reactions. A very simple example is the transition metal-catalyzed Diels–Alder reaction. Iron(0) complexes such as biscyclooctatetraeneiron(0) or a combination of ferric chloride and isopropylmagnesium chloride catalyze the reaction (418). Probably diene and acetylene ligand coupling occurs first and then a reductive elimination of the cyclic diene follows. Other catalysts capable of oxidative addition and elimination reactions would be expected to be useful for this reaction also.

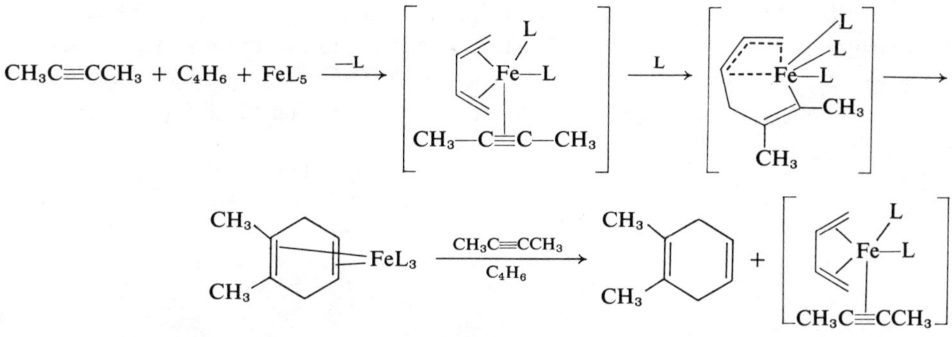

If the above example is attempted with norbornadiene instead of butadiene, a completely different reaction occurs. Initially a reverse Diels–Alder appears to take place with the norbornadiene because cyclopentadiene is one of the products formed. The acetylene from the norbornadiene ends up in a benzene ring with two of the added acetylenic molecules (418). Apparently, the iron(0) catalyst reverses the Diels–Alder reaction as a means for obtaining better ligands and then the acetylene trimerization reaction occurs, probably by way of a ferracyclopentadiene intermediate.

Alkyl- or aryl-substituted acetylenecarboxylic esters and butadiene react in the presence of nickel(0) catalysts to form vinylcyclohexadienes. Mixed ligand coupling probably occurs first. This step would be followed by an addition to a second acetylenic unit in either direction (giving a mixture of isomeric products) and a final reductive elimination of the organic ligands as a cyclohexadiene derivative (419) (Scheme VIII-6). These products undergo an interesting internal Diels–Alder reaction on heating to 180° (419). Less

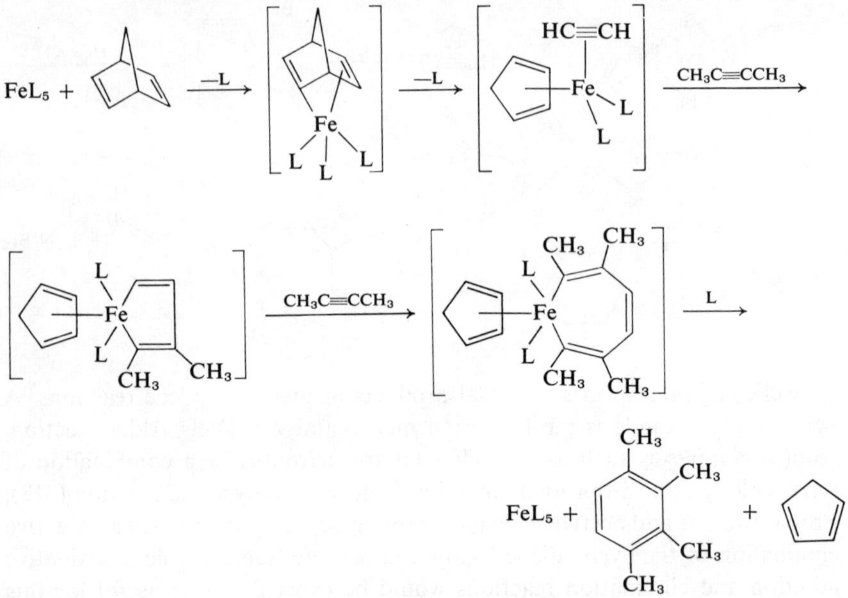

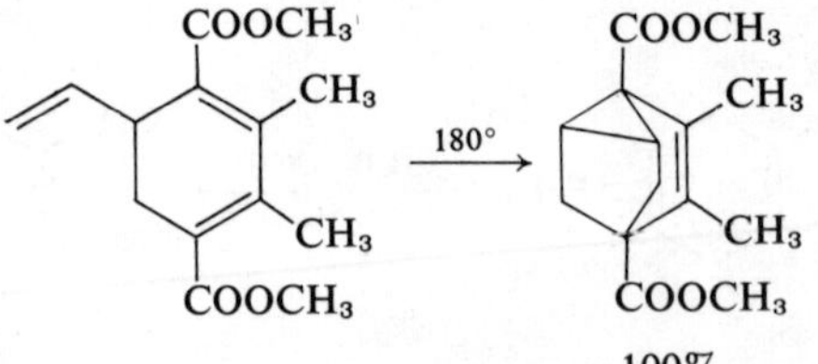

~ 100%

reactive acetylenes than the propiolate esters will react with conjugated dienes and nickel(0) catalysts to give cyclodecatriene derivatives composed of two diene units and one acetylene (420). The mechanism of the reaction appears to be related closely to the two-diene–one-olefin cyclic coupling reactions described in Section A, with an acetylene replacing the olefin in the octadiene-diylnickel(II) intermediate. The cyclodecadiene products readily undergo the Cope rearrangement on warming. The divinylcyclohexadienes so formed can then be aromatized with potassium *t*-butoxide in dimethyl sulfoxide (DMSO) providing a useful synthesis for some tetrasubstituted benzene derivatives

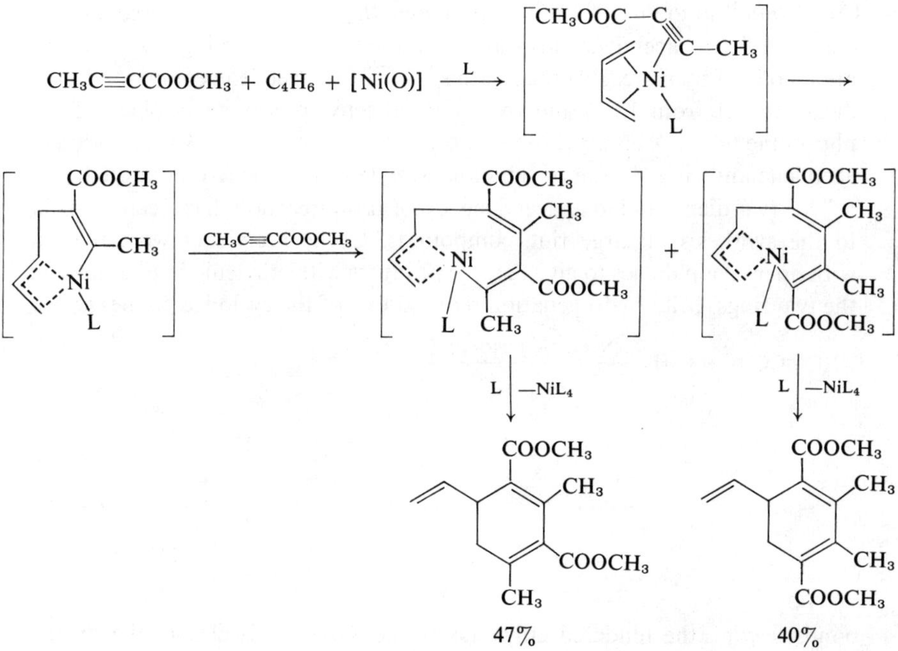

Scheme VIII-6.

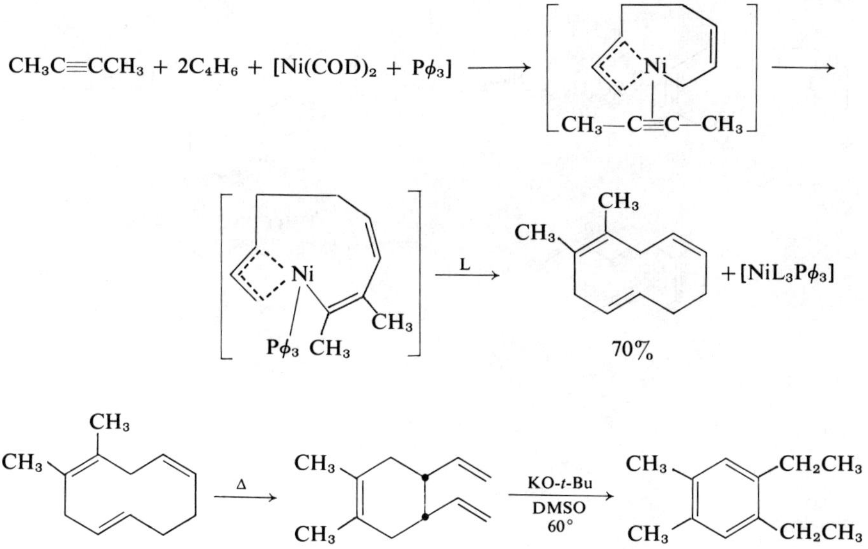

(420). Small amounts of twelve-membered ring products composed of two
dienes and two acetylenes also are produced in the coupling (421). Yields
are increased by increasing the relative amount of acetylene compared to the
diene (to 1:1 from 10:1) and by using triphenyl phosphite in place of the
phosphine (421). A competition seems to exist between insertion of a second
acetylene and ring closure of the monoacetylene–two-diene unit adduct.

The two-diene and one-acetylene cyclization reaction has been applied
to the synthesis of large ring compounds. Cyclic acetylenes were reacted
with open-chain dienes to give bicyclic products with ethylenic bridges joining
the two rings. Mild hydrogenation reduced two of the cyclodecatriene double

$$CH_3C{\equiv}CCH_3 + C_4H_6 \xrightarrow{\text{[Ni(COD)}_2 + \text{P(O}\phi)_3]}$$

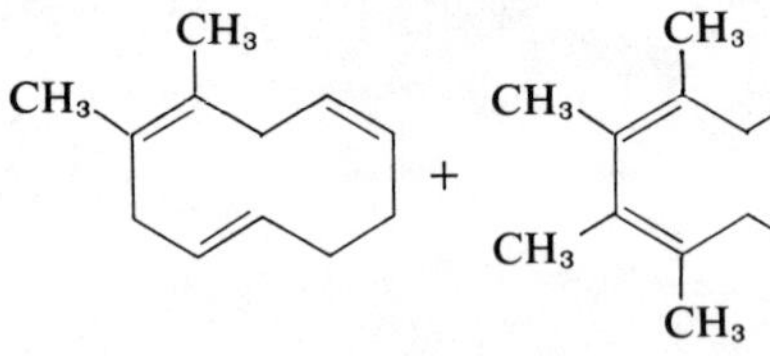

58%   19%

bonds leaving the hindered ethylenic bridge. Ozonolysis cleaves the bridge
to a larger ring diketone which then can be reduced by the Wolff–Kishner
reaction to the cyclic hydrocarbon. Cycloeicosane, for example, was prepared
from cyclododecyne and butadiene as follows (422).

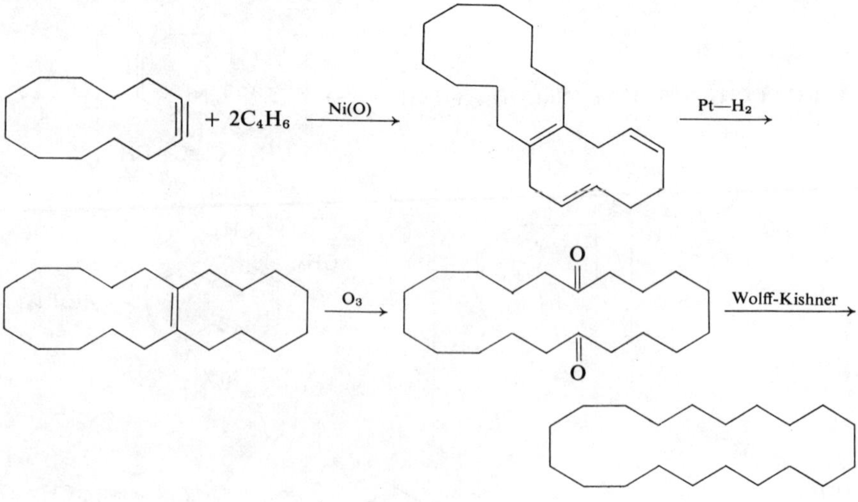

Reactions of Transition Metal Complexes with Carbon Monoxide and Organic Carbonyl Compounds

Some of the most useful transition metal reactions occur with carbon monoxide. The reaction products can be a wide variety of organic compounds such as aldehydes, ketones, alcohols, esters, lactones, amides, acids, and acid halides. Generally, acyl metal complexes are intermediates in the reactions. Reactions of organic carbonyl compounds with transition metal complexes are much rarer, although further study will more than likely turn up more examples.

A. CARBON MONOXIDE REACTIONS

1. Formation Reactions of Acyl Metal Complexes

a. The 1:2 Shift Reaction

The 1:2 shift is the basic step in most carbon monoxide reactions. The mechanism and some variations of it have already been discussed in some detail in Chapter II, Section B,4,a. The question of whether true coordinately unsaturated intermediates are produced in the shift (first-order reaction) or whether solvent or other potential ligands present assist the shift (second-order reaction) has only been answered in a few cases. Generally, it is assumed that if solvent effects on the 1:2 shift are small and entropies of activation are near zero a coordinately unsaturated intermediate is involved. An instructive example is the reaction of *cis*-dichlorodicarbonyl(dimethylphenylarsine)alkyliridium(III) complexes with another ligand to form the acyl complex with the new ligand ultimately coordinating with the metal

(423). The structures of the starting materials and products were deduced mainly from NMR and IR data. The following transformations were observed.

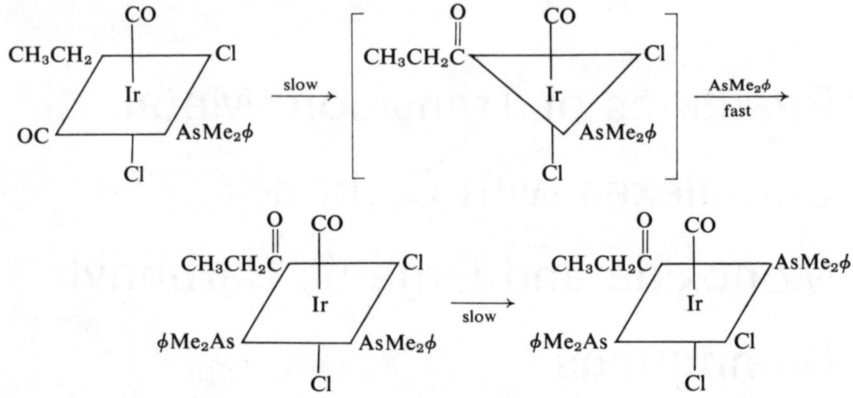

The ethyl group apparently shifted from iridium to carbon monoxide, producing a five-coordinate trigonal-bipyramidal intermediate which then reacted with the new ligand. The new ligand entered trans to the acyl group because of the large trans effect of the acyl ligand. A slow isomerization then produced the thermodynamically favored cis isomer. Kinetics of the reaction have been measured in several solvents of different dielectric constants. Differences were small and in a direction opposite to that expected based on the coordinating ability of the solvents (424). The reaction rates were independent of the reacting ligand concentration. Entropies of activation were measured for the related triphenylarsine reactions and found to be -5.0 e.u. for the ethyl compound and -6.7 e.u. for the methyl derivative, about what would be expected for first-order reactions.

The reaction of pentacarbonylphenylmanganese with $C_2H_5C(CH_2O)_3P$ in chloroform solution, on the other hand, shows second-order kinetics which become first-order in solvents of higher dielectric constant and better coordinating ability, indicating solvent participation in the phenyl shift (425). The triphenylarsine(methyl)iridium complex is about 3.5 times less reactive than the corresponding dimethylphenylarsine derivative in chlorobenzene solution. The triphenylarsine(ethyl)iridium complex is about 6.3 times more reactive than the corresponding methyl compound (424).

The 1:2 shift reaction readily forms acyl metal derivatives, but has never been observed to form α-ketoacyl metal complexes where addition to two carbon monoxides has taken place. In the few examples known where organic 1,2-diketonic products are formed in transition metal reactions, the diketones are probably produced by reductive elimination of two acyl groups or by some other mechanism than a double CO insertion. Carbonylation of *cis*-bistriethylphosphinedimethylplatinum(II), for example, yields first the unstable diacetyl derivative and then biacetyl (59).

$$
\begin{array}{c}
Et_3P \diagdown \quad \diagup CH_3 \\
\quad Pt \quad + 2CO \\
Et_3P \diagup \quad \diagdown CH_3
\end{array}
\longrightarrow
\left[
\begin{array}{c}
Et_3P \diagdown \quad \diagup COCH_3 \\
\quad Pt \\
Et_3P \diagup \quad \diagdown COCH_3
\end{array}
\right]
\xrightarrow{\;CO\;}
$$

$$
[(Et_3P)_2Pt(CO)_2] + CH_3COCOCH_3
$$

The 1:2 shift of an organic group from the metal to a coordinated carbonyl group and the reverse reaction proceed with retention of stereochemistry about the carbon, at least in the few examples studied. The carbonylation and decarbonylation of optically active pentacarbonyl-1-phenyl-2-propyl-manganese(I), for example, occur with retention of configuration, and without loss of optical activity (426).

$$
\phi CH_2 \overset{*}{C}HMn(CO)_5 + CO \;\rightleftharpoons\; \phi CH_2 \overset{*}{C}HCOMn(CO)_5
$$

with CH_3 substituents on the starred carbons.

Similarly, reaction of π-cyclopentadienyldicarbonyl-*threo*-1,2-dideutero-3,3-dimethylbutyliron(II) with triphenylphosphine produces the phosphine-substituted acyl derivative (two isomers) with retention of stereochemistry of the *threo*-deuterium arrangement. Two isomeric products are obtained because the iron atom becomes asymmetric with the phosphine substitution and it may have the same or opposite configuration compared with the two asymmetric deuterium-substituted carbons (427).

$$
(CH_3)_3CC\text{--}C\text{--}Fe\text{--}\langle\text{Cp}\rangle + P\phi_3 \longrightarrow
$$

$$
(CH_3)_3CC\text{--}C\text{--}C\text{--}Fe\text{--}\langle\text{Cp}\rangle + (CH_3)_3CC\text{--}C\text{--}C\text{--}Fe\text{--}\langle\text{Cp}\rangle
$$

In most useful applications of the addition of metal alkyls to carbon monoxide, the acyl intermediate and often the precursor alkyl are not isolated. Three examples are discussed below to illustrate the reaction and show three ways that the acyl groups may be removed from a metal.

A catalytic synthesis of esters from primary and secondary halides occurs with carbon monoxide and an alcohol using tetracarbonylcobaltate anion as catalyst in the presence of a hindered amine (94c). Initially, the cobaltate anion reacts with the halide (S_{N^2}) to form the alkyl, which then adds to carbon monoxide producing the acyl metal derivative. The acyl group is lost from the metal by a simple alcoholysis which gives the ester and hydrido-tetracarbonylcobalt(I). The hindered amine converts the acidic hydride back into the anion and completes the cycle. A hindered amine is used to minimize side reactions which otherwise would produce amides or alkylated amines.

$$RI + Co(CO)_4{}^- \longrightarrow RCo(CO)_4 + I^-$$

$$RCo(CO)_4 + CO \longrightarrow RCOCo(CO)_4$$

$$RCOCo(CO)_4 + R'OH \longrightarrow RCOOR' + HCo(CO)_4$$

$$HCo(CO)_4 + R_3{}''N \longrightarrow R_3{}''NH^+Co(CO)_4{}^-$$

In another example symmetrical ketones are obtained catalytically from diaryl- or dialkylmercury compounds and carbon monoxide with octacarbonyldicobalt(0) and ultraviolet irradiation as catalysts in THF solution (428). The tetrahydrofuran first disproportionates the octacarbonyldicobalt into tetracarbonylcobaltate anion and a positively charged cobalt species. The anion is then alkylated (arylated) by the mercurial. Both organic groups in the mercurial are used giving bis(tetracarbonylcobalt)mercury(II). Irradiation converts this compound back into octacarbonyldicobalt and mercury metal. The ketonic products are obtained by the second-order decomposition of the alkylcobalt complex, a known reaction (429). The cobalt presumably is initially converted into heptacarbonyldicobalt which reacts with the carbon monoxide present to reform the catalyst.

$$Co_2(CO)_8 + THF \rightleftharpoons [(THF)Co(CO)_4{}^+ Co(CO)_4{}^-]$$

$$R_2Hg + (THF)Co(CO)_4{}^+ Co(CO)_4{}^- \longrightarrow [RCo(CO)_4] + [RHgCo(CO)_4] + THF$$

$$2[RHgCo(CO)_4] \longrightarrow R_2Hg + Hg[Co(CO)_4]_2$$

$$Hg[Co(CO)_4]_2 \xrightarrow{h\nu} Hg + Co_2(CO)_8$$

$$2[RCo(CO)_4] \longrightarrow RCOR + [Co_2(CO)_7]$$

$$[Co_2(CO)_7] + CO \longrightarrow Co_2(CO)_8$$

The carbon monoxide insertion step probably involves a 1:2 shift of the organic group to form a tricarbonylacylcobalt(I) species which then can undergo an oxidative addition reaction with an unreacted tetracarbonylalkyl- or arylcobalt complex. A reductive elimination of the ketone could then follow and form heptacarbonyldicobalt. (See Eq. at top of p. 205.)

A third example of the use of the metal to carbon monoxide 1:2 shift of an alkyl group occurs in an aldehyde synthesis from alkyl halides and carbon monoxide. The reaction requires three steps (all done in the same reaction vessel) using tetracarbonylferrate dianion as a stoichiometric reagent (430).

$$RCo(CO)_4 \rightleftharpoons RCOCo(CO)_3$$

$$RCo(CO)_4 + RCOCo(CO)_3 \longrightarrow \left[\begin{array}{c} R \\ | \\ RCOCoCo(CO)_4 \\ \diagup \ | \ \diagdown \\ OC \ | \ CO \\ CO \end{array} \right] \longrightarrow RCOR + Co_2(CO)_7$$

In the first step the dianion is reacted with one equivalent of an organic halide to form a monoalkyliron complex. One mole of triphenylphosphine then is added to convert the alkyl into the triphenylphosphine derivative of the acyliron complex and, finally acetic acid is added to protonate the acyl anion. The hydride thus formed probably transfers hydrogen to the acyl carbon with reductive elimination producing aldehyde and tricarbonyltriphenylphosphineiron which will ultimately form tricarbonyltriphenylphosphineiron trimer. The same reaction can be carried out with acid halides in

$$RX + Na_2Fe(CO)_4 \longrightarrow [RFe(CO)_4]^- Na^+ \xrightarrow{P\phi_3}$$

$$[RCOFe(CO)_3P\phi_3]^- Na^+ \xrightarrow[-NaOAc]{HOAc} \begin{array}{c} H \\ \diagdown \\ Fe(CO)_3P\phi_3 \\ \diagup \\ RCO \end{array} \longrightarrow RCHO + \{[Fe(CO)_3P\phi_3]_3\}$$

place of the alkyl halides, in which case one CO is evolved (431). The triphenylphosphine is not necessary for this reaction to occur. Ketones are produced by the reaction if two moles of alkyl or acyl halide per mole of tetracarbonylferrate anion are used. Mixed ketones are produced if two different alkylating or acylating agents are reacted successively (432).

b. Acylation of Anionic Complexes

Nucleophilic transition metal anionic complexes react readily with acid halides and often with acid anhydrides to form the acyl metal derivatives (433).

$$(P\phi_3)(CO)_3Co^- + CH_3COCl \longrightarrow (P\phi_3)(CO)_3CoCOCH_3 + Cl^-$$

$$(CO)_5Mn^- + (CF_2CO)_2O \longrightarrow (CO)_5MnCOCF_3 + CF_3COO^-$$

An interesting α-ketoacylmanganese derivative has been prepared by the acid chloride method. The pentacarbonylpyruvylmanganese(I) prepared lost carbon monoxide on heating, but at a rate 21 times slower than the related acetylmanganese compound did. The reaction could not be reversed even at 258 atm of CO and 80° in 9 hr (434). It is curious that the pyruvyl complex is so stable, but yet cannot be made by a double CO insertion.

$$CH_3COCOCl + NaMn(CO)_5 \xrightarrow{-NaCl} CH_3COCOMn(CO)_5 \longrightarrow$$

$$CO + CH_3COMn(CO)_5 + CH_3Mn(CO)_5$$

c. Formation of Acyl Metal Complexes by Oxidative Addition Reactions

Acyl metal complexes generally can be prepared from complexes reactive in oxidative addition reactions by addition of acid halides. For example, tetrakistriphenylphosphineplatinum(0) and acetyl chloride react to form the chlorobisphosphine(acyl)platinum complex (435). Oxidative addition of acid

$$Pt(P\phi_3)_4 + CH_3COCl \longrightarrow cis(?)\text{-}(P\phi_3)_2Pt(Cl)COCH_3$$

chlorides to chlorotristriphenylphosphineiridium(I), however, is accompanied by decarbonylation. In the absence of another ligand, the initial complex apparently converts the acyl group into two ligands by shifting the alkyl group to the metal (436). This reaction is unusual in that both 2-methyl-propanoyl and butanoyl chlorides add to produce only the n-propyliridium complex. The sterically less-crowded chlorocarbonylbiscycloocteneiri-

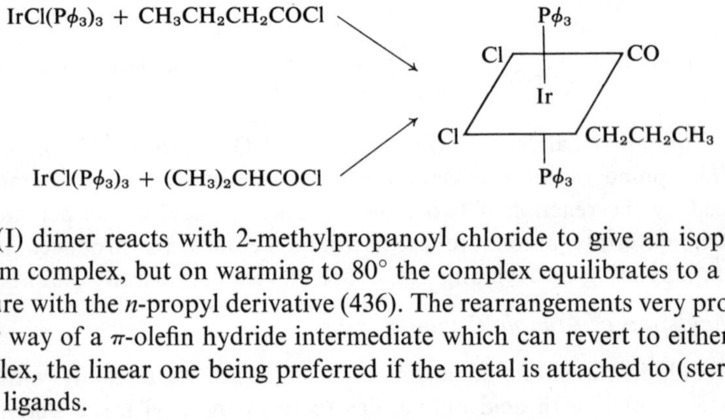

dium(I) dimer reacts with 2-methylpropanoyl chloride to give an isopropyl-iridium complex, but on warming to 80° the complex equilibrates to a 50:50 mixture with the n-propyl derivative (436). The rearrangements very probably go by way of a π-olefin hydride intermediate which can revert to either alkyl complex, the linear one being preferred if the metal is attached to (sterically) large ligands.

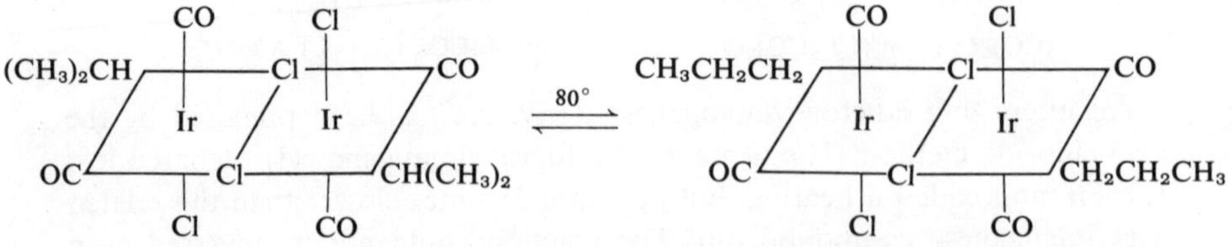

A similar equilibration of isomers to a 50:50 mixture occurs with tetra-carbonylbutanoyl- and tetracarbonyl-2-methylpropanoylcobalt(I) at 25° (437) (see Scheme IX-1). As expected, the isomerization is inhibited by excess carbon monoxide because it reacts with the necessary coordinately unsaturated intermediates.

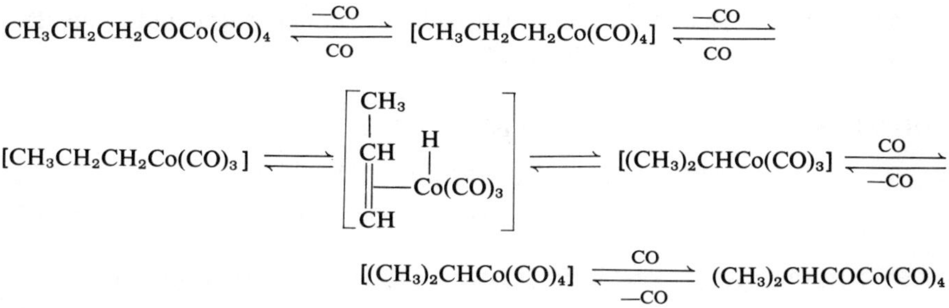

$$CH_3CH_2CH_2COCo(CO)_4 \underset{CO}{\overset{-CO}{\rightleftharpoons}} [CH_3CH_2CH_2Co(CO)_4] \underset{CO}{\overset{-CO}{\rightleftharpoons}}$$

$$[CH_3CH_2CH_2Co(CO)_3] \rightleftharpoons \begin{bmatrix} CH_3 \\ | \\ CH \quad H \\ | \quad | \\ \| - Co(CO)_3 \\ CH \end{bmatrix} \rightleftharpoons [(CH_3)_2CHCo(CO)_3] \underset{-CO}{\overset{CO}{\rightleftharpoons}}$$

$$[(CH_3)_2CHCo(CO)_4] \underset{-CO}{\overset{CO}{\rightleftharpoons}} (CH_3)_2CHCOCo(CO)_4$$

Scheme IX-1.

Few useful synthetic reactions employ the oxidative addition reaction of acid chlorides because the acyl metal complexes usually can be prepared in the reaction mixture by carbonylation of an alkyl group which often is available from a more readily available starting material.

A typical catalytic example of the preparation of an acyl chloride by oxidative addition is the synthesis of 3-butenoyl chloride from allyl chloride and carbon monoxide with chloro-π-allylpalladium dimer as catalyst (438). The kinetics of the reaction show the rate to depend on the first power of the allyl chloride and palladium complex with a second-order dependence on the carbon monoxide concentration. A mechanism has been proposed in which two carbon monoxide molecules first coordinate with the allylpalladium complex. The π-allyl group becomes a σ-allyl group and then shifts to one of the carbon monoxides. A reductive elimination of 3-butenoyl chloride follows, probably assisted by more carbon monoxide or other ligands present in the reaction mixture. The palladium(0) product then undergoes oxidative addition with allyl chloride reforming the catalyst (439).

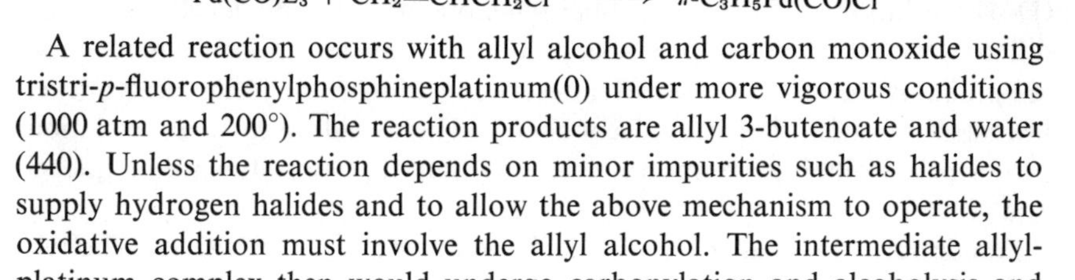

$$[\pi\text{-}C_3H_5PdCl]_2 + CO \longrightarrow [\pi\text{-}C_3H_5Pd(CO)Cl] \xrightarrow{CO} \begin{bmatrix} & CO \\ & | \\ CH_2=CHCH_2PdCl \\ & | \\ & CO \end{bmatrix} \xrightarrow{L}$$

$$[CH_2=CHCH_2COPd(CO)LCl] \xrightarrow{L} CH_2=CHCH_2COCl + Pd(CO)L_3$$

$$Pd(CO)L_3 + CH_2=CHCH_2Cl \xrightarrow{-L} \pi\text{-}C_3H_5Pd(CO)Cl$$

A related reaction occurs with allyl alcohol and carbon monoxide using tristri-p-fluorophenylphosphineplatinum(0) under more vigorous conditions (1000 atm and 200°). The reaction products are allyl 3-butenoate and water (440). Unless the reaction depends on minor impurities such as halides to supply hydrogen halides and to allow the above mechanism to operate, the oxidative addition must involve the allyl alcohol. The intermediate allyl-platinum complex then would undergo carbonylation and alcoholysis and form the ester and a hydrido metal hydroxide. The last product could eliminate water and react again with allyl alcohol.

$$(R_3P)_3Pt + CH_2=CHCH_2OH \longrightarrow [C_3H_5Pt(PR_3)_2OH] \xrightarrow{CO}$$

$$[CH_2=CH\,CH_2COPt(PR_3)_2OH] \xrightarrow{C_3H_5OH}$$

$$CH_2=CHCH_2COOCH_2CH=CH_2 + [HPt(PR_3)_2OH]$$

$$[HPt(PR_3)_2OH] + CH_2=CHCH_2OH \longrightarrow H_2O + [C_3H_5Pt(PR_3)_2OH]$$

The solvent employed in oxidative addition reactions can be quite important since hydroxylic solvents may react with the acyl intermediate before reductive elimination can occur. Iodobenzene and tetracarbonylnickel(0), for example, form benzil in 80% yield in THF solution, whereas in methanol, methyl benzoate is produced in 60% yield (441) (see Scheme IX-2).

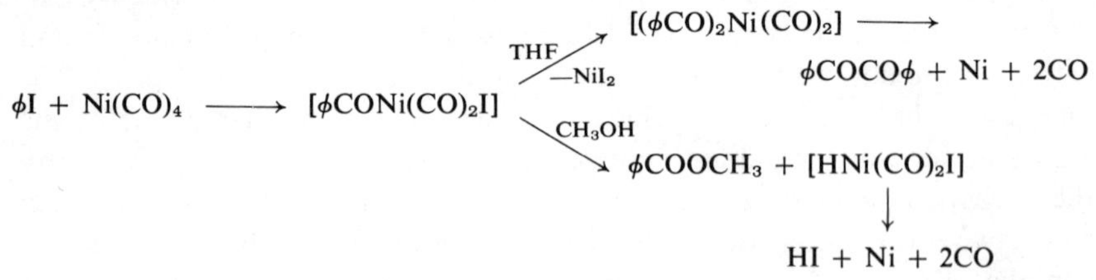

$$\phi I + Ni(CO)_4 \longrightarrow [\phi CONi(CO)_2I] \begin{array}{c} \xrightarrow[\text{—NiI}_2]{\text{THF}} [(\phi CO)_2Ni(CO)_2] \longrightarrow \phi COCO\phi + Ni + 2CO \\ \\ \xrightarrow{CH_3OH} \phi COOCH_3 + [HNi(CO)_2I] \\ \downarrow \\ HI + Ni + 2CO \end{array}$$

Scheme IX-2.

Maitlis has pointed out the effect of the basicity of the medium on the course of the reaction of an isolable chelation-stabilized alkylpalladium complex with carbon monoxide. In neutral or acidic methanol, carbon monoxide insertion with cyclization by addition of the acyl group to a chelating double bond occurs, whereas in basic methanol solution (NaOCH₃ added) the acyl insertion product undergoes alcoholysis forming an open-chain ester (442). The methoxide ion apparently attacks the acylpalladium intermediate more rapidly than the acyl complex cyclizes.

Benzyl bromide reacts with tetracarbonylnickel(0) in dimethylformamide (DMF) solution to form 93% dibenzylketone and 4% dibenzyl (443). The differences in the types of products produced from the iodobenzene reaction are probably the result of the easier decarbonylation of the phenylacetylnickel intermediate rather than the solvent change. High yields of carboalkoxylation products are obtained from aryl and vinyl halides with tetracarbonylnickel and alkoxides (444).

Amides are obtained from vinyl and aromatic halides with tetracarbonylnickel(0) in methanol solution in the presence of amines, trans-β-bromostyrene and pyrrolidine, for example, give an 82% yield of the *trans*-cinnamamide derivative.

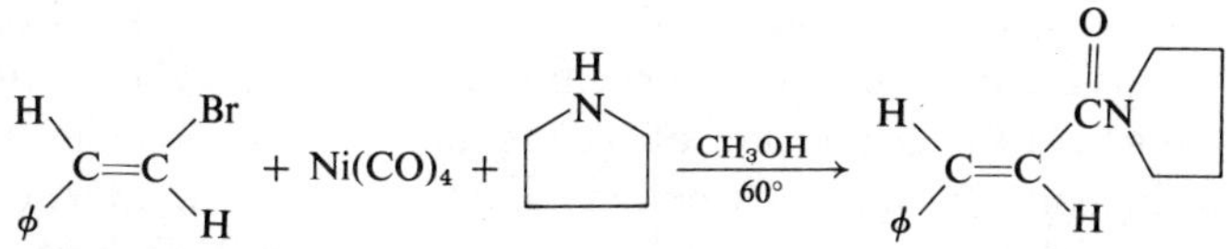

A commercial application of the oxidative addition–carbonylation reaction has recently been developed to prepare acetic acid from methanol and carbon monoxide with a rhodium(III) oxide–methyl iodide catalyst (445). The true catalyst has not been identified, but a good possibility is the iododicarbonyl-rhodium(I) dimer. The reduced rhodium catalyst likely reacts by oxidative addition with the methyl iodide catalyst. The methylrhodium compound then shifts the methyl to coordinated carbon monoxide and undergoes hydrolysis with the water formed in a subsequent reaction (or with the small amount present initially) to form acetic acid and a rhodium hydride. The hydride will lose hydrogen iodide which then can react with more methanol to reform methyl iodide and water and complete the cycle (see Scheme IX-3).

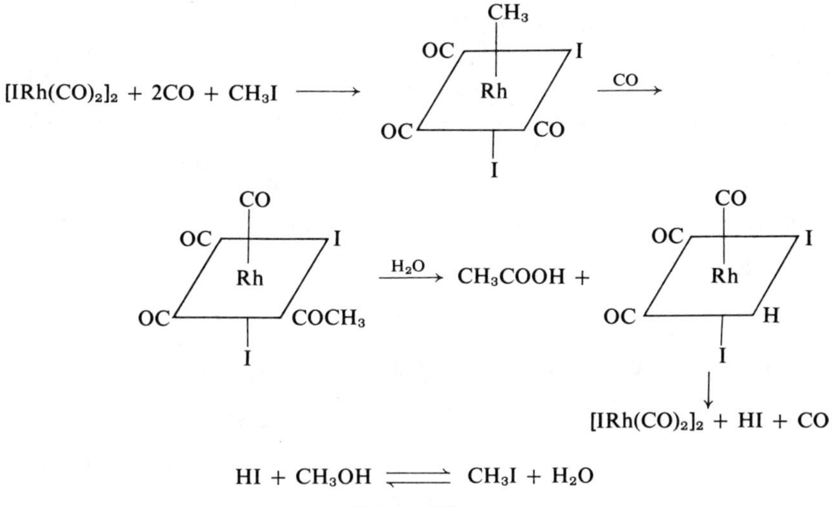

$$[IRh(CO)_2]_2 + 2CO + CH_3I \longrightarrow$$

$$\xrightarrow{H_2O} CH_3COOH +$$

$$[IRh(CO)_2]_2 + HI + CO$$

$$HI + CH_3OH \rightleftharpoons CH_3I + H_2O$$

Scheme IX-3.

d. Formation of Acyl Metal Complexes by Metal Hydride Addition and Carbonylation

Catalytic reactions very often involve metal hydride additions to olefins, acetylenes, dienes, or organic carbonyl compounds to form alkyl metal derivatives. When carried out in the presence of carbon monoxide, acyl complexes are formed directly which then can give products by a variety of reactions of the types discussed in Chapter III, Section B. Typical of these are the various olefin carboxylation reactions. Octacarbonyldicobalt(0), for example, catalyzes the conversion of olefins into acids in aqueous acetone solution with about 100 atm of carbon monoxide at about 165°. Cyclohexene under these conditions gave an 86% yield of cyclohexanecarboxylic acid (446). Unsymmetrical olefins give mixtures of acids. The mechanism of the

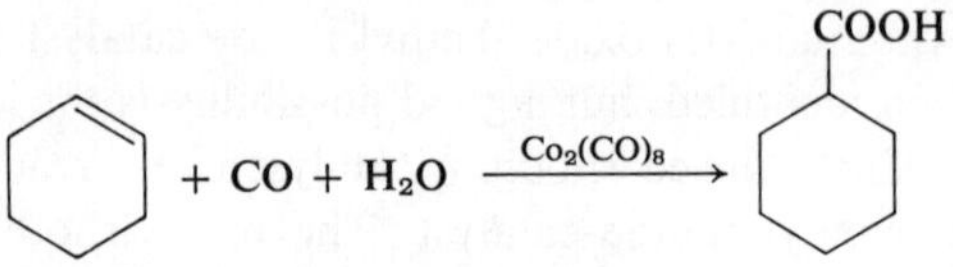

$$\text{(cyclohexene)} + CO + H_2O \xrightarrow{Co_2(CO)_8} \text{(cyclohexanecarboxylic acid, COOH)}$$

reaction almost certainly involves formation of hydridotetracarbonylcobalt(I) from the octacarbonyl and water. Then, an addition of the hydride to the olefin probably occurs, followed by a carbon monoxide insertion and hydrolysis. The proposed steps all can be demonstrated separately. The addition of a concentrated solution of hydridotetracarbonylcobalt(I) to an excess of 1-pentene produces a 50:50 mixture of the 1- and 2-pentylcobalt complexes. These were isolated as the acyltriphenylphosphine derivatives formed by reaction of the initially formed products with one mole of the phosphine (447). Tetracarbonylacylcobalt complexes can be obtained if the hydride addition is carried out under carbon monoxide (Scheme IX-4). The hydrolysis

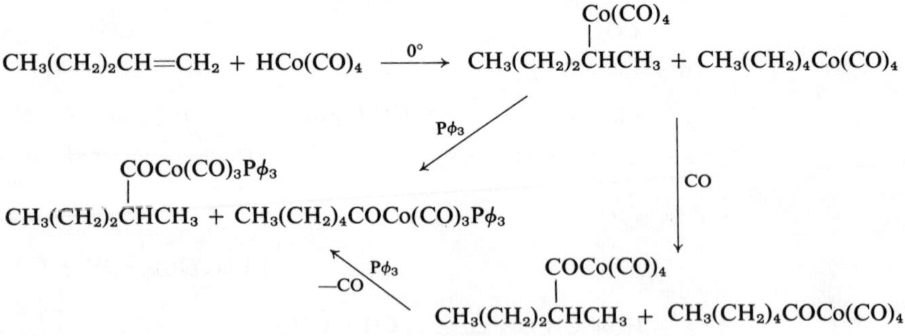

$$CH_3(CH_2)_2CH{=}CH_2 + HCo(CO)_4 \xrightarrow{0°} \underset{\overset{|}{Co(CO)_4}}{CH_3(CH_2)_2CHCH_3} + CH_3(CH_2)_4Co(CO)_4$$

$$\underset{\overset{|}{COCo(CO)_3P\phi_3}}{CH_3(CH_2)_2CHCH_3} + CH_3(CH_2)_4COCo(CO)_3P\phi_3$$

$$\underset{\overset{|}{COCo(CO)_4}}{CH_3(CH_2)_2CHCH_3} + CH_3(CH_2)_4COCo(CO)_4$$

Scheme IX-4.

of isolated acylcobalt compounds has not been carried out directly, but alcoholysis was demonstrated in the alkyl halide carboxylation described in Section A,1,a of this chapter.

Nickel(II) iodide is also employed as a catalyst for the olefin carboxylation (448). The true catalyst is likely a hydride probably formed in two steps.

The first step could be a reduction of the nickel(II) iodide with carbon monoxide and water to tetracarbonylnickel and the second step an oxidative addition of hydrogen iodide to tetracarbonylnickel with loss of two carbon monoxides. The hydride would then add to the olefin, a carbonylation could occur, and hydrolysis would produce the carboxylic acids. The carboxylation

$$NiI_2 + 5CO + H_2O \rightleftharpoons 2HI + CO_2 + Ni(CO)_4$$

$$Ni(CO)_4 + HI \rightleftharpoons 2CO + [HNi(CO)_2I]$$

$$[HNi(CO)_2I] + CH_2{=}CH_2 \rightleftharpoons CH_3CH_2Ni(CO)_2I \underset{-CO}{\overset{CO}{\rightleftharpoons}}$$

$$[CH_3CH_2CONi(CO)_2I] \xrightarrow{H_2O} CH_3CH_2COOH + [HNi(CO)_2I]$$

of norbornene with CH_3CH_2OD and tetracarbonylnickel using a little deuteroacetic acid as a catalyst gave the cis-endo-deuteroethyl ester and a little exo,exo-dinorbornylketone (449). The cis addition is consistent with a

82%   16%

covalent 1:2 addition of the hydride. Several other transition metal catalysts are known to produce the olefin carboxylation also and presumably their reaction mechanisms are similar to the above.

e. Formation of Acyl Complexes by Metalation and Carbonylation

Metalation is not a very general reaction and, when it does occur, it is often not selective. These restrictions limit the usefulness of the metalation and carbonylation combinations to only a few special cases. The metalation reaction occurs easily and is selective only when coordinating groups are present. Presumably the effect is mainly the result of lowering the entropy of the reaction by the coordinating group holding the metal close to the position reacting. The effect is best known for aromatic metalation, but saturated aliphatic carbons also may react in some instances (see Chapter II, Section B). The subsequent carbonylation of the often relatively stable, metalated products has been successfully employed in only a few cases, but could be expected to occur with other metalated products as well. The chloro-o-phenylazophenylpalladium(II) dimer from azobenzene and palladium chloride (450), for example, undergoes carbonylation in methanol solution to give a 17% yield of 2-phenyl-1H-indazolone as the only isolable product (450, 451). This reaction possibly proceeds by way of an o-carbomethoxy intermediate which also produces a hydridopalladium complex capable of reducing the azo group. Once reduced, cyclization could occur. The possible intermediate can be isolated in the case of the carbonylation of tricarbonyl-2-phenylazophenylcobalt(I) in methanol (451). A catalytic version

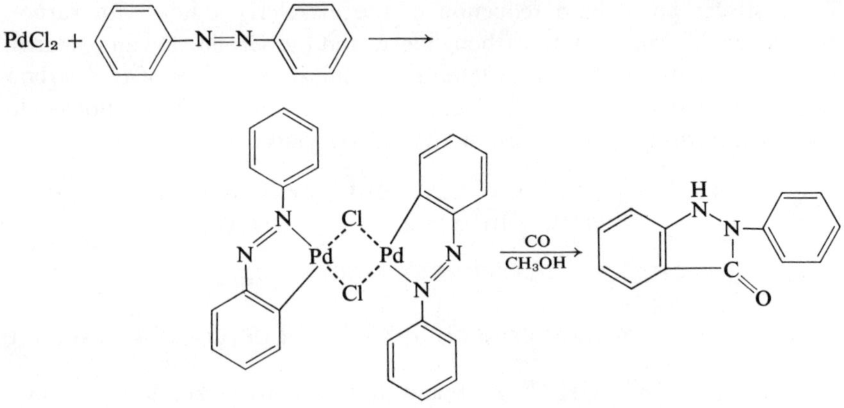

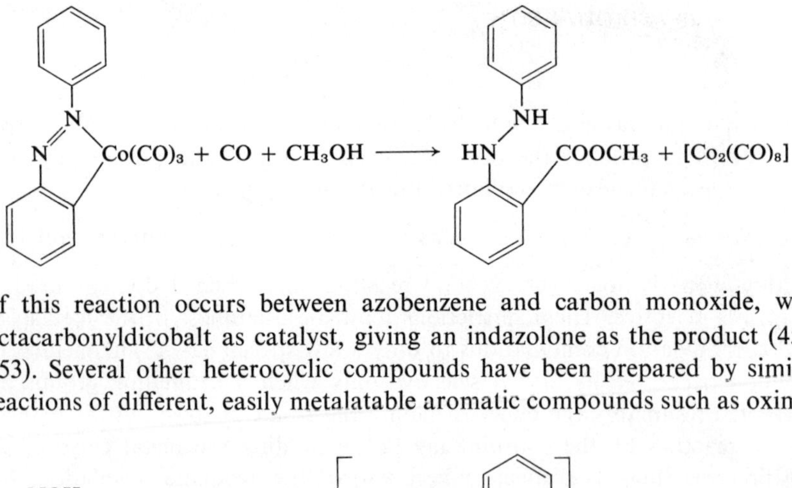

of this reaction occurs between azobenzene and carbon monoxide, with octacarbonyldicobalt as catalyst, giving an indazolone as the product (452, 453). Several other heterocyclic compounds have been prepared by similar reactions of different, easily metalatable aromatic compounds such as oximes

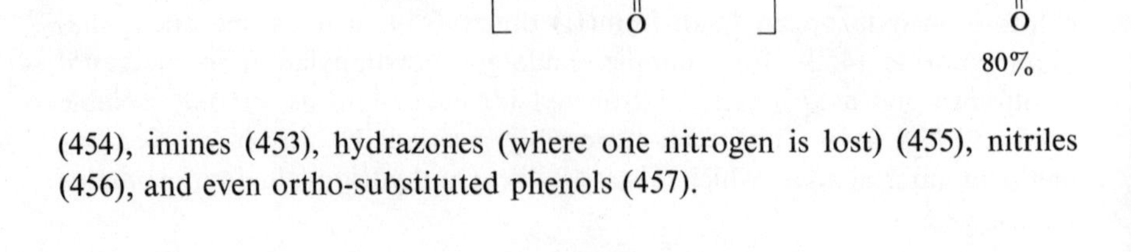

(454), imines (453), hydrazones (where one nitrogen is lost) (455), nitriles (456), and even ortho-substituted phenols (457).

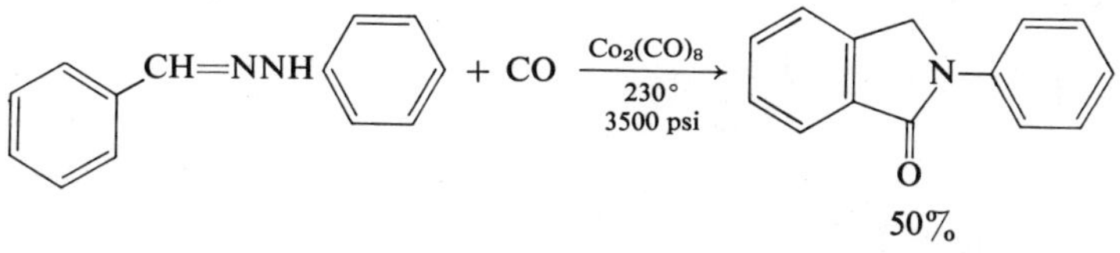

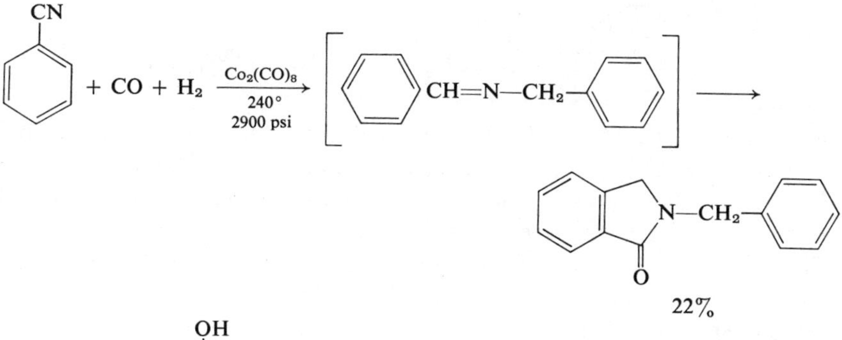

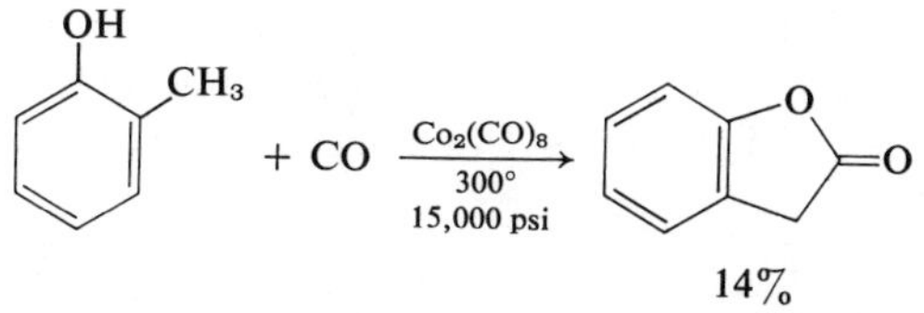

50%

22%

14%

Nonspecific metalation of aromatics without a coordinating group is also possible under vigorous conditions with palladium (Chapter II, Section B,5), but the reaction apparently has not been reported to occur in the presence of carbon monoxide.

f. Alkoxy-, Acyloxy-, and Chlorometalation with Carbonylation

The 1:2 addition of alkoxy, acyloxy, and halo metal groups to unsaturated organic compounds gives metal–carbon-bonded complexes which can react with carbon monoxide to form the acyl derivatives. Palladium compounds are the most useful in these reactions. Only in cases where chelating groups are present can intermediate adducts be isolated. A good example is the reaction of palladium chloride in methanol with 1,5-cyclooctadiene. The isolable methoxypalladation adduct with the second double bond chelating to the metal can be carbonylated in methanol solution in the presence of sodium acetate to neutralize the hydrogen chloride produced. The product is the *trans*-2-methoxy ester indicating that the alkoxypalladation was trans since carbonylation and alcoholysis occur with retention (458).

A catalytic example occurs when ethylene, carbon monoxide, acetic acid, and oxygen are reacted in the presence of catalytic amounts of palladium acetate, lithium chloride, cupric chloride, and acetic anhydride. The reaction produces 3-acetoxypropionic acid in 85% yield (315). The palladium acetate apparently first adds to the ethylene and the adduct reacts with carbon monoxide. Reaction with acetic acid or reductive elimination could give hydridopalladium acetate and a mixed anhydride which exchanges with more acetic acid giving free 3-methoxypropionic acid and acetic anhydride. The palladium metal is then reoxidized to palladium(II) acetate with oxygen and the lithium chloride–cupric chloride catalyst. Water is a by-product of the reoxidation and it reacts with the acetic anhydride (see Scheme IX-5). If significant amounts of water were allowed to accumulate, acetaldehyde probably would be produced instead of 3-acetoxypropionic acid (see Chapter V, Section D,2).

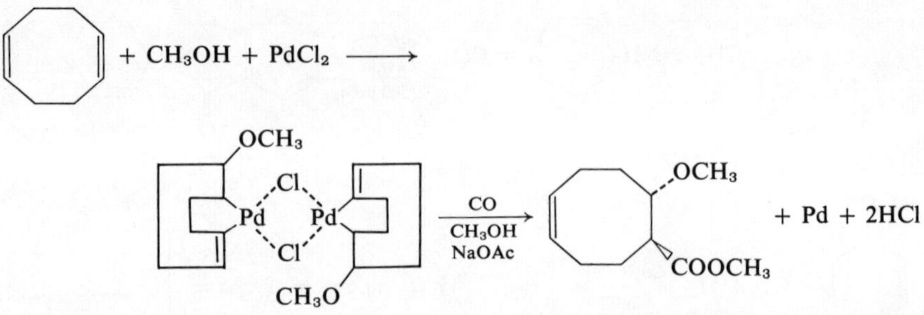

$$CH_2{=}CH_2 + Pd(OAc)_2 \rightleftharpoons^{L} AcOCH_2CH_2\underset{L}{\overset{L}{PdOAc}} \xrightarrow{CO}$$

$$AcOCH_2CH_2CO\underset{L}{\overset{L}{PdOAc}} \xrightarrow[-L]{HOAc} \left[AcOCH_2CH_2\overset{O}{\overset{\|}{C}}O\overset{O}{\overset{\|}{C}}CH_3 \right] + Pd$$

$$\left[AcOCH_2CH_2\overset{O}{\overset{\|}{C}}O\overset{O}{\overset{\|}{C}}CH_3 \right] + HOAc \rightleftharpoons AcOCH_2CH_2COOH + Ac_2O$$
$$85\%$$

$$Pd + \tfrac{1}{2}O_2 + HOAc \xrightarrow[LiCl]{CuCl_2} Pd(OAc)_2 + H_2O$$

$$H_2O + Ac_2O \longrightarrow 2HOAc$$

Scheme IX-5.

g. Formation of Acyl Metal Complexes from Metal Carbonyls and Organic Anions

Coordinated carbon monoxide is susceptible to nucleophilic attack and this reaction can be used to prepare some types of carboalkoxy or acyl metal complexes which may then react further to produce totally organic products. Cationic complexes are particularly reactive. Tetracarbonylbistriphenylphosphinemanganese(I) cation reacts with methoxide ion, for example, to form the carbomethoxymanganese complex (459). Neutral metal carbonyls

$$Mn(CO)_4(P\phi_3)_2{}^+ + CH_3O^- \longrightarrow CH_3OCOMn(CO)_3(P\phi_3)_2$$

also react and form some useful reagents. Pentacarbonyliron(0) reacts with organolithium compounds to form tetracarbonylacyliron anions (431, 432, 460). Apparently, these are the same compounds that are obtained from the reaction of tetracarbonylferrate dianion with acid chlorides. A similar reagent

$$Fe(CO)_5 + RLi \longrightarrow [RCOFe(CO)_4]^- Li^+$$

is obtained from tetracarbonylnickel(0) and organolithium compounds. This reagent readily adds to α,β-unsaturated carbonyl compounds, producing after hydrolysis 1,4-diketones (461) (see Section 2,b below). A review has been written on carbamoyl and alkoxycarbonyl complexes, many of which are prepared by anionic attack on metal carbonyls (462).

2. Carbonylation Reactions of Olefins

a. Hydroformylation and Related Reactions

The hydroformylation reaction was the first important homogeneous transition metal-catalyzed reaction to be discovered. Roelen discovered it in 1938, but publication of the results was delayed by World War II until 1948 (463). The basic reaction is the conversion of olefins, carbon monoxide, and hydrogen under pressure with cobalt catalysts to aldehydes containing one more carbon than the starting olefin. Unsymmetrical olefins generally yield mixtures of mainly two aldehydes arising from addition of the elements of formaldehyde in both possible directions to the original double bond.

$$CH_3CH{=}CH_2 + CO + H_2 \xrightarrow{\text{Co}} \overset{\displaystyle CHO}{\underset{\displaystyle |}{CH_3CHCH_3}} + CH_3CH_2CH_2CHO$$

The reaction is applicable to a wide variety of mono- and disubstituted olefins and it usually proceeds in high yield under the proper conditions. It is still an important commercial reaction.

The mechanism of the hydroformulation reaction was not unraveled until 1960, some 22 years after its discovery (447, 464). This was the first catalytic transition metal carbonyl reaction to be explained in detail and has formed the basis for much of the subsequent work in the area. The reaction was found to involve hydridotetracarbonylcobalt as the catalyst. Kinetic studies showed that the rate of aldehyde formation depended directly on the olefin, hydrogen, and cobalt concentrations, but inversely on the carbon monoxide concentration above about 10 atm, where the reaction is generally carried out (465, 466). A study of reactions of possible intermediates suggested that three basic steps were involved: (1) an addition of the $HCo(CO)_4$ to the olefin (two isomers may be formed from unsymmetrical olefins); (2) a 1:2 shift of the alkyl group from cobalt to coordinated carbon monoxide; and (3) a hydrogenolysis of the acylcobalt group. The carbon monoxide inhibition of the reaction arises because the reactive forms of the intermediates are coordinately unsaturated and carbon monoxide competes with the other reactants for the available coordination sites. The addition of $HCo(CO)_4$ to 1-pentene discussed in Section A,1,d above is inhibited by as little as 1 atm of carbon monoxide, indicating hydridotricarbonylcobalt(I) is the actual reactant. Presumably, olefin coordination with the hydride is necessary before addition can occur. The dissociated carbon monoxide then must reassociate and form the tetracarbonylalkylcobalt intermediates. A shift of the alkyl to a coordinated carbonyl produces an unsaturated tricarbonylacylcobalt(I) complex. Carbon monoxide now competes with hydrogen for this unsaturated intermediate. If the tetracarbonylacyl species is formed, it apparently cannot be reduced until it dissociates again since small amounts of carbon monoxide greatly decrease the rate of reduction of isolated tetracarbonylacylcobalt(I) complexes with hydrogen. The hydrogen probably reduces the tricarbonylacyl complex by first oxidatively adding to the cobalt and then one hydrogen shifts from cobalt to the acyl carbon, a reaction which also regenerates the active catalyst $HCo(CO)_3$. Another source of CO inhibition exists in the catalyst formation reaction. Hydridotetracarbonylcobalt(I) readily loses hydrogen in a second-order decomposition, forming

$$Co_2(CO)_8 \rightleftharpoons Co_2(CO)_7 + CO$$

$$Co_2(CO)_7 + H_2 \rightleftharpoons [HCo(CO)_3] + HCo(CO)_4$$

$$[HCo(CO)_3] + CO \rightleftharpoons HCo(CO)_4$$

$$[HCo(CO)_3] + CH_2{=}CH_2 \rightleftharpoons \left[\begin{array}{c} CH_2{=}CH_2 \\ | \\ HCo(CO)_3 \end{array} \right] \rightleftharpoons [CH_3CH_2Co(CO)_3]$$

$$[CH_3CH_2Co(CO)_3] + CO \rightleftharpoons CH_3CH_2Co(CO)_4 \rightleftharpoons [CH_3CH_2COCo(CO)_3]$$

$$[CH_3CH_2COCo(CO)_3] + CO \rightleftharpoons CH_3CH_2COCo(CO)_4$$

$$[CH_3CH_2COCo(CO)_3] + H_2 \rightleftharpoons \left[\begin{array}{c} H \\ | \\ CH_3CH_2COCo(CO)_3 \\ | \\ H \end{array} \right] \longrightarrow$$

$$CH_3CH_2CHO + [HCo(CO)_3]$$

octacarbonyldicobalt(0). The reverse reaction is inhibited by CO (467, 468). Which of the possible sources of CO inhibition is most important in a specific hydroformylation depends on reaction conditions, reagent concentrations, and olefin structure.

Acylcobalt complexes can also be reduced to aldehydes by $HCo(CO)_4$ (464). Coordinately unsaturated species, probably the acyltricarbonyl, are apparently involved here also since CO inhibits the reduction. Presumably, an oxidative addition mechanism is involved where the hydride adds to the acylcobalt intermediate and a 1:2-hydrogen shift occurs to give aldehyde and $[Co(CO)_8]$. It is unlikely that this reaction occurs to a significant extent

$$
\begin{array}{c}
\text{O} \\
\parallel \\
\text{RCCo(CO)}_3
\end{array}
+ HCo(CO)_4 \longrightarrow
\left[
\begin{array}{c}
\text{O H} \\
\parallel\ | \\
\text{RCCo(CO)}_3 \\
| \\
\text{Co(CO)}_4
\end{array}
\right]
\longrightarrow RCHO + [Co_2(CO)_7]
$$

under hydroformylation conditions because it is second-order in cobalt complexes and the concentration of $HCo(CO)_4$ should be very low. Not only is the total concentration of cobalt low, but the $HCo(CO)_4$ is much lower because most of the cobalt will likely be present in the form of the other more stable and less reactive intermediate complexes.

The stereochemistry of the hydroformylation reaction has been determined by reacting 3,4-di-O-acetyl-D-xylal with carbon monoxide and deuterium using $Co_2(CO)_8$ as catalyst. By carrying out the reaction at higher temperatures than required for the hydroformylation, the initially formed aldehydes are reduced to alcohols. The product contained deuterium and the deuterated carbinol group cis to each other indicating a cis addition of the hydride and CO insertion with retention (469).

The reduction of aldehydes by hydrogen in the presence of carbon monoxide and $Co_2(CO)_8$ as catalyst to alcohols is commonly carried out during the hydroformylation reaction by employing higher temperatures than are necessary for the hydroformylation. The mechanism of this reduction probably involves addition of hydridotricarbonylcobalt(I) to the aldehyde (94a, 470) in either of two possible directions, followed by hydrogenolysis, by way of an oxidative addition of hydrogen, of the cobalt–oxygen- or cobalt–carbon-bonded intermediate (Scheme IX-6). Hydrogenolysis with

$HCo(CO)_4$ appears less likely than with hydrogen in view of the probable low concentration of $HCo(CO)_4$ in catalytic reaction mixtures.

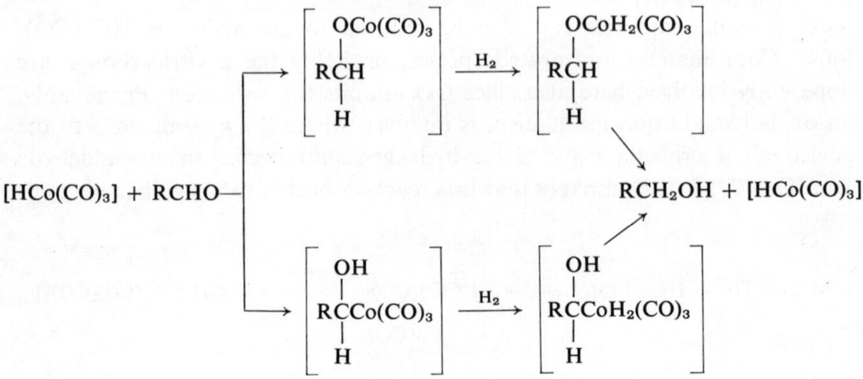

$$[HCo(CO)_3] + RCHO \longrightarrow \begin{cases} \begin{bmatrix} OCo(CO)_3 \\ | \\ RCH \\ | \\ H \end{bmatrix} \xrightarrow{H_2} \begin{bmatrix} OCoH_2(CO)_3 \\ | \\ RCH \\ | \\ H \end{bmatrix} \searrow \\ \qquad\qquad RCH_2OH + [HCo(CO)_3] \\ \begin{bmatrix} OH \\ | \\ RCCo(CO)_3 \\ | \\ H \end{bmatrix} \xrightarrow{H_2} \begin{bmatrix} OH \\ | \\ RCCoH_2(CO)_3 \\ | \\ H \end{bmatrix} \nearrow \end{cases}$$

Scheme IX-6.

Reduction by way of the intermediate with the cobalt–oxygen bond is probably preferred, since formate esters are minor side products in the hydroformylation reaction and can be produced from aldehydes under hydroformylation conditions (471). Very probably, the intermediate in the formate ester reaction is the $[HCo(CO)_3]$-aldehyde adduct which subsequently is carbonylated and hydrogenated. The products expected from carbonylation

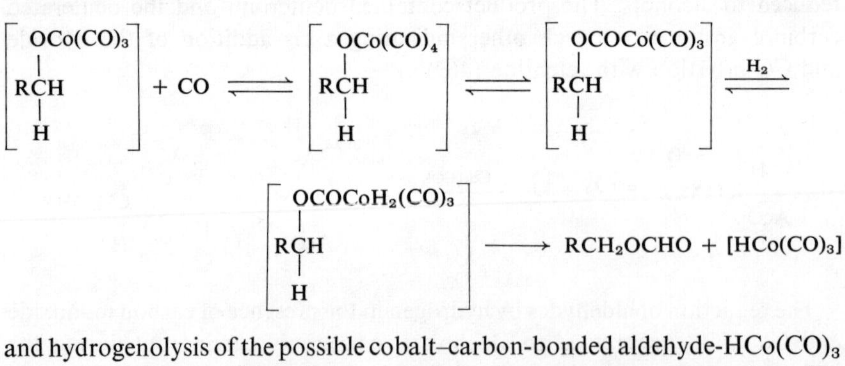

$$\begin{bmatrix} OCo(CO)_3 \\ | \\ RCH \\ | \\ H \end{bmatrix} + CO \rightleftarrows \begin{bmatrix} OCo(CO)_4 \\ | \\ RCH \\ | \\ H \end{bmatrix} \rightleftarrows \begin{bmatrix} OCOCo(CO)_3 \\ | \\ RCH \\ | \\ H \end{bmatrix} \xrightarrow[\rightleftharpoons]{H_2}$$

$$\begin{bmatrix} OCOCoH_2(CO)_3 \\ | \\ RCH \\ | \\ H \end{bmatrix} \longrightarrow RCH_2OCHO + [HCo(CO)_3]$$

and hydrogenolysis of the possible cobalt–carbon-bonded aldehyde-$HCo(CO)_3$ adducts, α-hydroxy aldehydes or 1:2-diols, have not been reported as side products. These results do not necessarily mean that that type of intermediate is not formed, however. It may only show that that intermediate does not undergo further carbonylation and hydrogenolysis under the usual conditions. The mechanism of the aldehyde reduction has been discussed in more detail in Chapter IV, Section D.

A carboalkoxycobalt complex of the type suggested as an intermediate in the formate formation reaction has been isolated by reaction of tetra-carbonylcobaltate anion with t-butyl hypochlorite followed by treatment with triphenylphosphine (60).

$$Co(CO)_4^- + (CH_3)_3COCl \longrightarrow [(CH_3)_3COCo(CO)_4] \xrightarrow{\text{P}\phi_3} (CH_3)_3CO\overset{\displaystyle O}{\overset{\|}{C}}Co(CO)_3P\phi_3$$

As might have been anticipated from the discussion in Section A,1,c and earlier sections, the hydroformylation reaction may produce aldehydes not derivable by simply adding the elements of formaldehyde to the double bond. Double-bond isomerization by hydride addition–elimination mechanisms and isomerization of the acylcobalt intermediates as indicated in Section A,1,c may occur. The extent of the isomerization is generally quite dependent on reaction conditions. Most often isomeric aldehydes appear to be formed directly through π- to σ-complex rearrangements under the usual hydroformylation conditions rather than through prior formation of isomerized olefins. This point is clearly illustrated in studies of the hydroformylation of optically active 3-methyl-1-hexene. Reaction of the olefin with carbon monoxide and hydrogen gives as expected the two optically active aldehydes formed by formylation in both possible directions of the double bond, but also small amounts of other isomers are formed. Particularly interesting is the product with the formyl group on the original 3-methyl group. In order for this product to form, the cobalt group would have to move past the chiral center. This aldehyde was found to be optically active and, therefore, could not have been formed by way of a free, isomerized and necessarily optically inactive olefin (472). The use of the 3-deutero olefin and determination of the fate of the deuterium in the rearranged aldehyde showed that the reaction mechanism was consistent with a hydride addition–elimination sequence in which the olefin never left the metal during rearrangement. The cobalt-containing intermediates were all asymmetric and, therefore, gave optically active products. The 3-deutero group was transferred to the (initial) 2-position in the product as expected from an elimination of metal deuteride followed by a readdition in the reverse direction (473) (see Scheme IX-7). The absence of aldehyde formed by carbonylation of the *tertiary* alkylcobalt intermediate is not surprising since it appears that under the reaction conditions elimination of hydride from such compounds is much more rapid than the shift of the tertiary alkyl group to a coordinated carbonyl group and products arise only from more stable intermediates (447).

The equilibration of possible alkylcobalt intermediates in the hydroformylation reaction may be complete or only partial, depending on reaction conditions and the olefin being reacted. Under fairly vigorous conditions 1- and

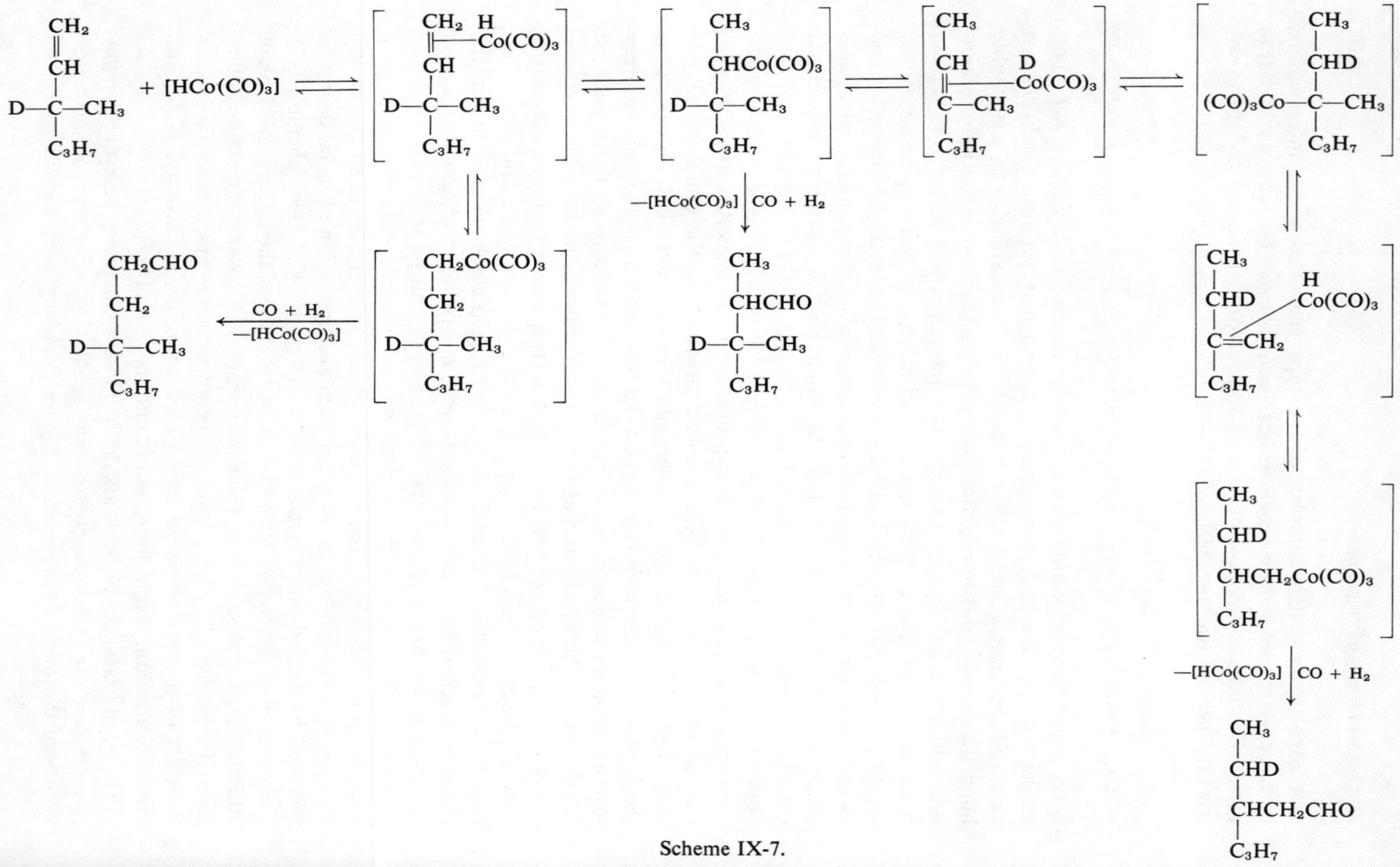

Scheme IX-7.

2-pentene give exactly the same mixture of hydroformylation products (474), whereas under milder conditions the same products are obtained but in different amounts, indicating equilibration is then incomplete (475). The cobalt group may move over many carbon atoms in these reactions. The hydroformylation of oleic acid, for example, gives 20% 18-formylstearic acid, where the cobalt in the original intermediate must have moved over eight carbon atoms (476).

From the large number of hydroformylation reactions carried out it is clear that with linear olefins, 1- and 2-formyl-substituted products are usually much preferred over the other possible isomers. This probably occurs because the terminal olefin-$HCo(CO)_3$ π complex is favored over all the possible internal π complexes. The smaller steric interactions of the terminal olefins with the coordinated carbonyl groups probably account for the higher stability of the terminal complexes.

The direction of addition of the metal hydride to the olefin from within the π complex depends on both steric and electronic factors (cf. Chapter V, Section A). These factors are probably working in opposite directions in the hydridotetracarbonylcobalt addition. Sterically, the larger, ligated metal group would prefer to be on the least substituted carbon of the double bond, while electronically the cobalt group appears to prefer the more positive, more substituted carbon judging from the direction of addition observed with perdeuteropropylene as described in Chapter V, Section A,1. At 0° the addition of $HCo(CO)_4$ to isobutylene produces nearly exclusively the tertiary cobalt alkyl showing that electronic effects now dominate the reaction.

$$CH_3$$
$$\underset{\displaystyle |}{CH_3}$$
$$CH_3-C=CH_2 + HCo(CO)_4 \quad \xrightarrow{\quad P\phi_3 \quad}$$

$$\left[\begin{array}{c} CH_3 \\ | \\ CH_3CCOCo(CO)_3P\phi_3 \\ | \\ CH_3 \end{array} \right] \xrightarrow[CH_3OH]{I_2} (CH_3)_3CCOOCH_3$$

Under normal hydroformylation conditions, on the other hand, isobutylene yields 97% terminal formyl product and only 3% tertiary product (477). The tertiary cobalt alkyl is much less stable than the primary one. At higher temperatures its concentration is so low that very little product is formed from it, whereas at 0° both isomers are relatively stable.

The direction of addition of triorganophosphine-substituted hydrido-carbonylcobalt complexes to olefins is significantly affected by the size of the phosphine group (478, 479). The use of tri-*n*-butylphosphine as the ligand produces 85% primary and 15% 2-substituted products from linear 1-olefins rather than the 60:40 mixture typical of the reaction without the phosphine ligand present. Alcohols are produced rather than aldehydes since this catalyst reduces the initially formed aldehydes more easily than $HCo(CO)_4$ does. The effect on isomer distribution is probably mainly steric. The larger phosphine ligand makes the primary alkylcobalt adduct relatively more favorable than the secondary one because ligand interactions are smaller in the primary adduct.

A variety of catalysts other than $HCo(CO)_4$ are known which will catalyze the hydroformylation reaction. Some of these are $Fe(CO)_5$ (480), $[CpFe(CO)_2]_2$ (481), $HMn(CO)_5$ (482), $Ru_3(CO)_{12}$ (483), $Ir(CO)Cl(P\phi_3)_2$ (484), $Rh_2(CO)_8$ (485), and $HRh(CO)(P\phi_3)_3$ (486). The last compound is of particular interest since it catalyzes the hydroformylation at 50° and at pressures under 100 psi, producing almost entirely linear aldehydes from terminal olefins. The mechanism of this reaction has been studied and the steps shown in Scheme IX-8 proposed (487, 488). The proposed mechanism is quite similar to that

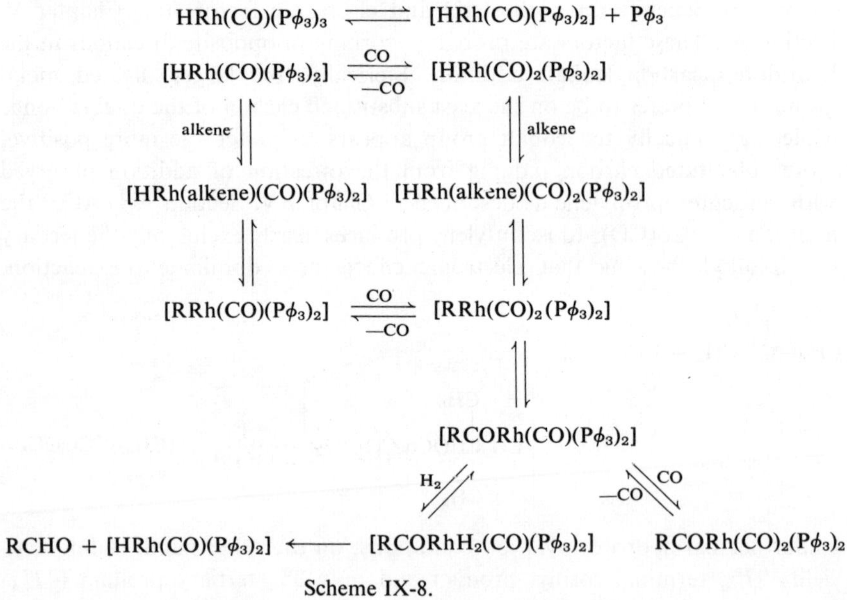

Scheme IX-8.

of the $HCo(CO)_4$-catalyzed reaction except that two paths are proposed, one involving addition of a coordinately saturated hydride to the olefin and the other addition of an unsaturated hydride analogous to the cobalt reaction. The associative pathway is proposed to account for the lower specificity in hydroformylation between 1- and 2-alkene than is observed in catalytic hydrogenation with the same catalyst. The hydrogenation presumably occurs only by the dissociative route.

The hydroformylation of unsaturated esters may lead to lactone products if five- or six-membered rings can be formed and the reaction is carried out under conditions where the formyl group is reduced to the alcohol. The hydroformylation of methyl methacrylate at 250° and 200 atm pressure with octacarbonyldicobalt, for example, gives 42% hydrogenation to methyl isobutyrate and 51% 2-methyl-4-butanolactone (489). The lactone very

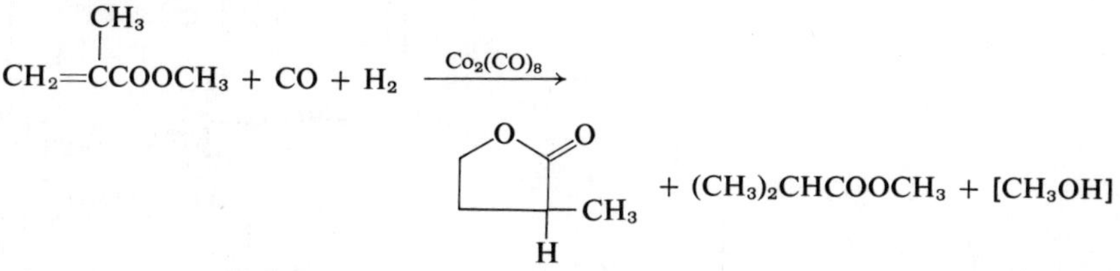

$$CH_2=\overset{\overset{\displaystyle CH_3}{|}}{C}COOCH_3 + CO + H_2 \xrightarrow{Co_2(CO)_8}$$

$+ (CH_3)_2CHCOOCH_3 + [CH_3OH]$

probably was formed from the 3-formyl ester by reduction and cyclization. A similar reaction, but with isomerization, occurs in the hydroformylation of methyl 3-methyl-2-butenoate (489). Carbonylation does not occur at the tertiary carbon, but cobalt moves and produces the terminal aldehyde which is then reduced and cyclized.

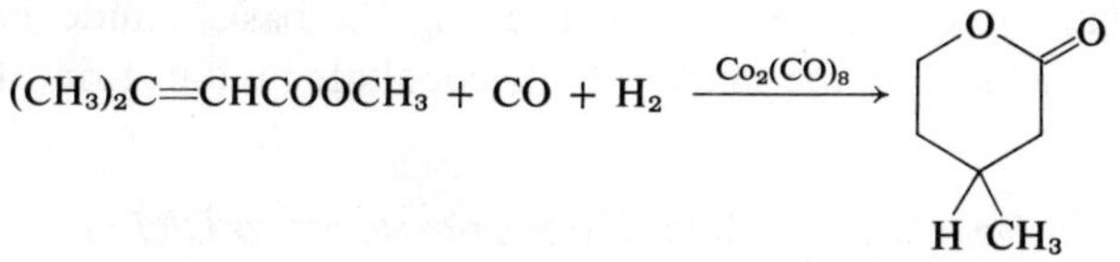

$$(CH_3)_2C=CHCOOCH_3 + CO + H_2 \xrightarrow{Co_2(CO)_8}$$

89%

Alken-3- and -4-ols normally rearrange mainly to aldehydes by a hydride addition–elimination sequence, or 1:3 hydrogen shifts, under hydroformylation conditions, but if this is blocked by gem disubstitution lactones can be obtained. The highest yields are obtained in the absence of hydrogen (490). Presumably, an intermediate hydroxyacylcobalt species undergoes an internal displacement of tetracarbonylcobaltate anion by the hydroxyl or alkoxide group. In the reaction of 2,2-dimethyl-3-buten-1-ol with CO and $Co_2(CO)_8$ as the catalyst, both five- and six-membered ring lactones are obtained because $HCo(CO)_3$ addition occurs both ways to the double bond (see Scheme IX-9). The source of the hydrogen required to form $HCo(CO)_4$ from $Co_2(CO)_8$ in this reaction is obscure. Presumably, dehydrogenation reactions occur and provide it.

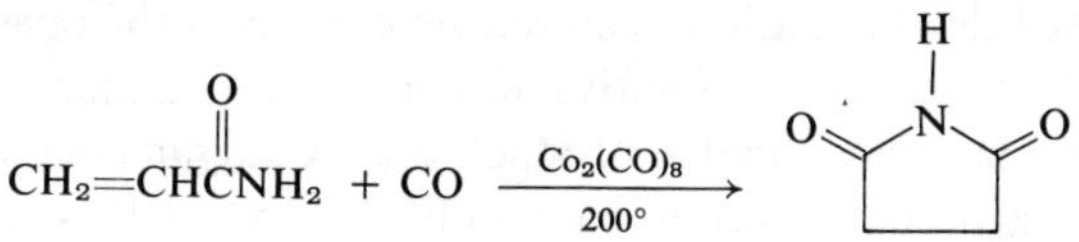

$$CH_2=CH\overset{\overset{\displaystyle O}{\|}}{C}NH_2 + CO \xrightarrow[200°]{Co_2(CO)_8}$$

81%

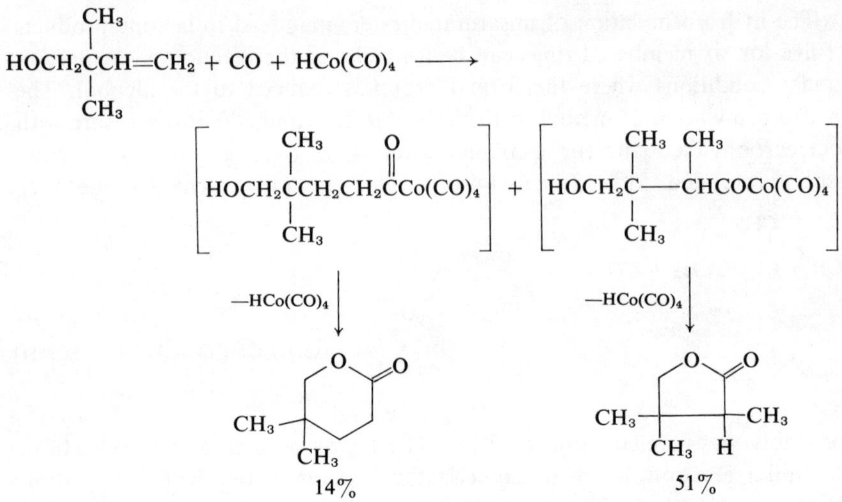

$$
\begin{array}{c}
CH_3 \\
| \\
HOCH_2CCH{=}CH_2 + CO + HCo(CO)_4 \\
| \\
CH_3
\end{array}
\longrightarrow
$$

$$
\left[
\begin{array}{c}
CH_3 \quad\quad O \\
| \quad\quad\quad || \\
HOCH_2CCH_2CH_2CCo(CO)_4 \\
| \\
CH_3
\end{array}
\right]
+
\left[
\begin{array}{c}
CH_3 \; CH_3 \\
| \quad\;\; | \\
HOCH_2C{\longrightarrow}CHCOCo(CO)_4 \\
| \\
CH_3
\end{array}
\right]
$$

$$
{-}HCo(CO)_4 \downarrow \qquad\qquad\qquad {-}HCo(CO)_4 \downarrow
$$

14%   51%

Scheme IX-9.

Unsaturated amides likewise may form cyclic imides under vigorous hydroformylation conditions (491). The slightly basic amide nitrogen is presumably displacing the tetracarbonylcobaltate anion from the intermediate.

b. Addition Reactions of Acyl Metal Complexes to Olefinic Compounds

Covalent 1:2 additions of acyl metal complexes to carbon–carbon double bonds occur in a variety of transition metal reactions. The reaction of o-diphenylphosphinostyrene with pentacarbonylmethylmanganese(I) is a good example (492, 493). The initial step must be a methyl shift from manganese to coordinated carbon monoxide, followed by coordination of phosphorus in the o-diphenylphosphinostyrene to the metal. Carbon monoxide dissociation probably then occurs and the olefinic double bond coordinates with the manganese. The acetyl group now shifts from mangenese to the coordinated olefinic group and addition occurs in both possible directions. The adduct with the terminal acyl group forms a coordinately saturated, isolable product by coordinating the acyl oxygen with the manganese. The adduct with the acetyl group on the benzyl carbon of the phosphine ligand very probably undergoes a hydride elimination next and readdition in the reverse direction to give a benzylmanganese derivative which can coordinate with the acyl group and form the other isolable product, a π-oxoallylmanganese complex. The last complex can be converted back to the "σ-oxoallyl" structure by CO under pressure at 70° (492) (Scheme IX-10).

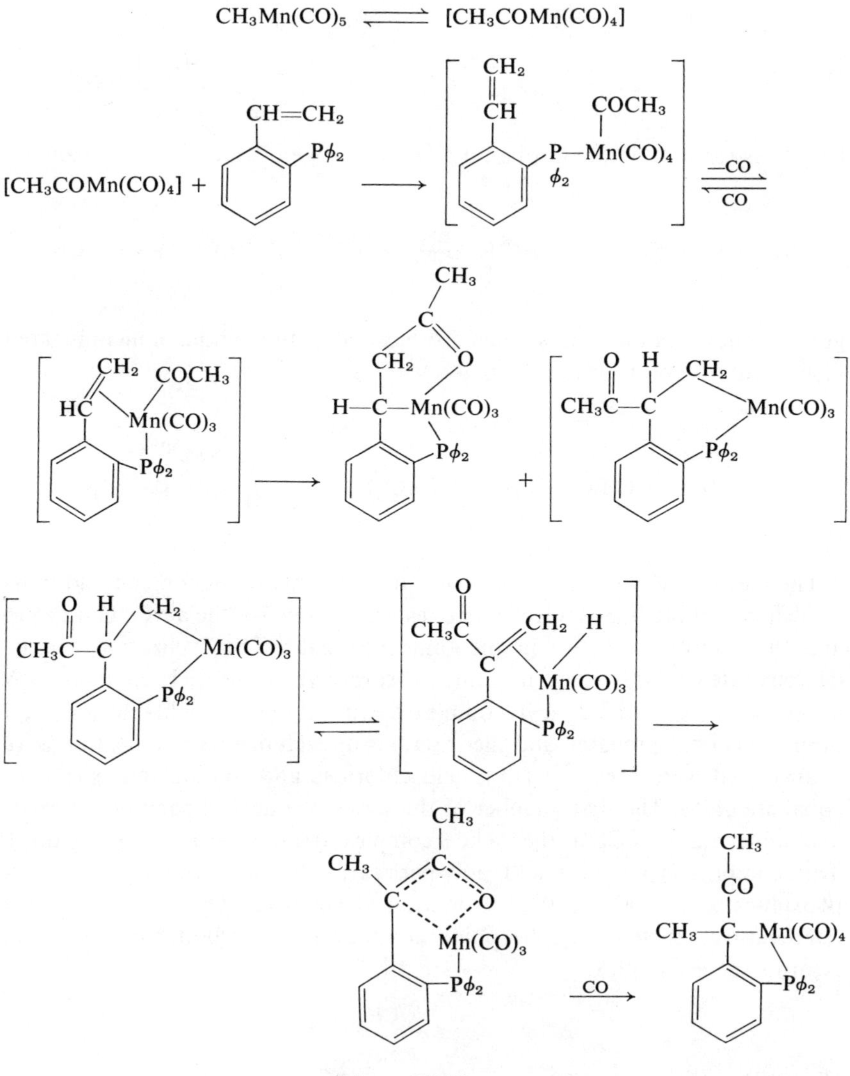

$$CH_3Mn(CO)_5 \rightleftharpoons [CH_3COMn(CO)_4]$$

Scheme IX-10.

Ketones are sometimes formed in small amounts in hydroformylation reactions. They may become major products if the hydrogen concentration is reduced to a low level. This reaction goes particularly well with ethylene. In one reported reaction a low concentration of hydrogen is provided by the slow *in situ* dehydrogenation of tetralin. This procedure produces diethyl ketone in 93% yield (494). A similar reaction has been carried out with

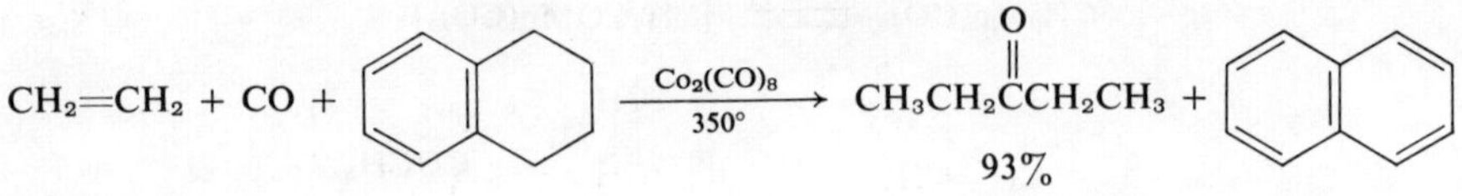

$$CH_2{=}CH_2 + CO + \underset{}{\bigcirc\!\!\bigcirc} \xrightarrow[350°]{Co_2(CO)_8} \underset{93\%}{CH_3CH_2\overset{\overset{\displaystyle O}{\|}}{C}CH_2CH_3} + \bigcirc\!\!\bigcirc$$

Rh_2O_3 as catalyst and isopropanol as the hydrogen source (495). An analo-

$$CH_2{=}CH_2 + CO + CH_3\overset{\overset{\displaystyle OH}{|}}{C}HCH_3 \xrightarrow[175°]{Rh_2O_3} \underset{90\%}{CH_3CH_2\overset{\overset{\displaystyle O}{\|}}{C}CH_2CH_3 + CH_3\overset{\overset{\displaystyle O}{\|}}{C}CH_3}$$

gous, cyclic reaction occurs when 1,4-hexadiene and similar nonconjugated olefins are hydroformylated (496, 497).

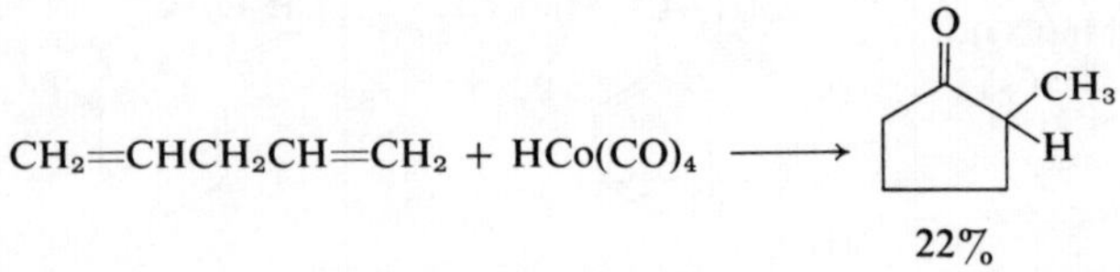

$$CH_2{=}CHCH_2CH{=}CH_2 + HCo(CO)_4 \longrightarrow$$

The mechanisms of these reactions appear to involve acyl metal additions to olefinic double bonds. Cyclic ketones are obtained if the acyl metal group and the double bond are in the same molecule. The cyclization can be demonstrated with isolated unsaturated carbonylacylcobalt(I) complexes. A series of ω-unsaturated acyl complexes with three to eight carbon acyl groups has been prepared and the reactions of each member on heating were studied. All were prepared from acid chlorides and sodium tetracarbonyl-cobaltate (498). The first member of the series, the acrylyl compound, exists at 0° under 1 atm of CO as the cyclic π complex, tricarbonyl-π-acrylylcobalt(I). This complex is unstable and polymerizes easily on warming. Triphenyl-phosphine reacts with it at 0° to form the acyl σ complex. On isolation at room temperature the triphenylphosphine complex loses carbon monoxide and reforms the π complex.

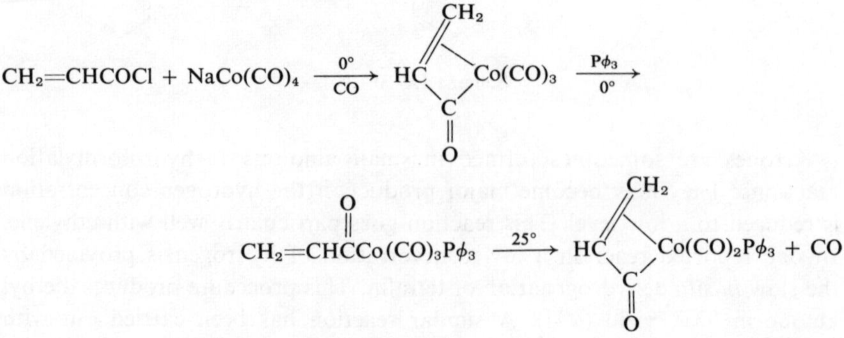

$$CH_2{=}CHCOCl + NaCo(CO)_4 \xrightarrow[CO]{0°} HC\overset{CH_2}{\underset{\underset{\displaystyle O}{\|}}{\underset{C}{\diagup}}}Co(CO)_3 \xrightarrow[0°]{P\phi_3}$$

$$CH_2{=}CH\overset{\overset{\displaystyle O}{\|}}{C}Co(CO)_3P\phi_3 \xrightarrow{25°} HC\overset{CH_2}{\underset{\underset{\displaystyle O}{\|}}{\underset{C}{\diagup}}}Co(CO)_2P\phi_3 + CO$$

The next member of the series exists only as an acyl σ complex at 0°. On warming, two carbon monoxide molecules are lost and tricarbonyl-π-allylcobalt(I) is formed. The third member of the series, the 4-pentenoyl

$$CH_2=CHCH_2\overset{\overset{\textstyle O}{\|}}{C}Co(CO)_4 \xrightarrow{25°} 2CO + \underset{CH_2}{\overset{CH_2}{HC}}\text{·····}Co(CO)_3$$

complex, cyclizes to the π complex at 0°, but the acyl group apparently cannot add to the internal double bond. There is probably too much strain for the olefin and acyl groups to get close enough to react. On warming, this complex also loses carbon monoxide and forms tricarbonyl-1-methyl-π-allylcobalt(I). An internal hydride elimination and readdition probably are involved.

$$\underset{CH_2}{\overset{CH_2}{\underset{H_2C}{\overset{HC}{\diagup}}\diagdown}}\overset{Co(CO)_3}{\underset{C=O}{|}} \longrightarrow \underset{CH}{\overset{CH_2}{HC}}\text{·····}Co(CO)_3 + CO$$
$$CH_3$$

The next homolog, the 5-hexenoylcobalt complex, exists at 0° under 1 atm of carbon monoxide as the σ complex. When the temperature is raised, internal addition of the acyl group to the double bond occurs, presumably by way of the olefin π complex. The cyclization produces both five- and six-membered ring products. These then undergo a disproportionation giving a mixture of saturated and unsaturated ketones and cobalt(0) carbonyls. Hydridotricarbonylcobalt elimination probably occurs and this hydride then reduces the unreacted, cyclized cobalt derivatives to form the saturated ketone. Approximately half of the product (54%) was the saturated ketone, 2-methylcyclopentanone, as this mechanism would predict. Part of the unsaturated ketone was lost. 2-Methylenecyclopentanone polymerized partially; 16% also rearranged to the more stable 2-methyl-2-cyclopentenone. Only 8% of a six-membered ring product, 2-cyclohexenone, was found. Tetracarbonyl-6-heptenoylcobalt also cyclizes on warming, but in lower yield, to give a mixture of ketones. Other products are π-allylic cobalt complexes.

The formation of cyclic ketones in the hydroformylation reaction of 1,4- or 1,5-dienes, therefore, probably occurs by way of a hydridotricarbonylcobalt addition to one of the double bonds. The alkyl group then shifts to coordinated carbon monoxide and a cyclization takes place with addition of the acylcobalt group to the remaining double bond. In the presence of hydrogen, the easily reduced unsaturated ketones would be hydrogenated.

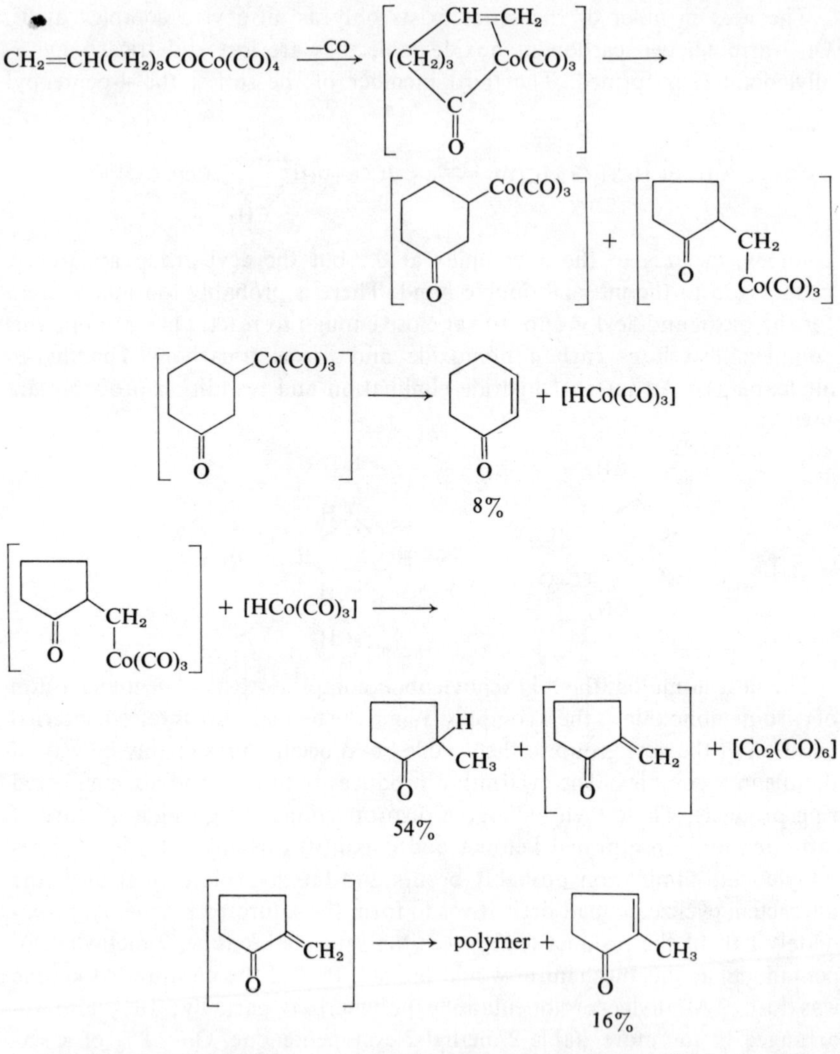

$$CH_2{=}CH(CH_2)_3COCo(CO)_4 \xrightarrow{-CO} \left[\begin{array}{c} CH{=}CH_2 \\ (CH_2)_3 \qquad Co(CO)_3 \\ C \\ \| \\ O \end{array} \right] \longrightarrow$$

$$\left[\text{(cyclohexanone with } Co(CO)_3) \right] + \left[\text{(cyclopentanone with } CH_2Co(CO)_3) \right]$$

$$\left[\text{(cyclohexanone with } Co(CO)_3) \right] \longrightarrow \text{(cyclohexenone)} + [HCo(CO)_3]$$

8%

$$\left[\text{(cyclopentanone with } CH_2\,Co(CO)_3) \right] + [HCo(CO)_3] \longrightarrow$$

$$\text{(cyclopentanone with } \overset{H}{\underset{CH_3}{\diagdown}}) + \left[\text{(cyclopentanone with } {=}CH_2) \right] + [Co_2(CO)_6]$$

54%

$$\left[\text{(cyclopentanone with } {=}CH_2) \right] \longrightarrow \text{polymer} + \text{(cyclopentenone with } CH_3)$$

16%

A related cyclization has been reported to occur when ethylene is heated with carbon monoxide and dichlorobistriphenylphosphinepalladium(II) as catalyst (499). The reaction produces a mixture of two isomeric unsaturated lactones which differ only in the position of the double bond. The reaction can be explained on the basis of a hydridopalladium addition to ethylene followed by an addition to CO forming a propionylpalladium complex. An addition of the propionylpalladium group to another ethylene followed by a second addition to CO and cyclization by addition of the acylpalladium group to the β-ketone carbonyl (placing acyl on oxygen and palladium on

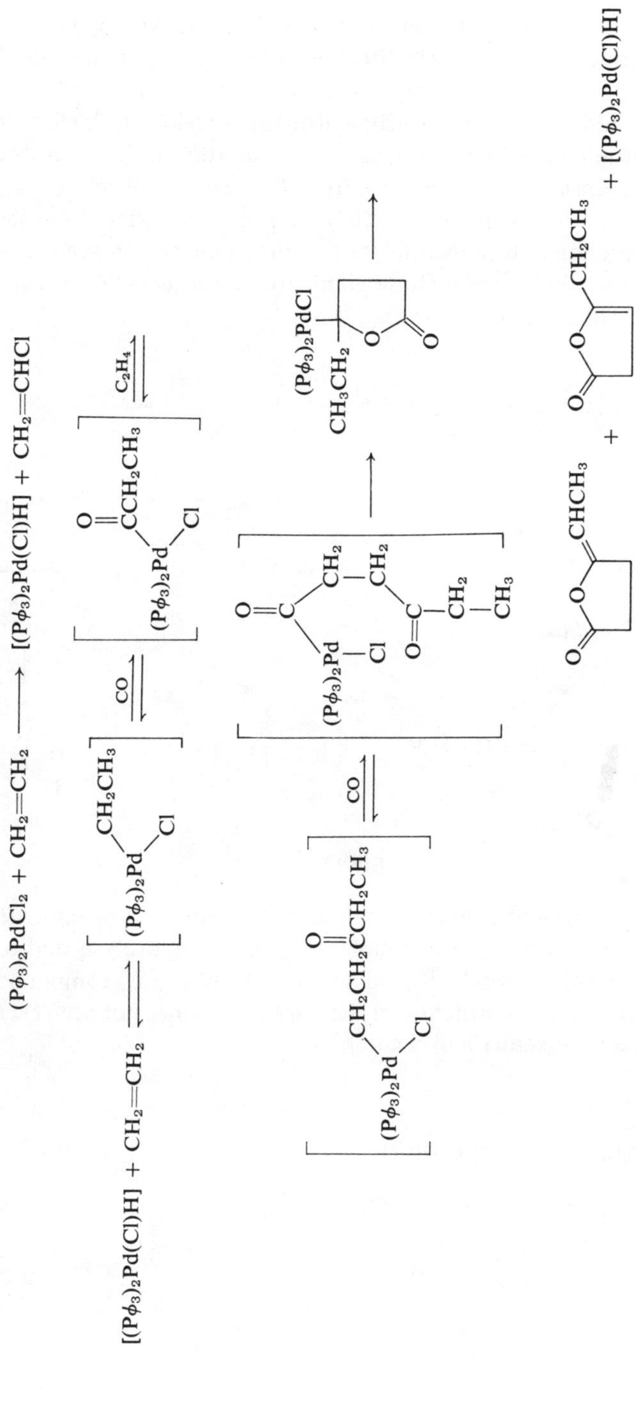

$$(P\phi_3)_2PdCl_2 + CH_2{=}CH_2 \longrightarrow [(P\phi_3)_2Pd(Cl)H] + CH_2{=}CHCl$$

$$[(P\phi_3)_2Pd(Cl)H] + CH_2{=}CH_2 \;\rightleftharpoons\; \left[(P\phi_3)_2Pd \underset{Cl}{\overset{CH_2CH_3}{\diagdown}} \right] \;\underset{CO}{\rightleftharpoons}\; \left[(P\phi_3)_2Pd \underset{Cl}{\overset{\overset{O}{\|}\;CCH_2CH_3}{\diagdown}} \right] \;\underset{C_2H_4}{\rightleftharpoons}$$

$$\left[(P\phi_3)_2Pd \underset{Cl}{\overset{CH_2CH_2CCH_2CH_3}{\diagdown}} \; \overset{O}{\|} \right] \;\underset{CO}{\rightleftharpoons}\; \left[\begin{array}{c} CH_2{-}CH_2 \\ O{=}C \qquad\quad CH_2 \\ (P\phi_3)_2Pd \quad C{=}O \quad CH_3 \\ Cl \end{array} \right] \longrightarrow$$

Scheme IX-11.

carbon) would give a palladium-substituted lactone. An elimination of the starting hydride in both possible directions would then give the two observed products and complete the catalytic cycle (Scheme IX-11).

A similar addition of iododicarbonylbenzoylnickel(II) to olefins probably is an important step in the formation of unsaturated ketones, saturated ketones, and unsaturated 4-lactones from the reaction of iodobenzene with olefins and tetracarbonylnickel(0) (500). The reaction differs from the above palladium reaction only in the way the first organometallic reagent is obtained and by the fact that more hydride elimination and reduction occurs in this case before cyclization (Scheme IX-12).

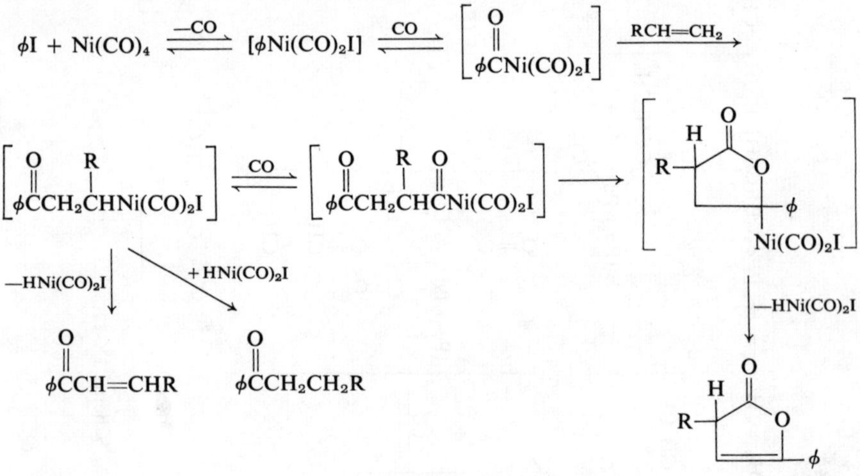

Scheme IX-12.

Anionic tricarbonylacylnickel complexes will also add to α,β-unsaturated carbonyl compounds to form complexes which on hydrolysis and oxidation of residual nickel carbonyl with iodine give 1,4-dicarbonyl compounds, often in high yield (461). The structure of the intermediate has not been established. Possibly it is a π-oxoallylnickel complex.

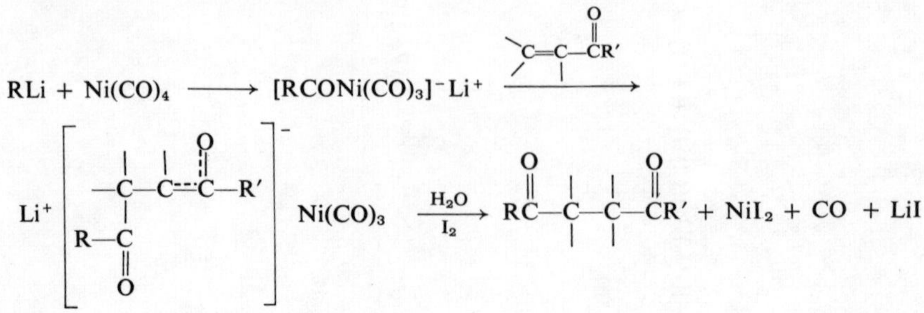

c. Dicarboxylation Reactions

Several types of alkoxycarbonyl metal complexes have been isolated, but none of these are known to add to olefins or acetylenes. An unisolable "methoxycarbonylpalladium acetate" prepared by an exchange reaction of methoxycarbonylmercuric acetate with palladium acetate does add, however. The adducts generally decompose rapidly by eliminating (probably) "hydrido-palladium acetate," thus producing a methoxycarbonyl derivative of the olefin (194). If this reaction is carried out in the presence of carbon monoxide

$$\text{Pd(OAc)}_2 + \text{CH}_3\text{OCOHgOAc} \xrightarrow{\text{L}} \left[\begin{array}{c} \text{L} \\ | \\ \text{CH}_3\text{OCO}\,\text{PdOAc} \\ | \\ \text{L} \end{array} \right] + \text{Hg(OAc)}_2$$

$$\left[\begin{array}{c} \text{L} \\ | \\ \text{CH}_3\text{OCOPdOAc} \\ | \\ \text{L} \end{array} \right] + \text{CH}_3(\text{CH}_2)_2\text{CH}=\text{CH}_2 \longrightarrow \left[\begin{array}{c} \text{L} \\ | \\ \text{LPdOAc} \\ | \\ \text{CH}_3(\text{CH}_2)_2\text{CHCH}_2\text{COOCH}_3 \end{array} \right] \xrightarrow{-\left[\begin{smallmatrix} \text{L} \\ | \\ \text{HPdOAc} \\ | \\ \text{L} \end{smallmatrix} \right]}$$

$$\text{CH}_3\text{CH}_2\text{CH}=\text{CHCH}_2\text{COOCH}_3 + \text{CH}_3\text{CH}_2\,\text{CH}_2\text{CH}=\text{CHCOOCH}_3$$
$$\qquad\qquad 43\% \qquad\qquad\qquad\qquad\qquad\qquad 30\%$$

at low pressure (30 psi) and methanol, 1,2-dicarboxylate esters are formed. The intermediate methoxycarbonylpalladium complex apparently reacts with carbon monoxide faster than it decomposes by hydride elimination and forms a carbomethoxyacylpalladium complex which then alcoholyzes to the 1,2-diester (501). The same reaction occurs in 25% yield without adding

$$\left[\begin{array}{c} \text{L} \\ | \\ \text{CH}_3\text{OCOPdCl} \\ | \\ \text{L} \end{array} \right] + \phi\text{CH}=\text{CH}_2 \longrightarrow \left[\begin{array}{c} \text{L} \\ | \\ \text{L}-\text{PdCl} \\ | \\ \phi\text{CHCH}_2\text{COOCH}_3 \end{array} \right] \xrightarrow{\text{CO}}$$

$$\left[\begin{array}{c} \text{L} \\ | \\ \text{L}-\text{PdCl} \\ | \\ \text{C}=\text{O} \\ | \\ \phi\text{CHCH}_2\text{COOCH}_3 \end{array} \right] \xrightarrow[-[\text{HPdL}_2\text{OAc}]]{\text{CH}_3\text{OH}} \begin{array}{c} \text{COOCH}_3 \\ | \\ \phi\text{CHCH}_2\text{COOCH}_3 \\ 60\% \end{array}$$

carbomethoxymercuric chloride to the reaction mixture. Apparently, the carbomethoxypalladium complex can be formed directly from palladium(II) chloride, CO, and methanol. The reaction is much improved if a molar

$$\text{PdCl}_2 + \text{CH}_3\text{OH} + \text{CO} \xrightarrow{\text{L}} \text{HCl} + \text{CH}_3\text{OCOPdL}_2\text{Cl}$$

equivalent of mercuric chloride (relative to the palladium chloride) is present. Although mercuric chloride (in contrast to the acetate), CO, and methanol do not react by themselves, in the presence of olefin and palladium chloride they apparently do and form the carbomethoxypalladium complex. Styrene under these conditions gives an 87% yield of the dicarboalkoxylated olefin along with 11% methyl cinnamate. A similar reaction with cyclohexene yields 52% *trans* methyl 2-methoxycyclohexanecarboxylate and only 6% *cis*-1,2-cyclohexanedicarboxylate. Methanol probably can attack coordinated carbon monoxide to form the carbomethoxy derivative or attack coordinated olefin and form a 2-methoxyethylpalladium derivative. Addition of the intermediates to carbon monoxide and alcoholysis would give the observed products. A cis addition of the carbomethoxypalladium group, CO insertion with retention, and methanolysis would give the observed cis isomers. Trans-methoxypalladation of the olefin would be expected by an external attack of methanol and the observed *trans*-methoxy ester would be obtained. Whether both of these reactions occur with a single complex such as dichlorocarbonyl-π-olefinpalladium(II) or with two different complexes, one containing CO and the other the olefin, is not known. A possible mechanism employing a single complex is shown in Scheme IX-13.

The dicarboxyalkylation of internal straight-chain olefins is not as specific, particularly with *cis*-olefins. While dicarbomethoxylation of *trans*-3-hexene yields 50% of the ($\pm$)-diester with little of the meso isomer or methoxy ester being formed, *cis*-3-hexene yields 16% methoxy ester, 5% *meso*-diester, and 30% ($\pm$)-diester (501).

The olefin dicarboalkoxylation may be carried out catalytically under pressure at 85° with cupric chloride and oxygen as reoxidants for the palladium (502).

3. Carbonylation Reactions of Dienes and Allenes

Carbonylation of 1,4- and 1,5-dienes often leads to cyclic products. The formation of cyclic ketones from 1,4- and 1,5-dienes and $HCo(CO)_4$ has already been mentioned in Section A,2. The high pressure reaction of 1, 4-pentadiene with carbon monoxide and diiodobistri-*n*-butylphosphinepalladium(II) in THF solution gave three cyclic products: 2-methyl-2-cyclopentenone and two isomeric bicyclic lactones, but only in 4% yield (503).

$$CH_2=CHCH_2CH=CH_2 + CO \xrightarrow[200°,\ 1000\ atm]{(n-Bu_3P)_2PdI_2}$$

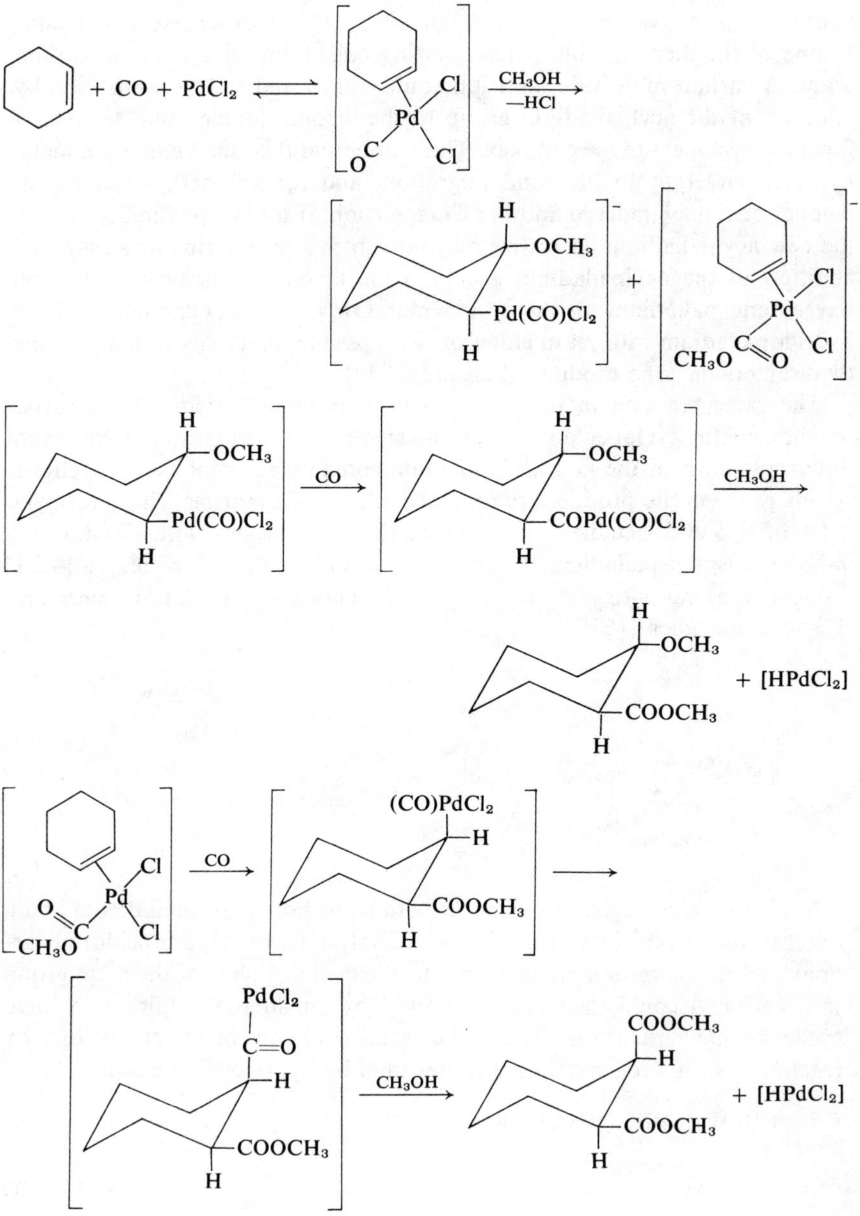
+ CO + PdCl2
Cl
Pd
C
O
Cl
CH3OH
—HCl
H
OCH3
Pd(CO)Cl2
H
+
Cl
Pd
CH3O
C
O
Cl
H
OCH3
Pd(CO)Cl2
H
CO
H
OCH3
COPd(CO)Cl2
H
CH3OH
H
OCH3
COOCH3
H
+ [HPdCl2]
Cl
Pd
O
C
CH3O
Cl
CO
(CO)PdCl2
H
COOCH3
H
PdCl2
C=O
H
COOCH3
H
CH3OH
COOCH3
H
COOCH3
H
+ [HPdCl2]

Scheme IX-13.

The true catalyst in this reaction is probably a palladium hydride such as hydridoiodobistri-*n*-butylphosphinepalladium. The hydride likely first adds to one of the diene double bonds, putting palladium on a terminal carbon atom. A carbon monoxide insertion could follow and then a cyclization by addition of the acylpalladium group to the second double bond to form a 2-palladacyclopentanone complex. This compound then may eliminate metal hydride, undergo double-bond migration, and form 2-methyl-2-cyclopentenone, or it may undergo another CO insertion. If the last possibility occurs, the new acylpalladium derivative may now form a second ring by an internal addition of the acylpalladium group to the ketone group placing acyl on oxygen and palladium on the cyclopentane ring. A final elimination of the hydridopalladium catalyst in either of two possible directions would give the two isomeric lactone products (Scheme IX-14).

The same reaction in methanol solution produces about 10% 2-carbomethoxymethylcyclopentanone and some methyl 5-hexenoate (504). Methanol intercepts most of the first acylpalladium complex before it cyclizes. Higher yields of the cyclic product are obtained with 1,5-hexadiene. The carbonylation of 1,5-cyclooctadiene even in methanol solutions with diiodobistri-*n*-butylphosphinepalladium(II) produces a 45% yield of a bicyclo[4.2.1] nonenone along with 30% dimethyl cyclooctanedicarboxylate isomers and 1,3- cyclooctadiene (505).

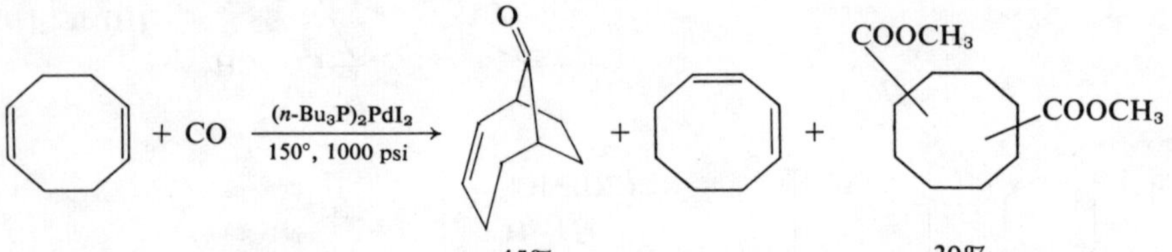

$$45\% \qquad\qquad 30\%$$

Similar catalytic cyclizations occur with some nonconjugated dienes and a tetracarbonylnickel–hydrogen chloride catalyst (506). Hydridochlorodicarbonylnickel(II) presumably first adds to the olefin. A shift of the alkyl group to a carbonyl could then occur followed by an internal addition to form cyclic ketone derivatives. The nickel groups are probably finally lost by reaction with hydrogen chloride rather than by hydride elimination.

$$CH_2\!=\!CHCH_2CH_2CH\!=\!CH_2 + Ni(CO)_4 + 2HCl \longrightarrow$$

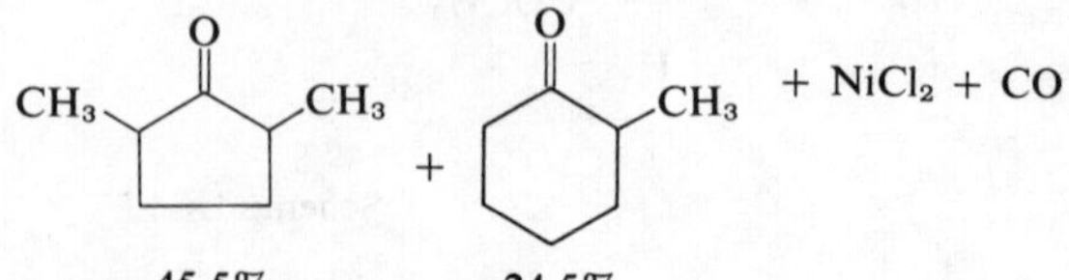

$$45.5\% \qquad\qquad 24.5\%$$

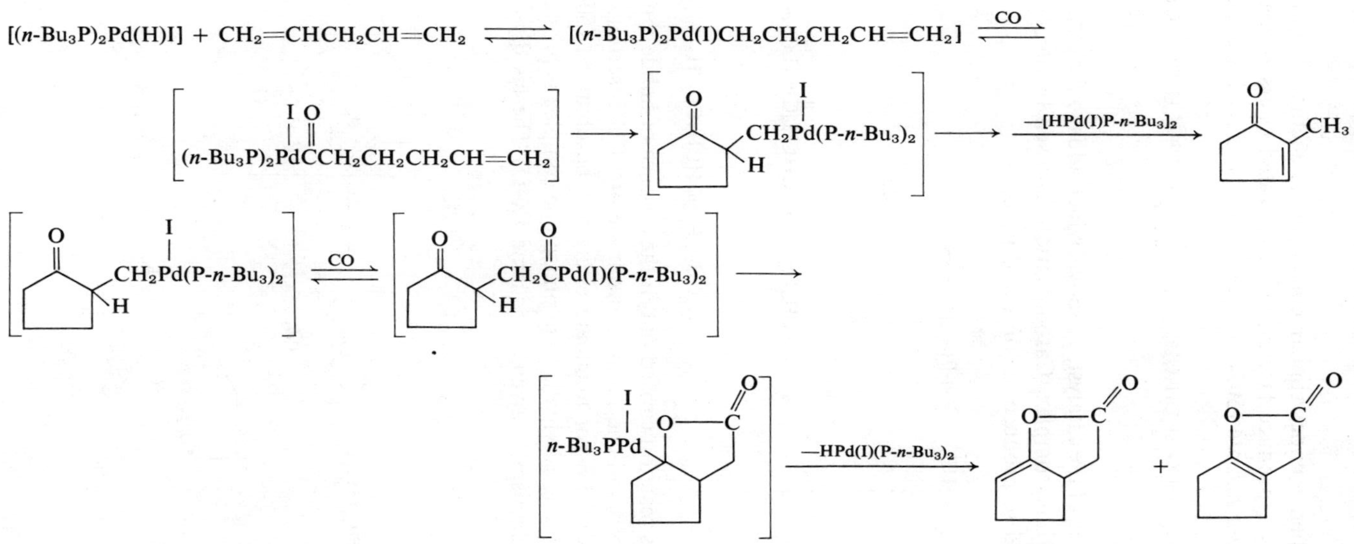

$$[(n\text{-}Bu_3P)_2Pd(H)I] + CH_2=CHCH_2CH=CH_2 \rightleftharpoons [(n\text{-}Bu_3P)_2Pd(I)CH_2CH_2CH_2CH=CH_2] \xrightarrow{CO}$$

Scheme IX-14.

Conjugated dienes under similar conditions in methanol solution with a palladium catalyst produce largely β,γ-unsaturated esters probably by way of a π-allylic intermediate (507). The reaction probably involves the mono-

$$CH_3CH{=}CHCH{=}CH_2 + CO + CH_3OH \xrightarrow[150°, 1000 \text{ atm}]{(n\text{-Bu}_3P)_2PdI_2} CH_3CH_2CH{=}CHCH_2COOCH_3$$

$$34\%$$

hydride derivative of the catalyst, as above, which adds to the diene to produce a π-allylicpalladium complex. Carbonylation followed by alcoholysis would then be expected to produce the observed product.

$$CH_3CH{=}CHCH{=}CH_2 + [HPd(I)(P\text{-}n\text{-}Bu_3)_2]^- \xrightarrow{\quad -n\text{-}Bu_3P \quad}$$

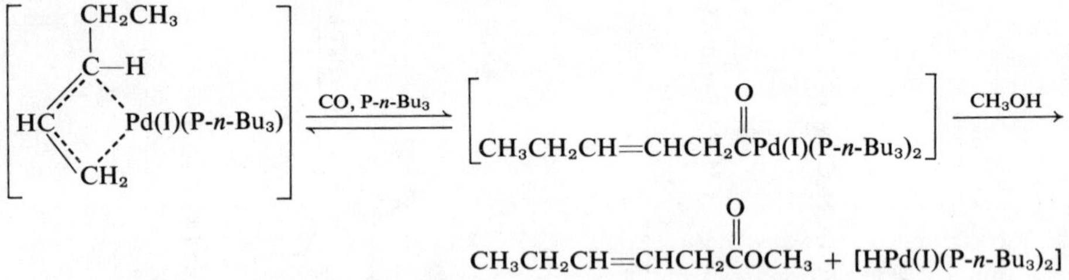

$$CH_3CH_2CH{=}CHCH_2\overset{O}{\overset{\|}{C}}OCH_3 + [HPd(I)(P\text{-}n\text{-}Bu_3)_2]$$

An analogous reaction carried out in the absence of halide ions allows the coupling of two butadiene units to occur before CO insertion takes place and high yields of 3,8-nonadienoate esters can be obtained (508, 509). An intermediate octadienediylpalladium phosphine complex is probably formed. A possible mechanism is given in Scheme IX-15. The cis β,γ-double bond

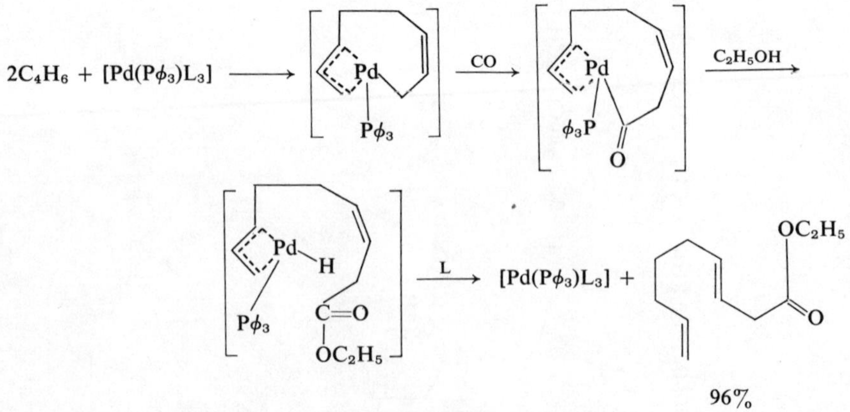

$$96\%$$

Scheme IX-15.

could be isomerized by an internal hydride addition and elimination reaction of the third intermediate but the stereochemistry has not been established.

Allene has been carbonylated in methanol solution at 140° and 1000 atm with a nonacarbonyldiruthenium catalyst and 18% methyl methacrylate and 79% dimethyl 2-methylene-4,4-dimethylglutarate were obtained (510). Under lower CO pressure (300 atm) 50% methyl methacrylate was produced. The

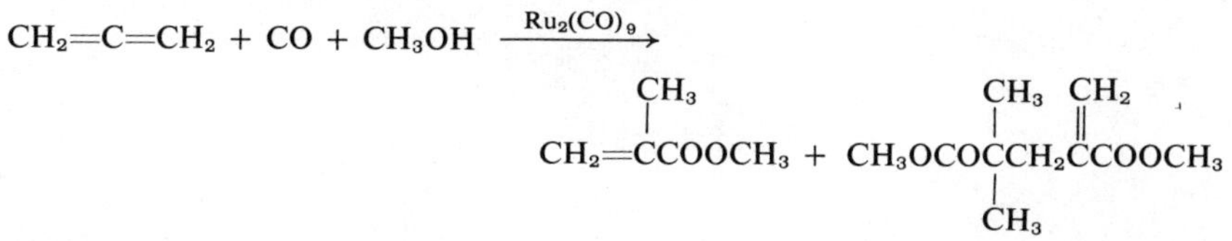

$$CH_2{=}C{=}CH_2 + CO + CH_3OH \xrightarrow{\ Ru_2(CO)_9\ }$$

$$
\underset{\substack{| \\ CH_3}}{CH_2{=}CCOOCH_3} \; + \; CH_3OCOCCH_2\overset{\substack{CH_3 \;\; CH_2 \\ | \;\;\;\; \|}}{C}COOCH_3 \atop \underset{CH_3}{|}
$$

mechanism of this reaction probably involves a ruthenium hydride addition to the allene, putting ruthenium on the central carbon, then CO insertion, and methanolysis to produce methyl methacrylate. The other product may be formed by an addition of the ruthenium hydride to the methyl methacrylate to give the tertiary ruthenium alkyl which then could add to allene and carbon monoxide and undergo alcoholysis. The formation and reaction of a tertiary alkyl under these conditions would be unusual and this mechanism must be considered speculative until more information is obtained. The exact catalyst reacting is also not clear (Scheme IX-16).

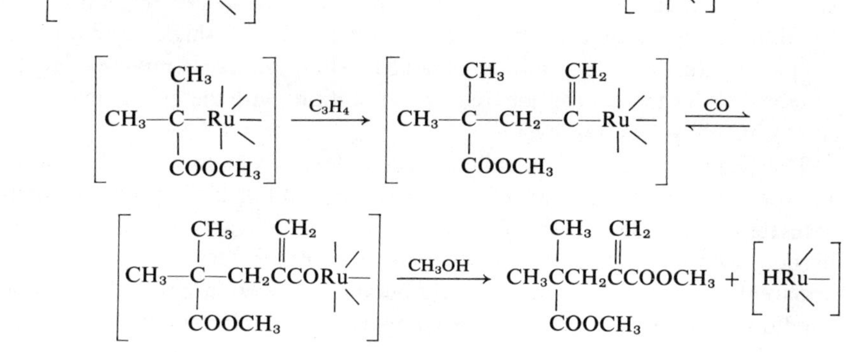

Scheme IX-16.

An interesting synthesis of muscone, a fifteen-carbon cyclic ketone, has been achieved from dodecatrienediylnickel, allene, and CO (511).

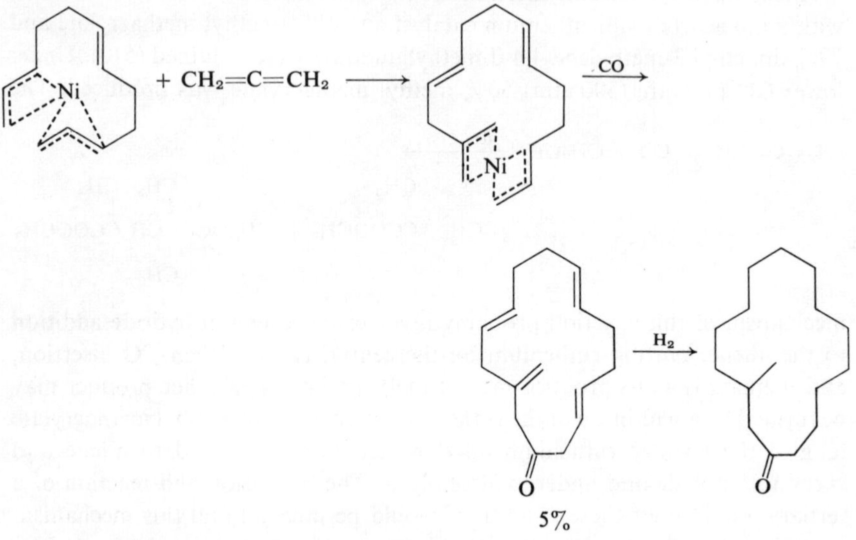

5%

4. Carbonylation Reactions of Acetylenes

a. Carbon Monoxide–Acetylene Reactions

A wide range of products has been obtained from transition metal complex reactions with acetylenes and carbon monoxide. The most versatile reactions have been found to occur with iron and cobalt carbonyls. Iron has yielded at least 20 types of complexes, while cobalt has given about ten (512). The structures of most of these complexes have been definitely established but still much remains to be learned about their mechanisms of formation. All the different types of complexes are not obtained in a single reaction, but simpler mixtures are commonly obtained. Many of the complexes can be obtained under the appropriate reaction conditions with the proper reactants, in reasonable yields, however.

The formation of cyclobutadiene π complexes from two acetylene units has already been mentioned (Chapter VII, Section C). If the probable intermediate in this reaction, the metalacyclopentadiene, reacts with coordinated carbon monoxide, a cyclopentadienone π complex results. This reaction apparently occurs when hexafluoro-2-butyne reacts with chlorodicarbonylrhodium(I) dimer and CO. The π complex is unstable and the tetrakistrifluoromethylcyclopentadienone is replaced by more acetylene and CO, allowing the reaction to proceed catalytically (513). A similar complex is

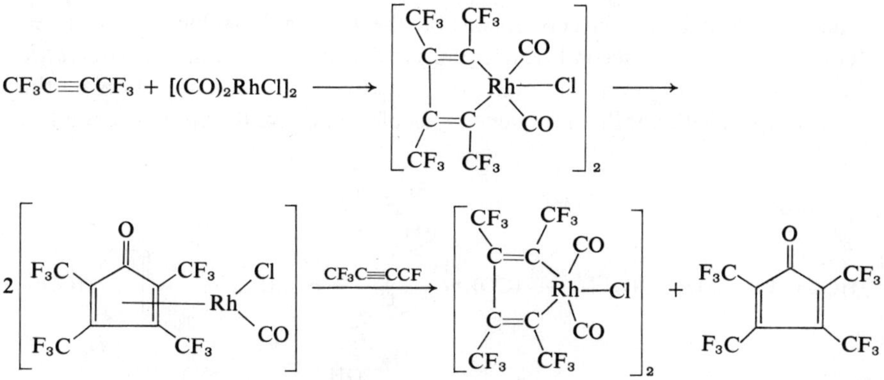

$$CF_3C \equiv CCF_3 + [(CO)_2RhCl]_2 \longrightarrow \left[\begin{array}{c} \text{structure} \end{array} \right]_2 \longrightarrow$$

$$2\left[\begin{array}{c} \text{structure} \end{array} \right] \xrightarrow{CF_3C \equiv CCF} \left[\begin{array}{c} \text{structure} \end{array} \right]_2 + \begin{array}{c} \text{structure} \end{array}$$

formed when diphenylacetylene (or another acetylene) is irradiated with pentacarbonyliron(0). Several other complexes are also produced at the same time in this reaction (514, 515).

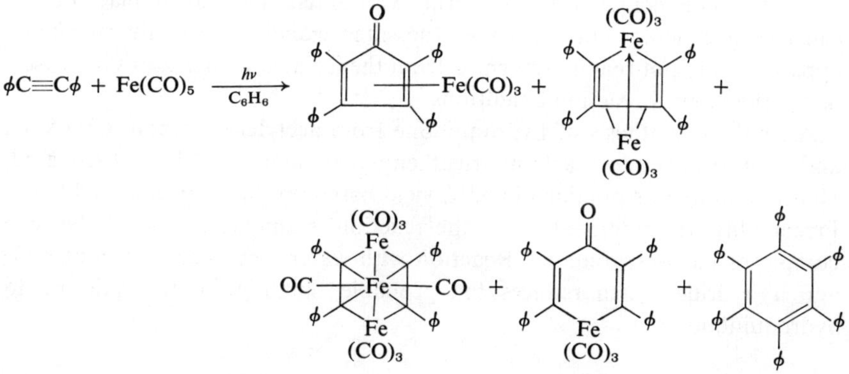

$$\phi C \equiv C\phi + Fe(CO)_5 \xrightarrow[C_6H_6]{h\nu} \begin{array}{c} \text{structure} \end{array} \text{—}Fe(CO)_3 + \begin{array}{c} \text{structure} \end{array} + $$

The fourth product above is also a possible precursor of the cyclopentadienone complex. This type of complex may be formed as an intermediate when 2-butyne is irradiated with $Fe(CO)_5$ and tricarbonyl-π-tetramethylquinoneiron(0) is obtained (516). Another type of iron complex could also

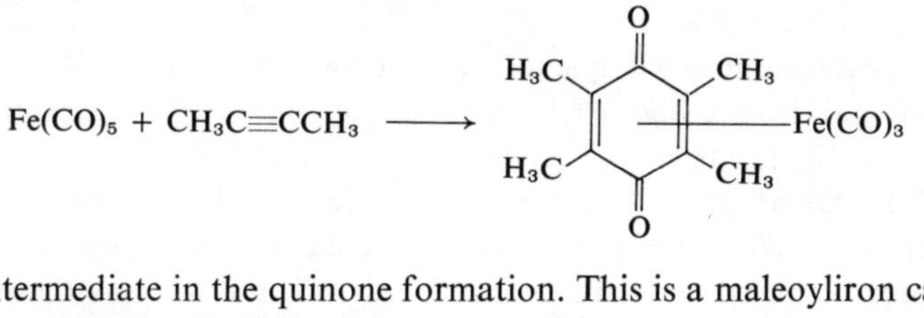

$$Fe(CO)_5 + CH_3C \equiv CCH_3 \longrightarrow \begin{array}{c} \text{structure} \end{array} \text{—}Fe(CO)_3$$

be an intermediate in the quinone formation. This is a maleoyliron carbonyl. One complex of this type has been obtained in two steps from the reaction of acetylene with $NaHFe(CO)_4$. The binuclear complex obtained has a succinoyl unit in the bisenol form and the two enol double bonds are π-complexed with the second iron atom. Oxidation of the complex with ferric ion yields the mononuclear maleoyl derivative (517–520). Reaction of the last complex with another acetylene molecule could give the quinone complex.

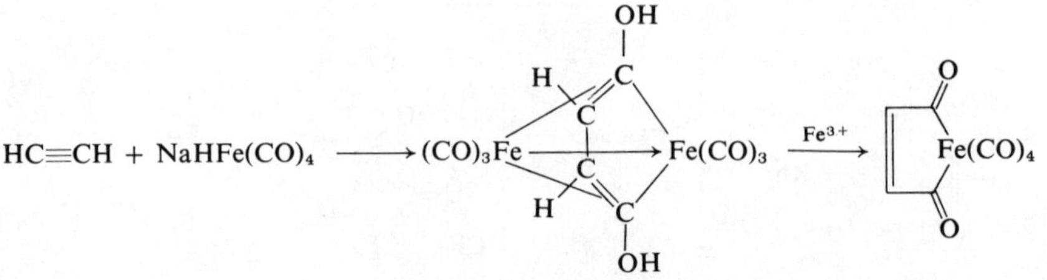

Often attempts to react isolated, presumed intermediate iron complexes further with acetylenes and CO to form other complexes fails. This does not necessarily mean that complexes of the type isolated are not intermediates but that CO dissociated forms or forms with different ligands in place of CO (such as acetylenic units) are the true intermediates. These intermediates apparently are not readily reformed from the isolated complex in such cases, under the normal reaction conditions.

A catalytic synthesis of hydroquinone from acetylene, carbon monoxide, and water using dodecacarbonyltriruthenium as catalyst has been discovered. Hydroquinone was obtained in 63% yield based on the acetylene used (521). Presumably, the intermediate in the reaction is similar to one of the iron complexes mentioned above. Reaction with water then occurs forming CO_2 and a hydride which reduces the probable, initially formed quinone to hydroquinone.

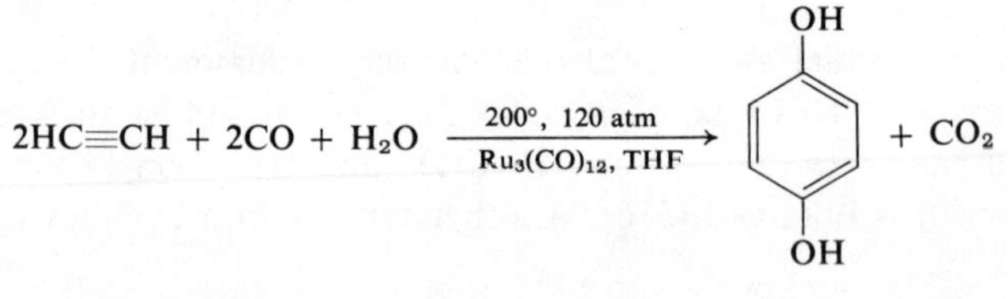

Cycloheptatrienone π complexes of iron have also been obtained from acetylenes and iron carbonyls. For example, nonacarbonyldiiron(0) and acetylene at 22 atm form tricarbonyl-π-cycloheptatrienoneiron(0) in 28% yield (522). Oxidation of the complex with ferric chloride produces cycloheptatrienone in 70% yield. In the cycloheptatrienone complex only two of the three olefinic bonds are π-complexed with the iron, as the inert gas rule would predict, since one double bond can be reduced catalytically to give an isomeric mixture of two cycloheptadienone iron complexes (522).

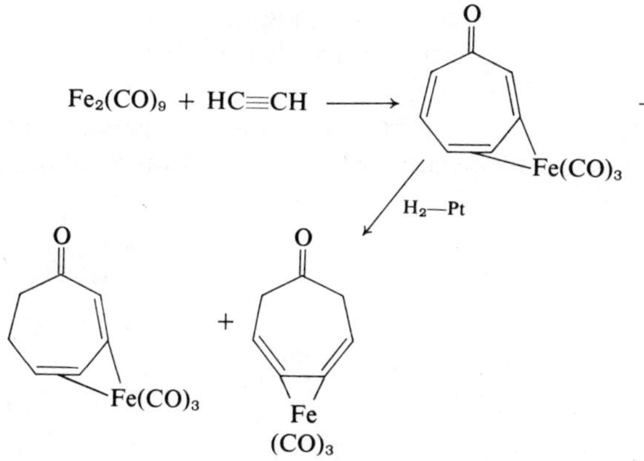

$$Fe_2(CO)_9 + HC\equiv CH \longrightarrow$$

The intermediate in the cycloheptatrienone formation may be of the type of the fourth complex obtained in the pentacarbonyliron–diphenylacetylene reaction mentioned above, where a ferracyclohexadienone complex is formed. Addition to the third acetylene and reductive elimination could produce the cycloheptatrienone complex.

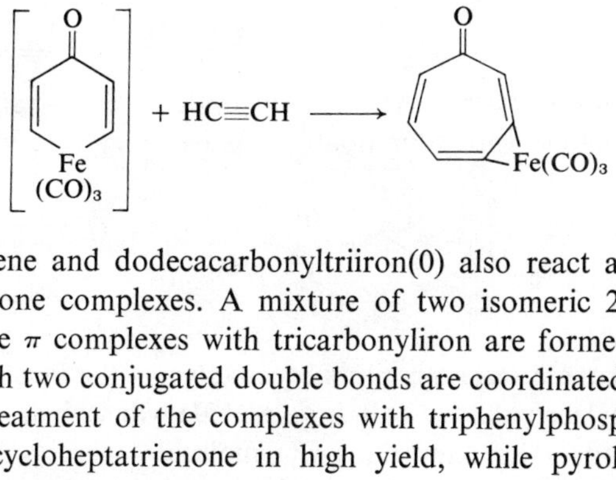

Phenylacetylene and dodecacarbonyltriiron(0) also react at 85° to form cycloheptatrienone complexes. A mixture of two isomeric 2,4,6-triphenyl-cycloheptatriene π complexes with tricarbonyliron are formed which differ only as to which two conjugated double bonds are coordinated with the iron atom (523). Treatment of the complexes with triphenylphosphine liberates 2,4,6-triphenylcycloheptatrienone in high yield, while pyrolysis generates

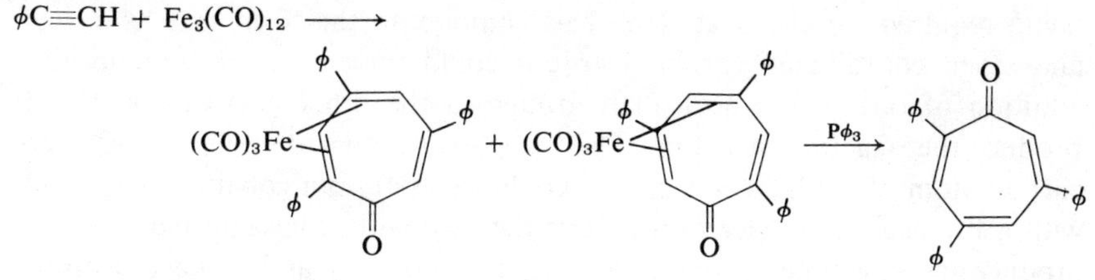

$$\phi C\equiv CH + Fe_3(CO)_{12} \longrightarrow$$

2,4,6-triphenylbenzene in 50% yield. The carbonyl group is probably abstracted in the pyrolysis because of the need for the iron atom to retain this ligand to compensate partially for the loss of the other three (olefinic) ligands.

Octacarbonyldicobalt reacts with acetylenes quite differently from penta-carbonyliron and readily forms hexacarbonylacetylenedicobalt(0) complexes (Chapter VII, Section A). At 75° and under 210 atm of carbon monoxide the last compound forms a binuclear complex with CO and γ-lactone-bridging groups (524, 525). At a higher temperature octacarbonyldicobalt catalyzes a

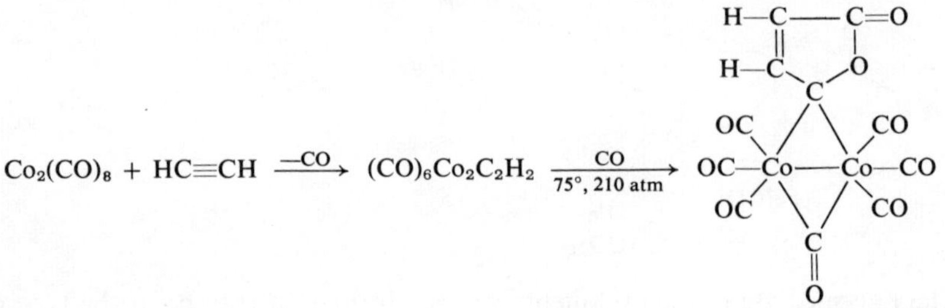

$$
Co_2(CO)_8 + HC{\equiv}CH \xrightarrow{-CO} (CO)_6Co_2C_2H_2 \xrightarrow[75°, 210 \text{ atm}]{CO}
$$

reaction between acetylene and carbon monoxide forming a cis-trans mixture of an eight-carbon dilactone with three double bonds, a dimer of the lactone ligand in the complex just mentioned, in 70% yield (526). When carried out

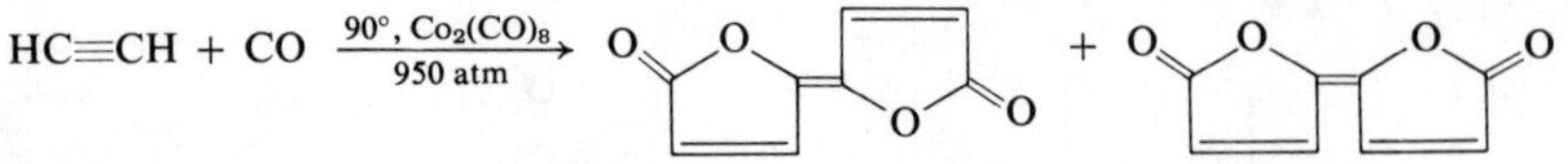

$$
HC{\equiv}CH + CO \xrightarrow[950 \text{ atm}]{90°, \, Co_2(CO)_8}
$$

with an excess of acetylene, the reaction produces a cis-trans mixture of a ten-carbon dilactone with four double bonds, in modest yield (527). The

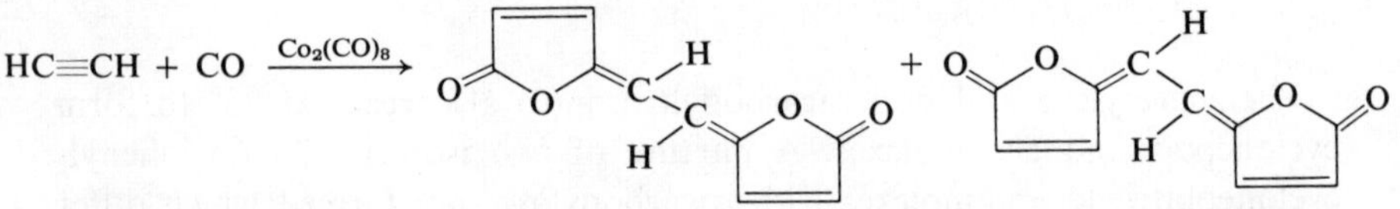

$$
HC{\equiv}CH + CO \xrightarrow{Co_2(CO)_8}
$$

mechanisms of these reactions probably involve as the first step conversion of hexacarbonylacetylenedicobalt into a maleoylbiscobalt derivative with carbon monoxide. The number of CO molecules absorbed in a stepwise manner could be from two to four with the three CO complex probably being required for the next step. This complex in the form with two acyl and seven coordinated carbonyl groups could undergo an intramolecular addition of one of the acylcobalt groups to the other acyl carbonyl and produce the γ-lactone structure. The two cobalt atoms are now on the same carbon atom of the lactone ring and could easily form a cobalt–cobalt bond with a bridged carbonyl group to give the dicobaltlactone intermediate. To produce the dilactone product, this intermediate must undergo a CO insertion, breaking the cobalt–cobalt bond again. A subsequent acetylene insertion and another CO insertion would form a new acylcobalt species with a β-carbonyl group which can cyclize again to form a bislactone derivative. The last complex then needs only to eliminate $Co_2(CO)_7$ in a cis or a trans manner to produce the two dilactone products. The $Co_2(CO)_7$ may then react again with acetylene (528) and start the catalytic cycle over or reversibly react with CO to form $Co_2(CO)_8$ (529) (Scheme IX-17).

The same type of reaction occurs with propyne, forming a dimethyl dilactone, where the additions to the triple bond occur in both instances to place the acyl group on the substituted acetylenic carbon (530).

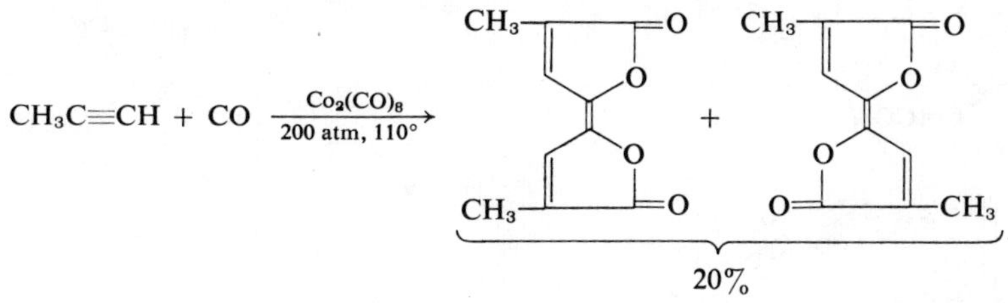

These reactions illustrate three general conclusions which can be drawn. (1) With comparable concentrations, additions of cobalt–carbon groups to carbon monoxide are faster than additions to acetylenes; (2) two carbon monoxide groups are never inserted next to each other; and (3) cyclization of acylcobalt complexes with β-carbonyl groups to form γ-lactone derivatives is a very favorable reaction. If this were not the case, alternating acetylene–carbon monoxide polymers would have been formed. These conclusions likely also apply to transition metals other than cobalt.

b. Carbonylation Reactions of Acetylenes in Hydroxylic Solvents

The acetylene–carbon monoxide reactions can be terminated by hydrolysis or alcoholysis of acyl intermediates if water or hydroxylic compounds are present in the reaction mixture. The octacarbonyldicobalt-catalyzed reaction of acetylene and carbon monoxide produces a 60% yield of succinic acid rather than the lactone products described above, if water is used as the solvent. Probably, the maleoylbiscobalt intermediate undergoes hydrolysis to maleic acid, which is then reduced to succinic acid by the hydridotetracarbonylcobalt(I) also formed in the hydrolysis. A minor amount of cyclopentanone is also produced. A π-cyclopentadienonecobalt intermediate from

$$Co_2(CO)_8 + HC{\equiv}CH \longrightarrow Co_2(CO)_6C_2H_2 \underset{}{\overset{4CO}{\rightleftharpoons}}$$

$$\left[(CO)_4Co\overset{O}{\overset{\|}{C}}CH{=}CH\overset{O}{\overset{\|}{C}}Co(CO)_4\right] \xrightarrow{H_2O} HOCOCH{=}CHCOOH + 2HCo(CO)_4 \longrightarrow$$

$$HOCOCH_2CH_2COOH + Co_2(CO)_8$$

two acetylenes and one carbon monoxide is probably being reduced by $HCo(CO)_4$ to form this product (531).

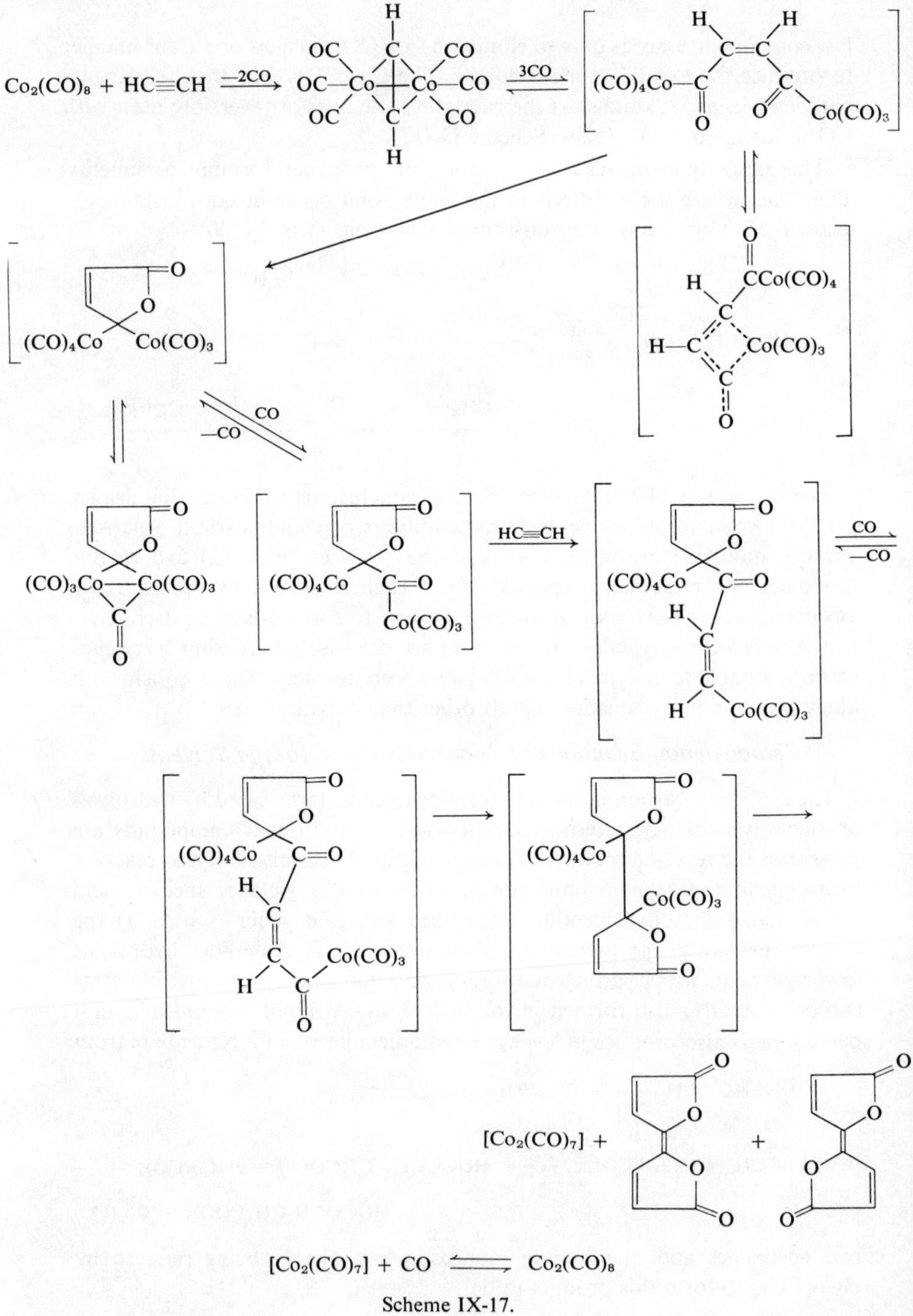

Scheme IX-17.

244

A much more complicated mixture of products is obtained when methanol is used instead of water. The major product is dimethyl succinate, but significant amounts of methyl acrylate, 2-cyclopentenone, dimethyl maleate, 1,1,2-tricarbomethoxyethane, and dimethyl 4-oxoheptanedioate are also produced (532). The succinate and maleate esters and the cyclopentenone are probably formed as in the hydrolysis reaction with alcoholysis replacing hydrolysis. An addition of hydridotetracarbonylcobalt to acetylene followed by carbonylation and methanolysis probably produces the methyl acrylate. The carbonylated hexacarbonylacetylenedicobalt intermediate may also react with a second acetylene and then undergo further carbonylation, methanolysis, and finally hydrogenation with $HCo(CO)_4$ to form the dimethyl 4-oxoheptanedioate. The 1,1,2-tricarbomethoxyethane may arise from an

$$\left[(CO)_4CoCCH{=}CHCCo(CO)_4 \right] + HC{\equiv}CH + CO \longrightarrow$$

$$(CO)_4CoCCH{=}CHCCH{=}CHCCo(CO)_4 \xrightarrow[-HCo(CO)_4]{CH_3OH}$$

$$CH_3OCOCH{=}CHCCH{=}CHCOOCH_3 \xrightarrow[-Co_2(CO)_8]{HCo(CO)_4} CH_3OCOCH_2CH_2CCH_2CH_2COOCH$$

$HCo(CO)_3$ addition to dimethyl maleate followed by carbonylation and methanolysis.

A synthesis of acrylic acid or esters occurs in much higher yield when acetylene, CO, and water or alcohols are reacted under pressure at elevated temperatures with a nickel iodide catalyst (533). Modifications of this reaction are still used for the commercial preparation of some acrylic compounds.

$$HC{\equiv}CH + CO + ROH \xrightarrow{NiI_2} CH_2{=}CHCOOR$$

Similar stoichiometric reactions occur under milder conditions when acetylenes and tetracarbonylnickel(0) are reacted in acetic acid solution. The carboxyl group adds to the substituted carbon of terminal acetylenes and cis addition of the hydride and carboxyl group is observed with disubstituted acetylenes (534, 535).

The mechanism of the nickel-catalyzed carboxylation very likely involves hydridonickel species which add to the acetylenes, undergo carbonylation, and then hydrolyze. The hydride in the nickel iodide case must be formed by reduction, whereas in the tetracarbonylnickel case it is probably formed by oxidative addition of acetic acid to coordinately unsaturated tricarbonylnickel(0).

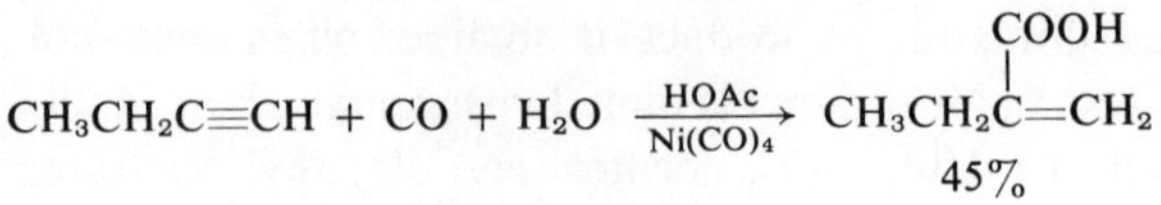

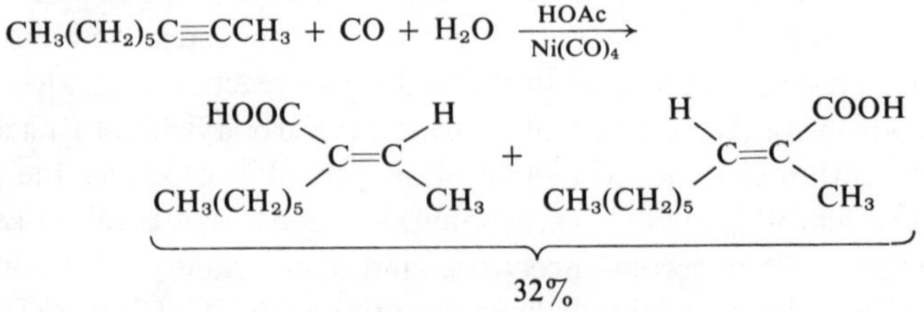

$$CH_3CH_2C{\equiv}CH + CO + H_2O \xrightarrow[Ni(CO)_4]{HOAc} \underset{45\%}{CH_3CH_2\overset{\overset{\textstyle COOH}{|}}{C}{=}CH_2}$$

$$CH_3(CH_2)_5C{\equiv}CCH_3 + CO + H_2O \xrightarrow[Ni(CO)_4]{HOAc}$$

$$32\%$$

$$NiI_2 + 3CO + 2CH_3OH \longrightarrow (CH_3)_2CO_3 + [HNi(CO)_2I] + HI$$

$$Ni(CO)_4 + HOAc \longrightarrow 2CO + [HNi(CO)_2OAc]$$

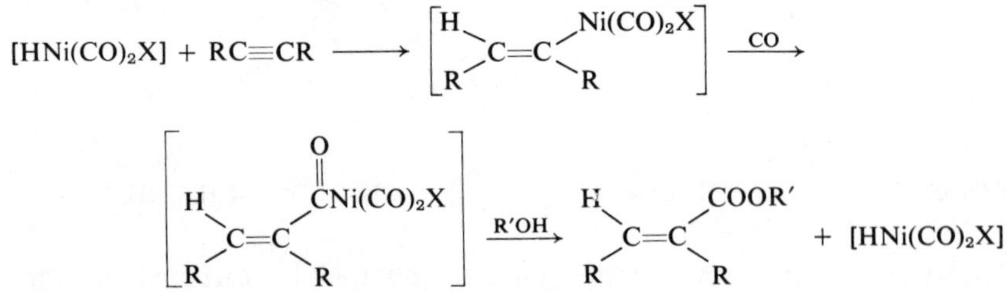

$$[HNi(CO)_2X] + RC{\equiv}CR \longrightarrow \left[\underset{R}{\overset{H}{}}C{=}C\underset{R}{\overset{Ni(CO)_2X}{}} \right] \xrightarrow{CO}$$

The use of palladium chloride as a catalyst in acetylene carbonylations in hydroxylic solvents yields maleic acid or ester derivatives. At atmospheric pressure and room temperature a mixture of acetylene and carbon monoxide reacts with methanolic palladium chloride to form dimethyl maleate along with smaller amounts of dimethyl fumarate and dimethyl muconate (mixture of isomers). The reaction is partially catalytic when thiourea is added as a ligand for the palladium and oxygen is passed into the reaction mixture (536). Some hydrogenation of the acetylene occurs also, which "reoxidizes"

$$HC{\equiv}CH + CO + CH_3OH \xrightarrow[NH_2CSNH_2]{PdCl_2, O_2}$$

$$CH_3OCOCH{=}CHCOOCH_3 + CH_3OCOCH{=}CHCH{=}CHCOOCH_3$$

part of the palladium. The formation of the muconate esters is an unusual example of the insertion of four unsaturated molecules in a series. The mechanism of the acetylene dicarbomethoxylation is apparently very similar to the olefin dicarboxyalkylation discussed in Section A,2,c of this chapter. Use of the same reaction conditions as for the olefins (25°–50° and 30 psi of carbon monoxide) allows various acetylenes to be dicarboxylated in a partially catalytic manner giving exclusively, or at least, predominantly *cis*-diesters (501).

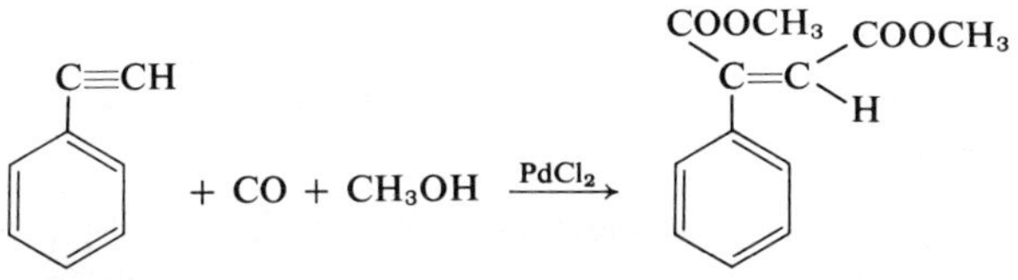

Propargyl bromide reacts with tetracarbonylnickel(0) in ethanol to form ethyl 3-bromo-3-butenoate (537). The mechanism may involve attack of tetracarbonylnickel on the triple bond with bromide displacement giving an allenylnickel complex. The bromide ion may then bond covalently to nickel with CO replacement or insertion, or insertion may occur without bromide participation. Ethanolysis of the acyl derivative would produce the allenic ester and hydrogen bromide. The hydrogen bromide then would add to the allenic ester forming the observed product.

$$HC\equiv CCH_2Br + Ni(CO)_4 \longrightarrow CO + (CO)_3\overset{+}{Ni}CH=C=CH_2 + Br^-$$

$$(CO)_3\overset{+}{Ni}CH=C=CH_2 + Br^- \rightleftharpoons (CO)_2\overset{+}{Ni}\overset{O}{\overset{\|}{C}}CH=C=CH_2 + Br^- \xrightarrow{EtOH}_{CO}$$

$$\overset{O}{\overset{\|}{EtOC}}CH=C=CH_2 + HBr + Ni(CO)_4$$

$$\overset{O}{\overset{\|}{EtOC}}CH=C=CH_2 + HBr \longrightarrow \overset{O}{\overset{\|}{EtOC}}CH_2\overset{Br}{\overset{|}{C}}=CH_2$$

c. Reactions of Acyl and Related Transition Metal Derivatives with Acetylenes and Carbon Monoxide

The cis addition of acyl transition metal complexes to acetylenes probably is general, although few clear examples are known. The reaction of pentacarbonylacetyl(or -methyl-)manganese(I) with phenylacetylene produces a 63% yield of an isolable adduct in which the acetyl group has added cis to the acetylene unit and the acetyl oxygen has coordinated with the manganese to fill the vacant position left by an initial CO dissociation. The initial dissociation is assumed necessary in order to allow the acetylenic group to coordinate before the addition can occur (358). An analogous reaction occurs with dicarbonylcyclopentadienylmethyliron(II) and 3-hexyne (538).

The reaction of tetracarbonylacetylcobalt(I) with 3-hexyne does not stop with the first addition, but a second CO inserts and a cyclization takes place. The cyclization apparently involves an addition of the acylcobalt group to

$$CH_3Mn(CO)_5 \rightleftharpoons [CH_3COMn(CO)_4] \underset{-CO}{\overset{CO}{\rightleftharpoons}} CH_3COMn(CO)_5$$

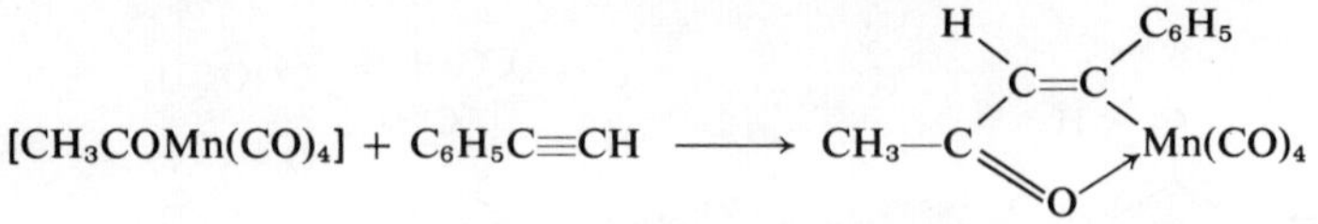

$$[CH_3COMn(CO)_4] + C_6H_5C{\equiv}CH \longrightarrow$$

the other ketone group to form a lactone complex with a π-allylic attachment to the cobalt (94b). The reaction seems to be analogous to the cyclization observed in the acetylene–CO reaction catalyzed by $Co_2(CO)_8$ (see Section A,4,a above). A catalytic application of this reaction to the preparation of

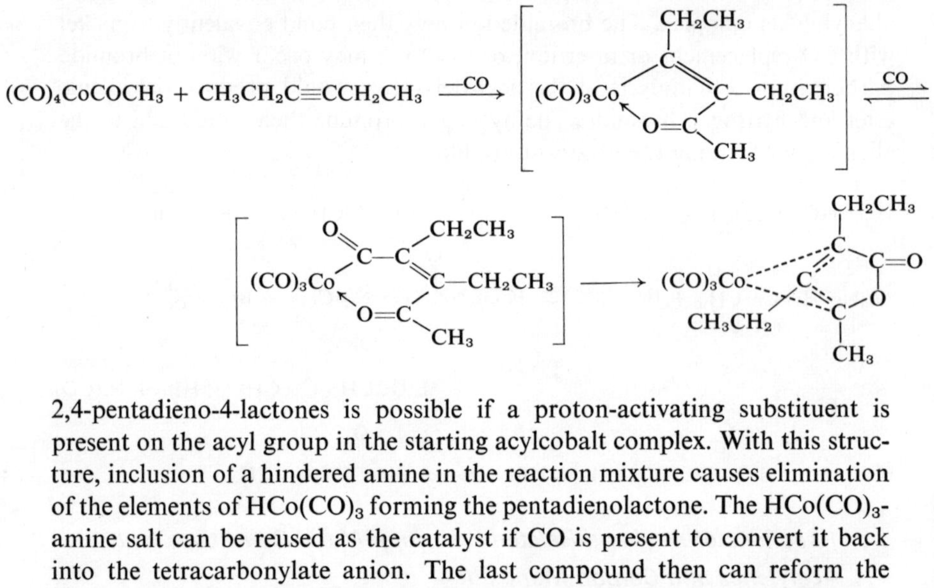

2,4-pentadieno-4-lactones is possible if a proton-activating substituent is present on the acyl group in the starting acylcobalt complex. With this structure, inclusion of a hindered amine in the reaction mixture causes elimination of the elements of $HCo(CO)_3$ forming the pentadienolactone. The $HCo(CO)_3$-amine salt can be reused as the catalyst if CO is present to convert it back into the tetracarbonylate anion. The last compound then can reform the starting acyl derivative by reaction with the appropriate halide (94b) (Scheme IX-18).

The cyclization of the acylcobalt derivatives to the lactone complex has an open chain analog. Tetracarbonylalkyl- or acylcobalt(I) complexes react readily with α,β-unsaturated aldehydes and ketones to form tricarbonyl-1-acyloxy-π-allylcobalt(I) derivatives (71).

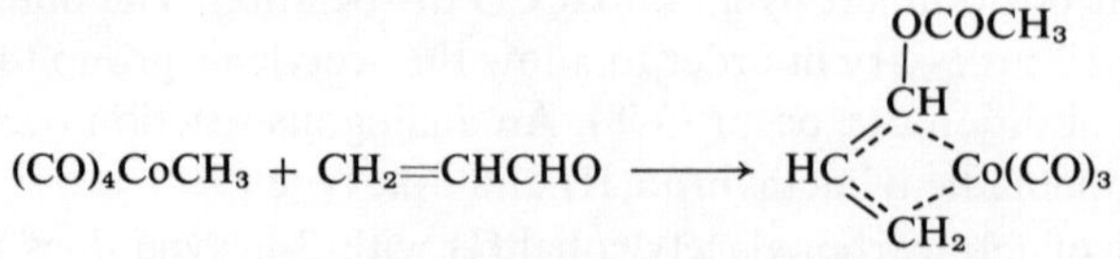

$$(CO)_4CoCH_3 + CH_2{=}CHCHO \longrightarrow$$

Acylnickel complexes also appear to add readily to acetylenes. The anionic acylnickel complex obtained from organolithium reagents and

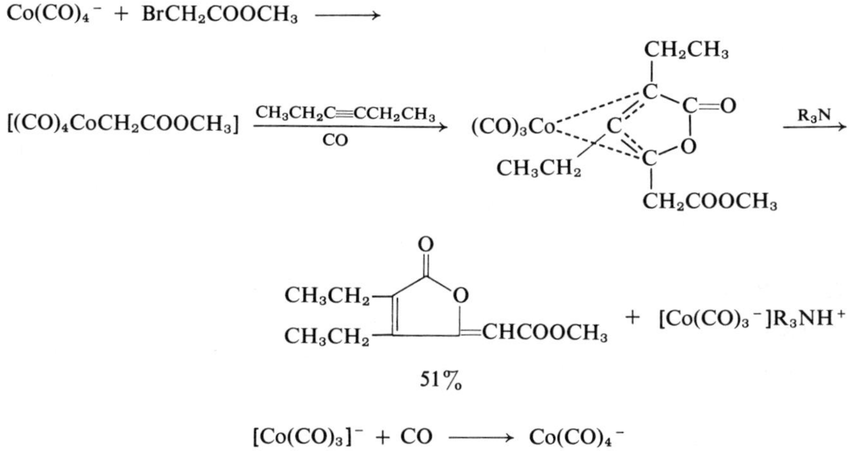

$$Co(CO)_4{}^- + BrCH_2COOCH_3 \longrightarrow$$

$$[(CO)_4CoCH_2COOCH_3] \xrightarrow[CO]{CH_3CH_2C\equiv CCH_2CH_3}$$

$$\xrightarrow{R_3N}$$

$$51\%$$

$$+ [Co(CO)_3{}^-]R_3NH^+$$

$$[Co(CO)_3]^- + CO \longrightarrow Co(CO)_4{}^-$$

Scheme IX-18.

tetracarbonylnickel(0) react with monosubstituted acetylenes to form two types of products: diacylated ethanes and 2,4-disubstituted-2-buteno-4-lactones (539). The mechanism of the reaction has not been investigated, but addition of two acyl groups to one acetylene must occur and there is a reduction which probably occurs at the time of hydrolysis on addition of aqueous acid. Although the details still require further investigation, a possible reaction sequence is given in Scheme IX-19.

An analogous reaction occurs when acetylenes are reacted with the anionic complex from tetracarbonylnickel(0) and lithium diethylamide with the difference that only the saturated dicarbonyl derivative is formed (540).

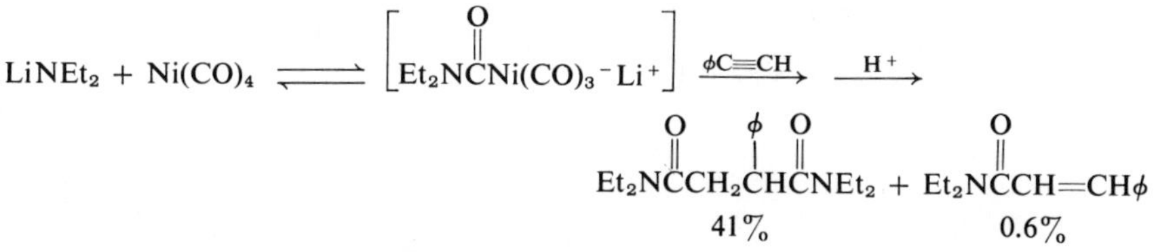

$$LiNEt_2 + Ni(CO)_4 \rightleftharpoons \left[Et_2NCNi(CO)_3{}^- Li^+ \right] \xrightarrow{\phi C\equiv CH} \xrightarrow{H^+}$$

$$\underset{41\%}{Et_2NCCH_2CHCNEt_2} + \underset{0.6\%}{Et_2NCCH=CH\phi}$$

The generation of acylnickel complexes from aryl iodides and tetracarbonyl-nickel also appears possible since iodobenzene, acetylene, CO, methanol, and tetracarbonylnickel(0) react to form methyl β-benzoylpropionate in high yield at 130° and 30 atm pressure (541). An oxidative addition of iodobenzene to nickel likely occurs first, followed by a phenyl shift to a coordinated carbonyl. The acyl complex then adds to acetylene, reacts again with CO, and finally alcoholyzes (Scheme IX-20).

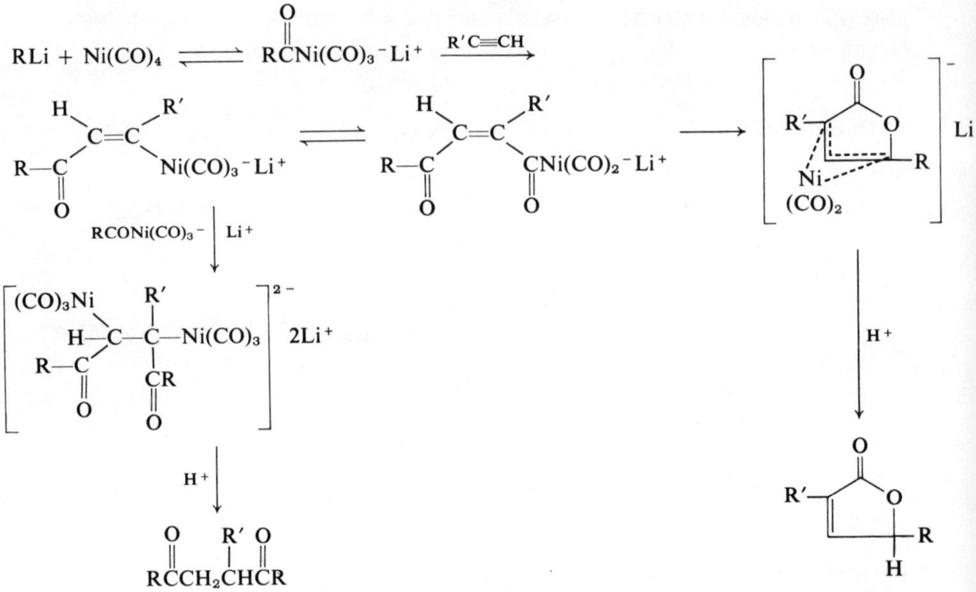

$$RLi + Ni(CO)_4 \rightleftharpoons R\overset{O}{\overset{\|}{C}}Ni(CO)_3{}^- Li^+ \xrightarrow{R'C\equiv CH}$$

Scheme IX-19.

Oxidative addition of acyl chlorides to tetracarbonylnickel(0) produces acylnickel complexes directly. Thus, acid chlorides, tetracarbonylnickel(0), and acetylene react in acetone solution to form 4-substituted-3-buteno-4-lactones and related derivatives with 2-isopropylidene groups arising from a reaction of an intermediate with the acetone solvent (542). The lactone is apparently formed analogously to the lactones in the anionic acylnickel carbonyl–acetylene reactions described above. The π-allylic intermediate

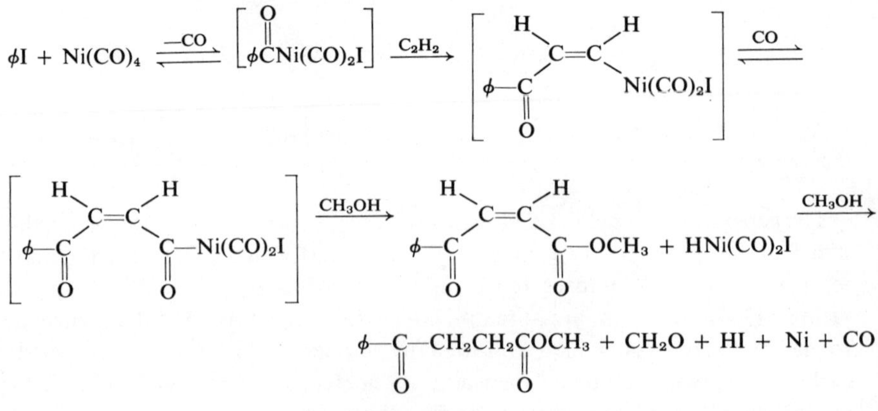

$$\phi I + Ni(CO)_4 \underset{-CO}{\rightleftharpoons} \left[\phi\overset{O}{\overset{\|}{C}}Ni(CO)_2 I \right] \xrightarrow{C_2H_2}$$

$$\xrightarrow{CH_3OH} \phi-\overset{O}{\underset{\|}{C}}-CH_2CH_2COCH_3 + CH_2O + HI + Ni + CO$$

$$\phi-\overset{O}{\underset{\|}{C}}-CH_2CH_2\overset{O}{\underset{\|}{C}}OCH_3 + CH_2O + HI + Ni + CO$$

Scheme IX-20.

must also react with acetone at the α-carbon in a manner analogous to the Grignard reaction. After hydrolysis the tertiary alcohol apparently dehydrates, since the isopropylidene product is isolated (see Scheme IX-21).

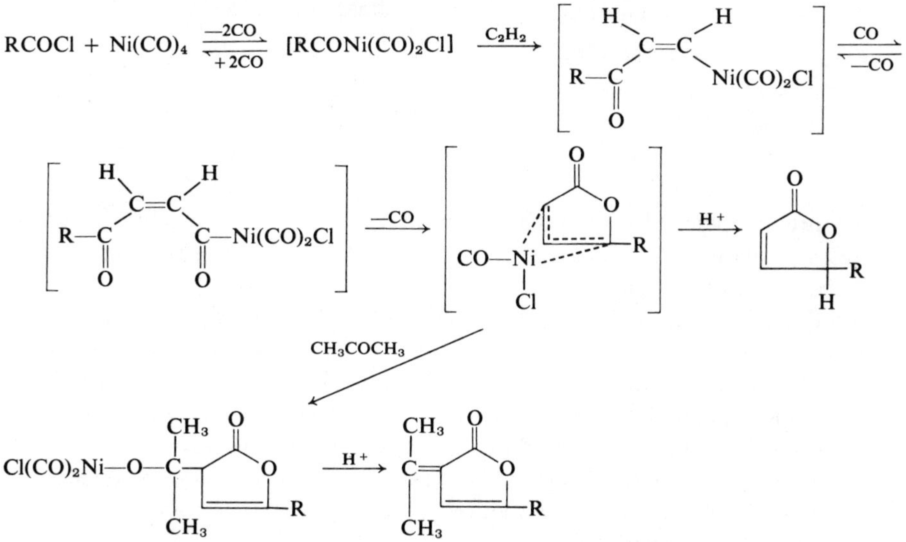

$$RCOCl + Ni(CO)_4 \underset{+2CO}{\overset{-2CO}{\rightleftharpoons}} [RCONi(CO)_2Cl] \overset{C_2H_2}{\longrightarrow}$$

Scheme IX-21.

The reaction of π-allylnickel halide dimers with acetone has been demonstrated independently with isolated bromo-π-1-carbomethoxyallylnickel(II) dimer. Two isomeric 2-hydroxy-2-propyl derivatives of methyl butenoate were formed. The major portion of the reaction took place at the α-carbon (542).

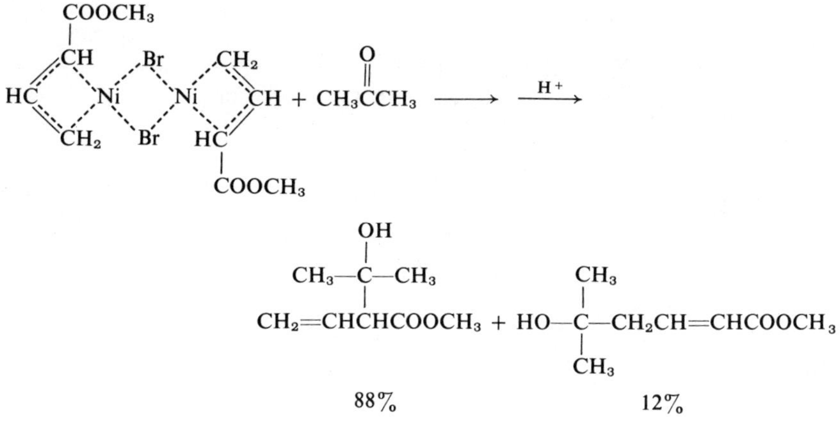

CH₂=CHCHCOOCH₃ + HO—C—CH₂CH=CHCOOCH₃

88%   12%

When a reaction similar to the above was carried out without acetone in THF solution with a continuous stream of acetylene to keep it in excess, a considerable amount of product consisted of the 6-substituted-3,5-hexadieno-6-lactone (543). Two consecutive acetylene additions apparently occurred before the addition to CO and the final cyclization.

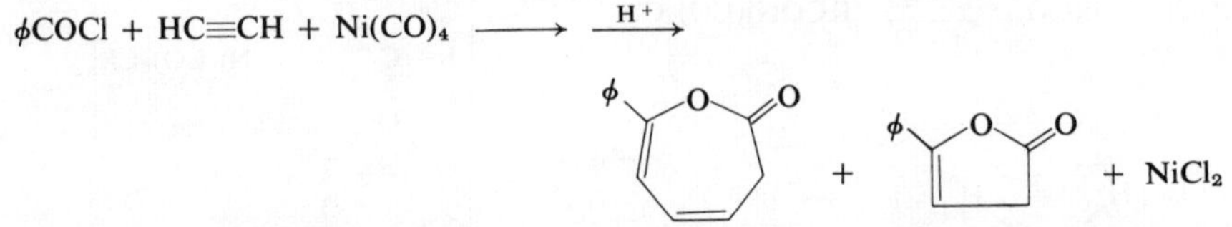

$$\phi COCl + HC\equiv CH + Ni(CO)_4 \longrightarrow \xrightarrow{\text{H}^+}$$

More complicated reactions were found when acrolein was added to the reaction mixture of an acid chloride, acetylene, and tetracarbonylnickel(0) (544). The intermediate acylnickel species appears to add to the acrolein in the same manner that tetracarbonylacylcobalt(I) complexes do and likely forms chlorocarbonyl-1-acyl-π-allylnickel(II) complexes. The π-allyl group may then revert to the σ form and add to the acetylene, and then to CO. The resulting complex is a 2,5-dienoylnickel derivative which can undergo an internal addition and produce a cyclopentenone derivative. The last complex may eliminate a hydridonickel group to form the acyloxymethylene-cyclopentenone product or be reduced by a hydridonickel complex to the acyloxymethylcyclopentenone product (Scheme IX-22).

d. Reactions of Allylic π Complexes with Acetylenes and Carbon Monoxide

Bromo-π-allylnickel(II) dimer reacts with carbon monoxide in ether solution at 0° to form 3-butenoyl bromide and tetracarbonylnickel(0), whereas if acetylene is present cis-2,5-hexadienoyl bromide and tetracarbonylnickel(0) are formed. One acetylene inserts faster than the CO does (545).

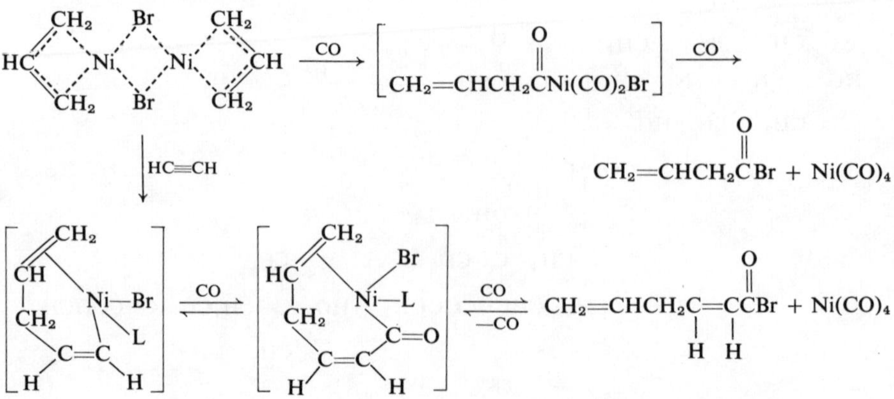

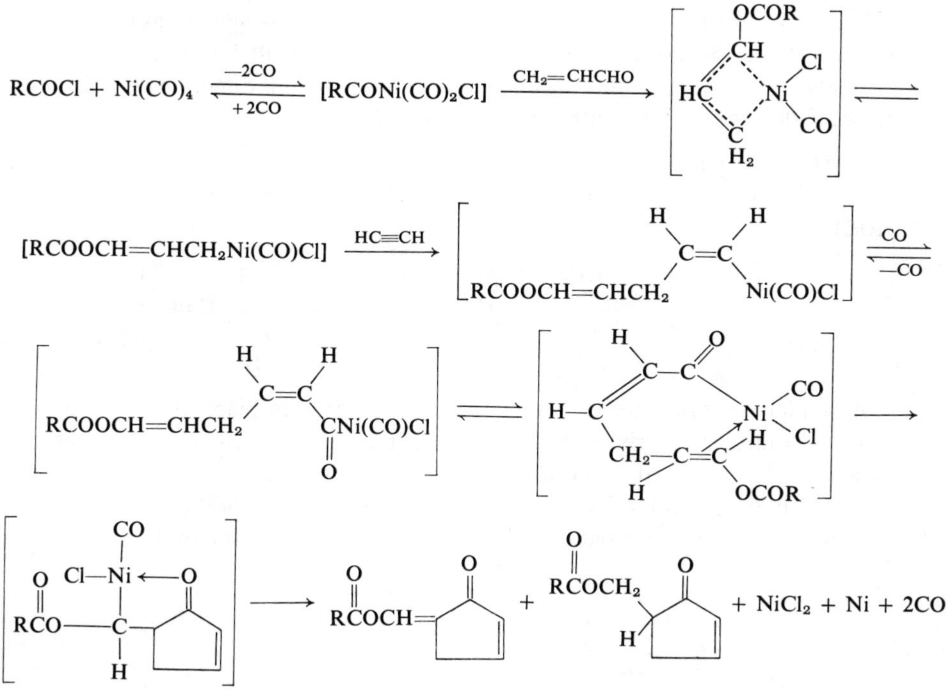

$$RCOCl + Ni(CO)_4 \underset{+2CO}{\overset{-2CO}{\rightleftarrows}} [RCONi(CO)_2Cl] \xrightarrow{CH_2=CHCHO}$$

$$\begin{bmatrix} \overset{OCOR}{\underset{}{}} \\ CH \\ HC \quad Cl \\ \| \quad Ni \\ C \quad CO \\ H_2 \end{bmatrix} \rightleftarrows$$

$$[RCOOCH=CHCH_2Ni(CO)Cl] \xrightarrow{HC\equiv CH} \begin{bmatrix} H \quad H \\ C=C \\ RCOOCH=CHCH_2 \quad Ni(CO)Cl \end{bmatrix} \overset{CO}{\underset{-CO}{\rightleftarrows}}$$

$$\begin{bmatrix} H \quad H \\ C=C \\ RCOOCH=CHCH_2 \quad CNi(CO)Cl \\ \| \\ O \end{bmatrix} \rightleftarrows \begin{bmatrix} H \quad O \\ C-C \\ H-C \quad Ni \quad CO \\ CH_2-C=C \quad H \quad Cl \\ H \quad OCOR \end{bmatrix} \longrightarrow$$

$$\begin{bmatrix} CO \\ O \quad Cl-Ni\leftarrow O \\ \| \\ RCO-C \\ H \end{bmatrix} \longrightarrow RCOCH= + \underset{H}{RCOCH_2} + NiCl_2 + Ni + 2CO$$

Scheme IX-22.

The faster rate of acetylene addition may be specific to the π-allylnickel system since apparently phenylnickel reacts more rapidly with CO (see Section A,4,c of this chapter).

The above reactions are probably involved in a reported catalytic synthesis of *cis*-methyl 2,5-hexadienoate from allyl chloride, acetylene, carbon monoxide, methanol, a manganese–iron alloy (reducing agent), and magnesium oxide (to neutralize the hydrogen chloride). The chloro-π-allylnickel dimer probably was generated in the reaction mixture from nickel chloride, the reducing agent (the manganese–iron alloy), and allyl chloride. Thiourea was also added as a ligand to solubilize nickel(0) (546).

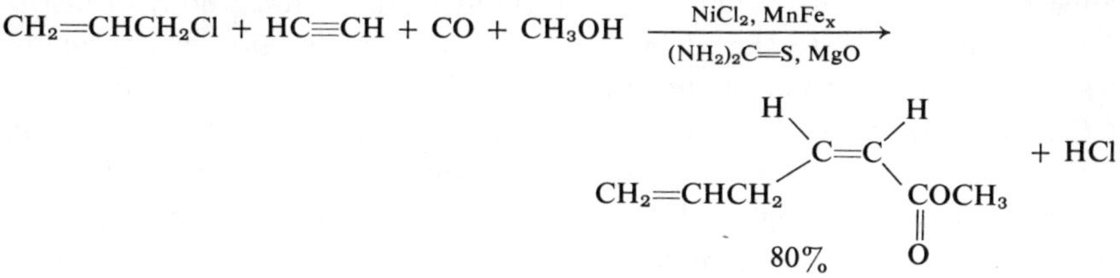

$$CH_2=CHCH_2Cl + HC\equiv CH + CO + CH_3OH \xrightarrow[(NH_2)_2C=S, MgO]{NiCl_2, MnFe_x}$$

$$\begin{array}{c} H \quad H \\ C=C \\ CH_2=CHCH_2 \quad COCH_3 \\ \| \\ O \end{array} + HCl$$

80%

In another example of this reaction, *trans*-1,4-dichloro-2-butene was formed from HCl and the diol in the reaction mixture and both allylic halide groups were added to acetylene and carbonylated. A 56% yield of dimethyl *cis,trans, cis*-2,5,8-decatrien-1,10-dioate was obtained (547).

$$
\begin{array}{c}
\text{H} \qquad \text{CH}_2\text{OH} \\
\diagdown \qquad \diagup \\
\text{C}{=}\text{C} \qquad + \text{ HCl } + \text{ CO } + \text{ CH}_3\text{OH } + \text{ HC}{\equiv}\text{CH } + \text{ Ni(CO)}_4 \longrightarrow \\
\diagup \qquad \diagdown \\
\text{HOCH}_2 \qquad \text{H}
\end{array}
$$

$$
\begin{array}{c}
\text{H}_2 \\
\text{CH}_3\text{OOC} \qquad \text{C} \qquad\qquad \text{H H} \qquad\qquad \text{H} \\
\diagdown \qquad \diagup \diagdown \qquad \diagup \diagdown \qquad \diagup \\
\text{C}{=}\text{C} \qquad \text{C}{=}\text{C} \qquad \text{C}{=}\text{C} \\
\diagup \qquad \diagdown \diagup \qquad \diagdown \qquad \diagdown \\
\text{H} \qquad\quad \text{H H} \qquad\qquad \text{C} \qquad\quad \text{COOCH}_3 \\
\qquad\qquad\qquad\qquad\qquad \text{H}_2
\end{array}
$$

A similar reaction carried out with phenylacetylene in place of acetylene, allyl chloride, tetracarbonylnickel and with aqueous acetone as solvent produced four related multiinsertion products (4). The major product, obtained in 64% yield, arises from a series of five specific addition reactions starting with chloro-π-allylnickel(II) dimer. The mechanism of formation

$$
C_6H_5C{\equiv}CH + CH_2{=}CHCH_2Cl + H_2O + Ni(CO)_4 \xrightarrow[20°]{\text{acetone}}
$$

of this and the other three complex products are now easily proposed on the basis of the various other nickel-promoted reactions already discussed. Initially, chloro-π-allylnickel dimer or a solvated monomeric form of it appears to be formed from the allyl chloride and tetracarbonylnickel(0). This complex adds to the phenylacetylene, placing the allyl group on the least-substituted carbon of the triple bond. This adduct then must add to CO and cyclize by addition of the acylnickel group to the terminal double bond. The cyclic product is mainly the five-membered ring derivative, but some six-membered ring product is also formed. The latter product appears to add to CO and undergo hydrolysis to form the third product listed above.

The five-ring intermediate also undergoes an addition to CO and some hydrolysis to form the second product listed. Only part of this complex is hydrolyzed; the rest adds to phenylacetylene. The adduct is probably cis and again has the organic (acyl) group on the least-substituted carbon of the acetylene. This reactive 2-acylvinylnickel complex then apparently may do two things: either react with HCl present in solution to form the major product isolated or react once more with CO, cyclize to a lactone π-allylnickel complex, and then react with HCl to form the last product of the group (Scheme IX-23).

5. Carbonylation Reactions of Epoxides and Trimethylene Oxides

a. Epoxides

Several cobalt-catalyzed carbonylation reactions of epoxides have been reported. All appear to involve either hydridotetracarbonylcobalt(I) or tetracarbonylcobaltate anion first opening the epoxide ring, forming carbon–cobalt σ bonds which then undergo CO insertion and various final steps to regenerate the catalyst. The basic steps of the hydride reactions have been demonstrated by isolation of intermediate cobalt complexes.

Hydridotetracarbonylcobalt(I) reacts rapidly with propylene oxide (and other epoxides) at 0° under 1 atm of carbon monoxide to form the 3-hydroxy-butanoylcobalt complex. The adduct has been isolated as the monotriphenyl-phosphine derivative (548). The direction of opening of the oxide is unusual and probably occurs the way it does for steric reasons. The large tetracarbonyl-

$$HCo(CO)_4 + CH_3CH\overset{O}{\overbrace{\hspace{1.2em}}}CH_2 + CO \xrightarrow{\ 0°\ } CH_3\overset{OH}{\underset{|}{C}}HCH_2COCo(CO)_4 \xrightarrow[-CO]{\ P\phi_3\ }$$

$$CH_3\overset{OH}{\underset{|}{C}}HCH_2COCo(CO)_3P\phi_3$$

cobaltate anion in the protonated epoxide–tetracarbonylcobaltate anion pair intermediate cannot attack the secondary carbon as easily as the primary one. The addition appears to be ionic since cyclohexene oxide yields only the *trans*-2-hydroxy-1-acylcobalt complex on reaction with $HCo(CO)_4$ and CO (548).

The tetracarbonylcobaltate anion also is able to open epoxides, but higher temperatures than for the hydride reaction are required. In inert solvents polymers are obtained, but in alcoholic solutions under CO, 3-hydroxy esters are formed in moderate yields catalytically at 50° (548, 549). The anionic opening of propylene oxide in methanol solution with CO and tetracarbonyl-cobaltate anion gives methyl 3-hydroxybutyrate (548). Thus, both $HCo(CO)_4$ and $Co(CO)_4{}^-$ open epoxides in the same direction.

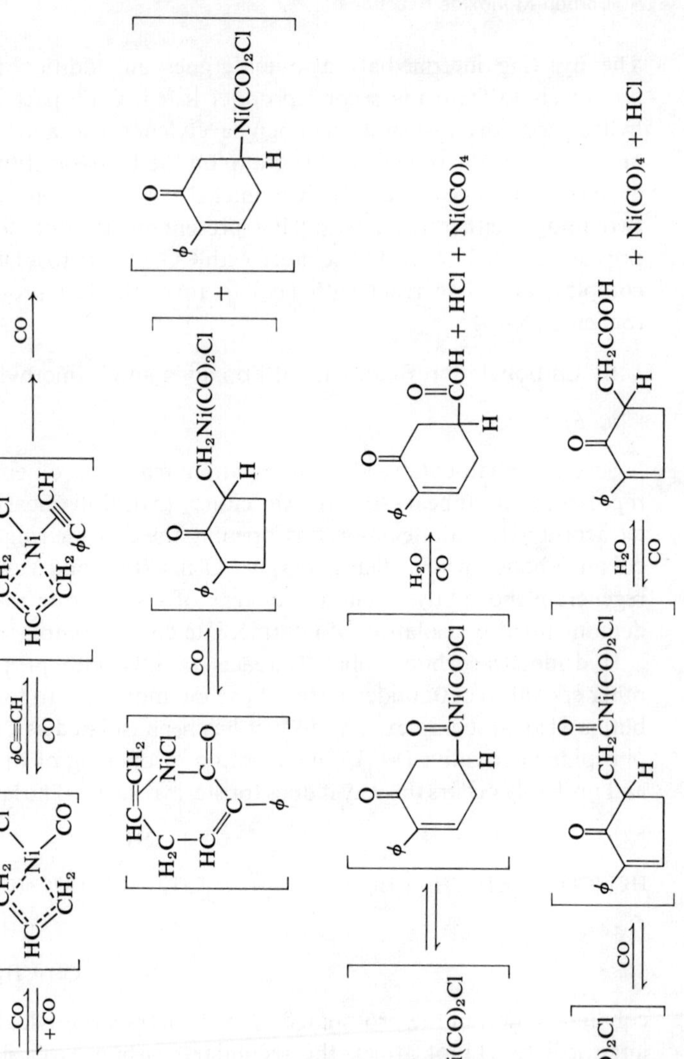

$CH_2=CHCH_2Cl + Ni(CO)_4$

$\dfrac{-CO}{+CO}$

$\phi C\equiv CH \Big/ CO$

CO

CO

$\dfrac{H_2O}{CO}$

$\dfrac{H_2O}{CO}$

CO

$+$

$COOH + HCl + Ni(CO)_4$

$CH_2COOH + Ni(CO)_4 + HCl$

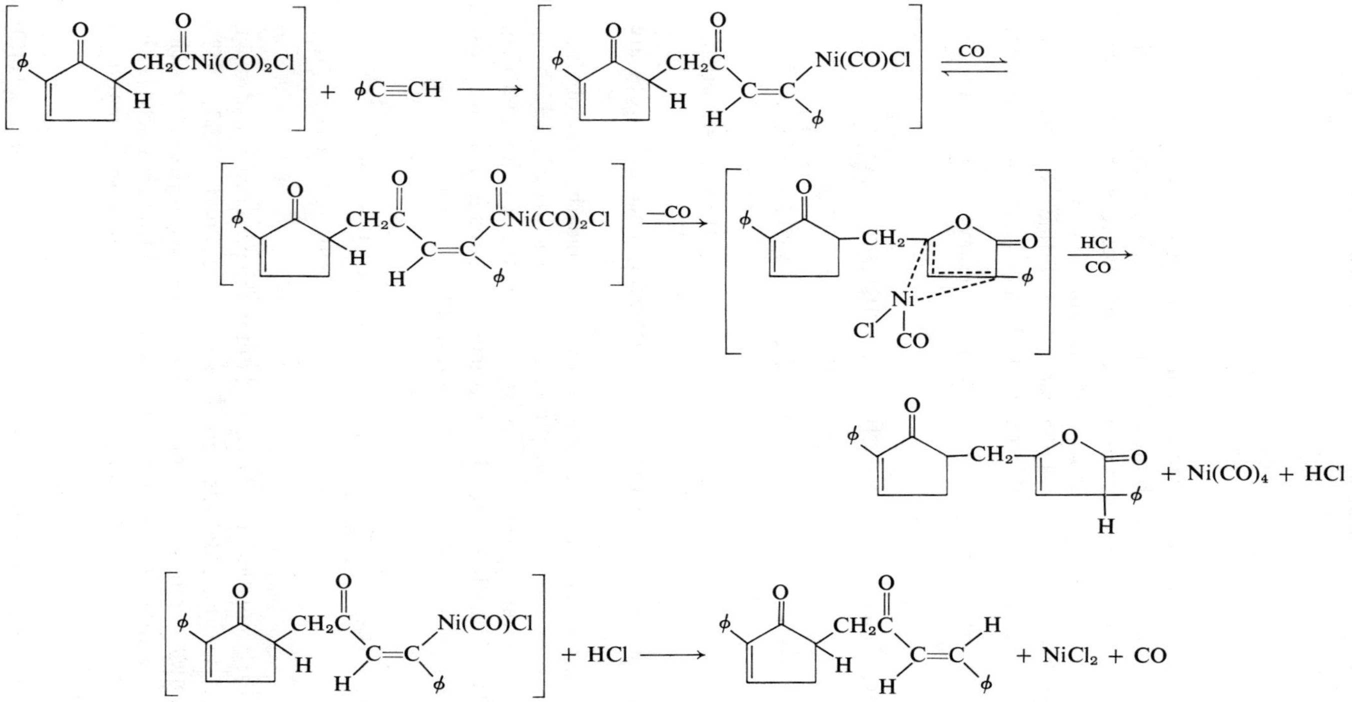

Scheme IX-23.

257

$$\underset{H_2C-CH_2}{\overset{O}{\triangle}} + CO + CH_3OH \xrightarrow[65°, 2000 \text{ psi}]{Co(CO)_4} HOCH_2CH_2COOCH_3 + CH_3CHO$$

$$55\%$$

Tetracarbonylcobaltate anion causes some rearrangement of epoxides to carbonyl compounds in the carbonylation reaction and thereby lowers the yield of product. The rearrangement is the major reaction for epoxides reacting with octacarbonyldicobalt in methanol solution in the absence of CO (550). Probably, tetracarbonylcobaltate anion, formed by disproportionation, is the true catalyst in this reaction. This isomerization contrasts

$$3Co_2(CO)_8 + 12CH_3OH \rightleftharpoons 2Co(CH_3OH)_6{}^{2+}(Co(CO)_4{}^-)_2 + 8CO$$

$$\underset{CH_3CH_2CH-CH_2}{\overset{O}{\triangle}} + Co(CO)_4{}^- \rightleftharpoons CH_3CH_2\underset{\overset{|}{O^-}}{CH}CH_2Co(CO)_4 \xrightarrow[CH_3O^-]{CH_3OH}$$

$$CH_3CH_2\underset{\overset{|}{OH}}{CH}CH_2Co(CO)_4 \rightleftharpoons CH_3CH_2\underset{\overset{|}{OH}}{C}=CH_2 + HCo(CO)_4 \xrightarrow{CH_3O^-}$$

$$CH_3CH_2\overset{\overset{O}{\|}}{C}CH_3 + Co(CO)_4{}^-$$

$$77\%$$

with the usual acid-catalyzed epoxide rearrangement, since aldehydes are produced rather than ketones.

The catalytic carbonylation of ethylene oxide with octacarbonyldicobalt as catalyst under vigorous conditions will produce acrylic acid in 81% yield (551). A small amount of water appears necessary to start the reaction and then dehydration of the initially formed hydroxyacid maintains it. It is not clear as yet whether $HCo(CO)_4$ or $Co(CO)_4{}^-$ is the catalyst.

$$\underset{H_2C-CH_2}{\overset{O}{\triangle}} + CO + H_2O \xrightarrow[160°, 6000 \text{ psi}]{Co_2(CO)_8} [HOCH_2CH_2COOH] \longrightarrow$$

$$CH_2=CH COOH + H_2O$$

$$81\%$$

The epoxide carbonylation may be carried out in the presence of hydrogen, in which case 3-hydroxyaldehydes are formed in moderate yields (552). Hydrogen doubtlessly reduces the intermediate 3-hydroxyacylcobalt species to aldehydes. The ease of condensation of hydroxyaldehydes likely is the reason that yields are not higher. The true catalyst in this transformation is probably $HCo(CO)_4$.

One reaction of a nickel complex with epoxides has been reported. The anionic complex obtained from phenyllithium and tetracarbonylnickel(0) reacts with styrene oxide producing, after treatment with acid, three products:

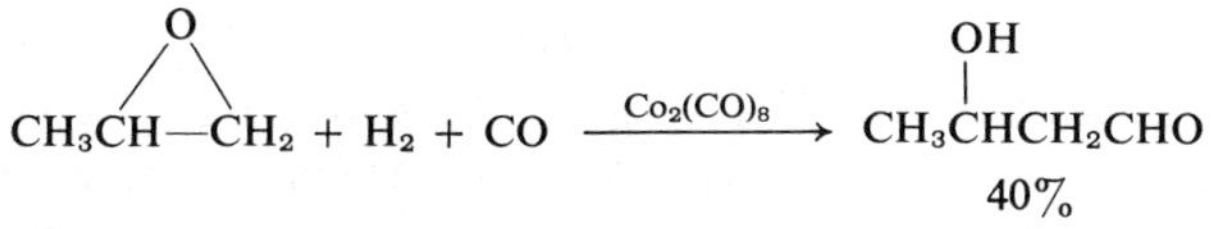

$$CH_3CH\overset{O}{\overset{\triangle}{-}}CH_2 + H_2 + CO \xrightarrow{Co_2(CO)_8} CH_3\overset{OH}{\underset{|}{C}}HCH_2CHO$$
$$40\%$$

2,3-diphenyl-4-butyrolactone (19%), benzoin (24%), and benzoic acid (9%) (553). The formation of the lactone is an unusual reaction and should be investigated further since its mechanism of formation is not obvious.

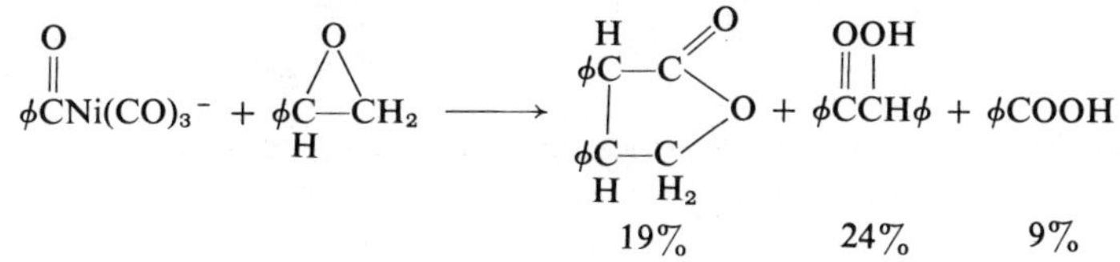

$$\phi C\overset{O}{\overset{||}{N}}i(CO)_3^- + \phi C\overset{O}{\overset{\triangle}{-}}CH_2 \longrightarrow \begin{matrix} H \\ \phi C-C \\ | \\ \phi C-C \\ H \quad H_2 \end{matrix} \overset{O}{\diagup}O + \phi C\overset{OOH}{\overset{||\,|}{C}}H\phi + \phi COOH$$
$$19\% \qquad 24\% \qquad 9\%$$

b. Trimethylene Oxides

Hydridotetracarbonylcobalt(I) reacts with trimethylene oxide and carbon monoxide at 0° to form tetracarbonyl-4-hydroxybutanoylcobalt(I) in good yield. The complex has been isolated as the triphenylphosphine derivative (548). The 4-hydroxybutyryl group cyclizes to 4-butyrolactone on treatment with a hindered amine.

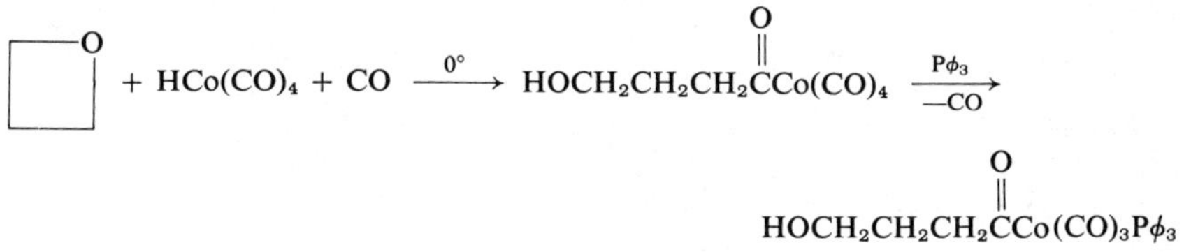

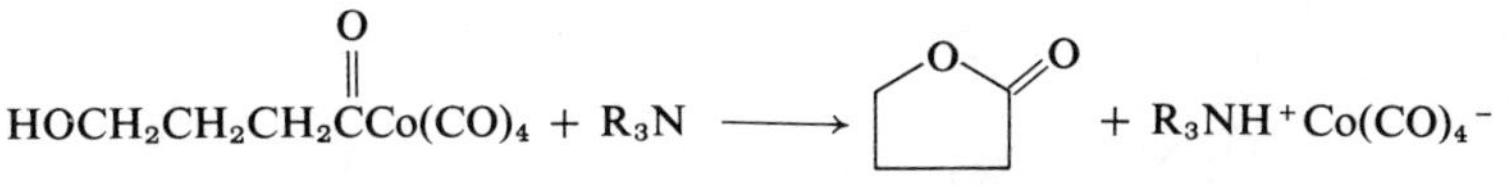

$$\overset{O}{\square} + HCo(CO)_4 + CO \xrightarrow{0°} HOCH_2CH_2CH_2\overset{O}{\overset{||}{C}}Co(CO)_4 \xrightarrow[-CO]{P\phi_3}$$

$$HOCH_2CH_2CH_2\overset{O}{\overset{||}{C}}Co(CO)_3P\phi_3$$

$$HOCH_2CH_2CH_2\overset{O}{\overset{||}{C}}Co(CO)_4 + R_3N \longrightarrow \underset{}{\square}\overset{O}{\diagdown}\overset{O}{\diagup} + R_3NH^+Co(CO)_4^-$$

The ring closure of 4- or 5-hydroxyacylcobalt complexes is probably general. 5-Hexenolactone was prepared similarly in a potentially catalytic reaction from cis-4-chloro-2-buten-1-ol, CO, and a hindered amine with tetracarbonylcobaltate anion as catalyst (548). This particular example was only partially catalytic, however, because unreactive tricarbonyl-π-1-hydroxy-methylallylcobalt was obtained in a side reaction.

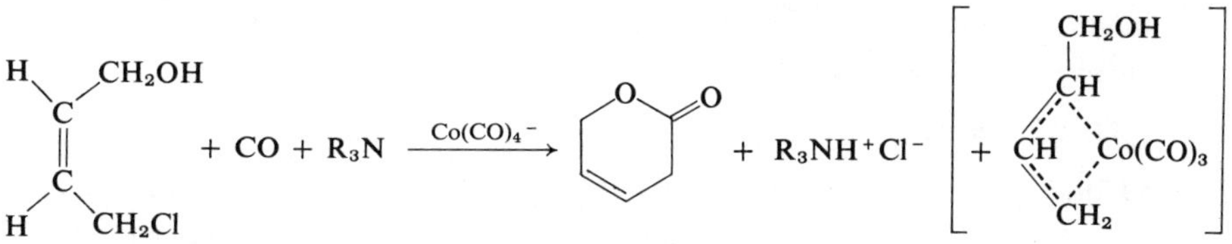

$$\begin{matrix} H & CH_2OH \\ \diagdown & \diagup \\ & C \\ & || \\ & C \\ \diagup & \diagdown \\ H & CH_2Cl \end{matrix} + CO + R_3N \xrightarrow{Co(CO)_4^-} \underset{}{\diagup O\diagdown}\overset{O}{\diagdown} + R_3NH^+Cl^- + \begin{bmatrix} CH_2OH \\ | \\ CH \\ \diagup \vdots \diagdown \\ CH \quad Co(CO)_3 \\ \diagdown \vdots \diagup \\ CH_2 \end{bmatrix}$$

3,3-Dimethyltrimethylene oxide has been catalytically carbonylated to 3,3-dimethyl-4-butyrolactone in 80% yield with cobalt(II) acetylacetonate as catalyst at 200° and 250 atm pressure (554). Presumably a small amount of HCo(CO)₄ is formed in the reaction mixture and it is the true catalyst.

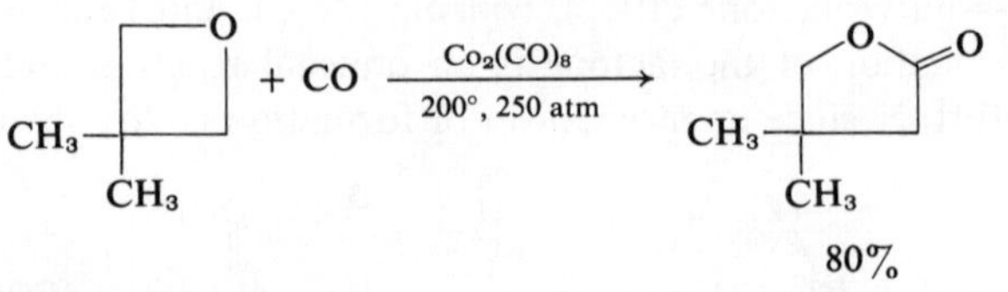

Tetracarbonylcobaltate anion does not open trimethylene oxide under mild conditions (548).

6. Carbonylation of Cyclopropanes and Cyclobutanes

a. Cyclopropanes

Cyclopropanes undergo oxidative additions with some transition metal complexes forming metalacyclobutane derivatives. As an example, dichloro-π-ethyleneplatinum(II) dimer reacts with phenylcyclopropane inserting the metal between the two unsubstituted cyclopropane carbons. The cyclopropane may be regenerated from the complex by reacting it with cyanide (555).

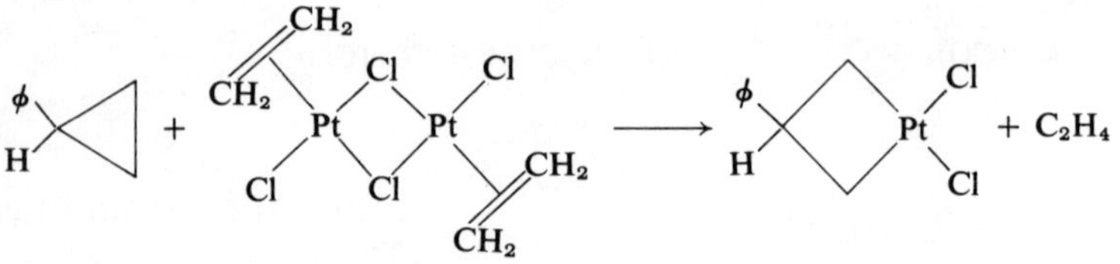

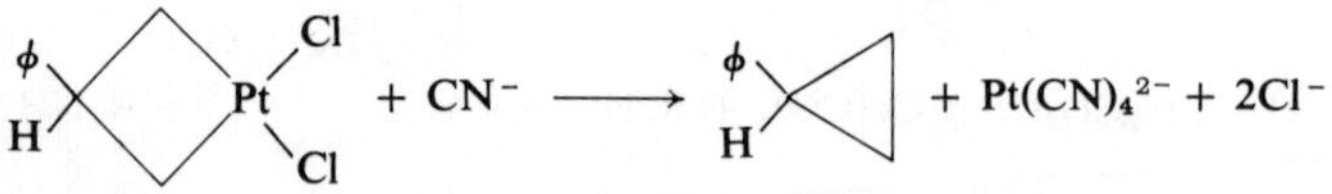

Phenylcyclopropane and chlorodicarbonylrhodium(I) dimer react forming a rhodiacyclopentanone complex. The oxidative addition in this case must have occurred between a substituted and unsubstituted carbon of the cyclopropane ring, in contrast to the preceding example, and then CO insertion followed. The CO apparently added to the primary carbon atom since reduction of the complex with NaBH₄ gave 4-phenyl-1-butanol (556).

Benzylcyclopropane in the same reaction undergoes opening between the unsubstituted carbons of the cyclopropane ring (556).

b. Cyclobutanes

Methylenecyclobutane reacts with palladium(II) chloride and with chlorodicarbonylrhodium(I) dimer with opening of the cyclobutane ring at an allylic carbon, forming π-allylic metal complexes. A second isomeric π-allylic complex is formed with rearrangement in the $PdCl_2$ reaction probably by way of a metal hydride shift from an alkyl formed by opening of the cyclobutane ring in the reverse direction from the reaction which gave the first isomer (557).

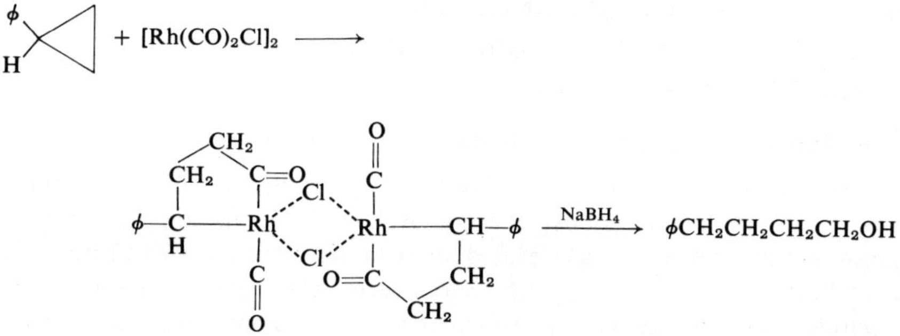

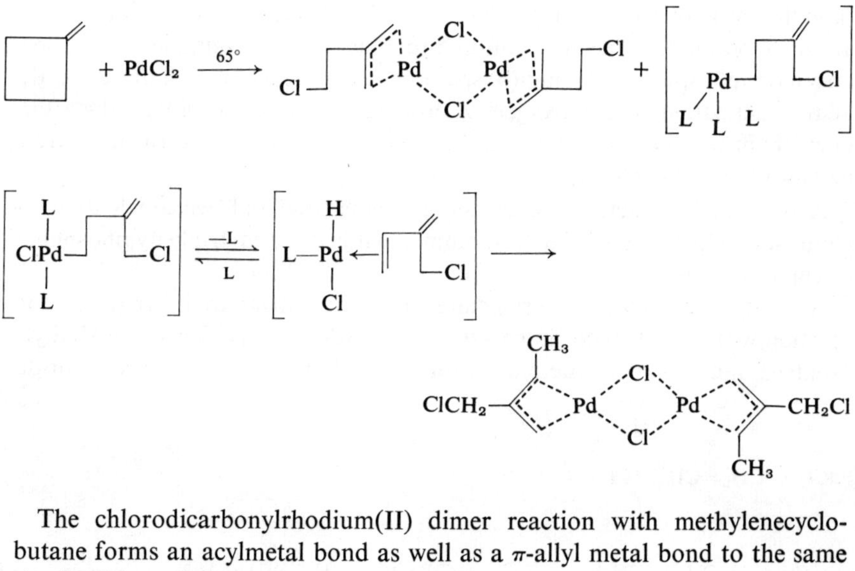

The chlorodicarbonylrhodium(II) dimer reaction with methylenecyclobutane forms an acylmetal bond as well as a π-allyl metal bond to the same rhodium atom (557).

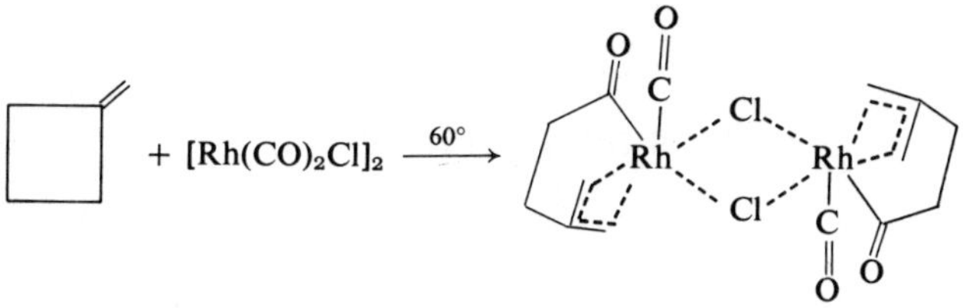

B. REACTIONS OF ORGANOTRANSITION METAL COMPLEXES WITH ORGANIC CARBONYL COMPOUNDS

1. Addition Reactions with Aldehydes and Ketones

The reaction of bromo-π-1-carbomethoxyallylnickel(0) dimer with acetone has already been noted above (Section A,4,c). Palladium appears able to promote similar reactions at elevated temperatures. The reaction of isolated π-allylic palladium complexes with organic carbonyl compounds has not been reported, but this reaction appears to occur when conjugated dienes and aldehydes are combined with a palladium acetylacetonate–triphenylphosphine catalyst. In this reaction two diene units are joined to one aldehyde group. A cyclic divinylpyran and an octatrienylcarbinol are obtained as products (558). The ratio of the two products formed depends on the phosphorus-to-palladium ratio used. With acetaldehyde and butadiene as reactants a 1:4 P to Pd ratio yields mainly the divinylpyran product, whereas with a 1:1 ratio the carbinol is the main product (558). Small amounts of 1,3,7-octatriene are also produced. The mechanism probably involves butadiene coupling on a monophosphinepalladium(0) species and then reaction with the aldehyde adding palladium to the oxygen. A hydrogen transfer would produce the open-chain product, whereas a reductive elimination could form the pyran derivative (Scheme IX-24).

A very similar reaction occurs between butadiene and benzaldehyde with chloro-π-allylpalladium dimer, sodium phenoxide, and triphenylphosphine as catalyst (559).

A π-allylic palladium intermediate does not appear to be required for reaction with an aldehyde, at least if the aldehyde is formaldehyde. 3-Methyl-1-butene and formaldehyde react with a palladium chloride–cupric chloride

$$
\text{PdCl}_2 + \underset{\begin{array}{c}\text{CH}_3\\|\end{array}}{\text{CH}_2=\text{CHCHCH}_3} \;\overset{L}{\underset{}{\rightleftharpoons}}
$$

$$
\underset{\substack{|\\L}}{\overset{\substack{L\quad\;\;\;Cl\\|\quad\;\;\;|}}{\text{ClPdCH}_2\text{CHCH(CH}_3)_2}} \;\xrightarrow{\text{H}_2\text{C}=\text{O}}\; \underset{\substack{|\\L}}{\overset{\substack{L\quad\qquad\;Cl\\|\qquad\quad\;\;|}}{\text{ClPdOCH}_2\text{CH}_2\text{CHCH(CH}_3)_2}} \;\xrightarrow{\text{H}_2\text{C}=\text{O}}
$$

$$
\underset{\substack{|\\L}}{\overset{\substack{L\qquad\qquad\qquad\;Cl\\|\qquad\qquad\qquad\;|}}{\text{ClPdOCH}_2\text{OCH}_2\text{CH}_2\text{CHCH(CH}_3)_2}} \;\xrightarrow{-L}\; \text{[pyran structure]} + \text{PdCl}_2
$$

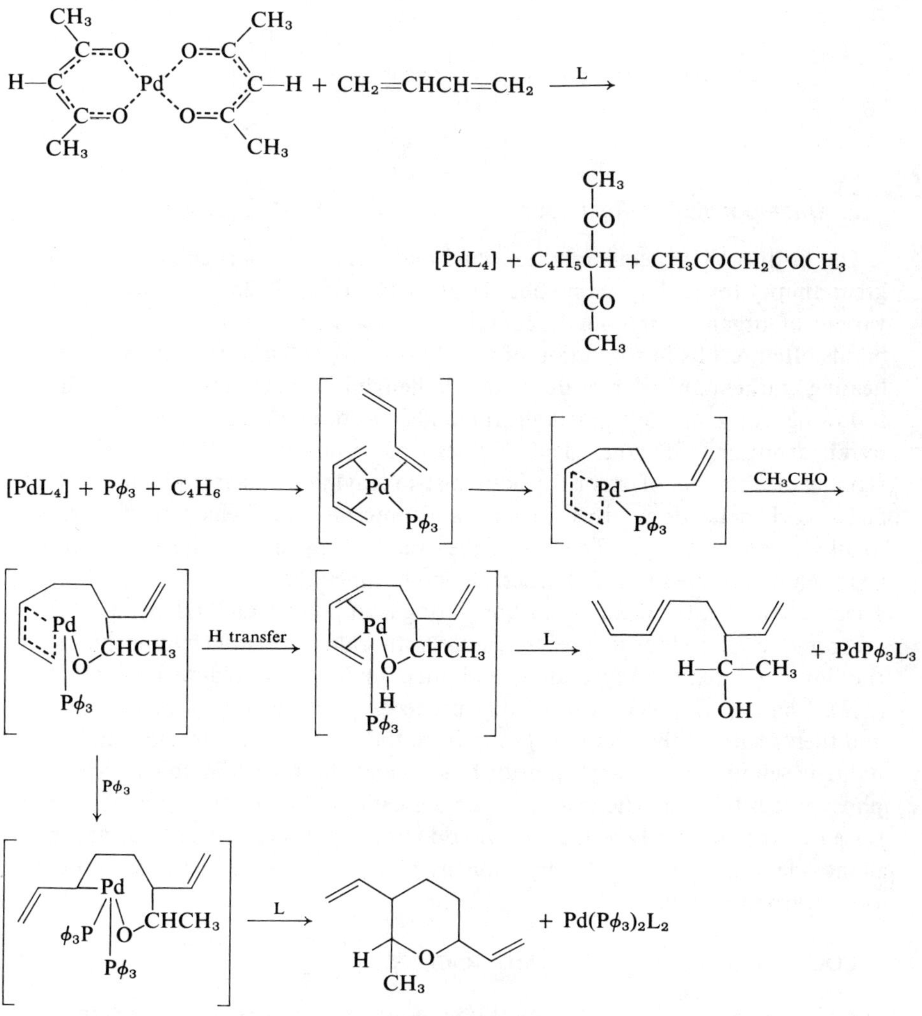

Scheme IX-24.

catalyst to produce a mixture of two isomeric six-membered ring formals (560). A possible mechanism for formation of one of the products is shown on p. 262. The second and major formal is probably obtained in a like manner from the isomeric olefin, 2-methyl-2-butene, which easily could be produced in the reaction mixture from 3-methyl-1-butene and [HPdL₂Cl] formed in side reactions.

$$2CH_2O + CH_3CH{=}\overset{\overset{\displaystyle CH_3}{|}}{C}CH_3 \xrightarrow[\text{CuCl}_2]{\text{PdCl}_2}$$

2. Decarbonylation Reactions

The Group VIII metals of the second and third row, particularly, have a great affinity for carbon monoxide. Heating chlorides of these metals with a variety of organic compounds containing carbon–oxygen single or double bonds often results in formation of metal carbonyl derivatives. For example, heating ruthenium trichloride with triphenylphosphine and any of the following compounds—dimethylformamide, acetophenone, cyclohexanone, tetrahydrofuran, dioxane, acid halides, or primary alcohols—produces significant amounts of chlorocarbonylbistriphenylphosphineruthenium (561, 562). Acyl metal derivatives are probably intermediates which then revert to alkyl metal carbonyls. The fate of the organic fragments in these reactions has been determined in a few cases. In one instance the reaction is useful for selectively and catalytically decarbonylating acid halides and aldehydes.

Aromatic acid chlorides react readily with chlorotristriphenylphosphine-rhodium first oxidatively adding and then undergoing a decarbonylation (563). The decarbonylation probably proceeds by a phosphine dissociation and then a shift of the aromatic group from the acyl carbonyl to the metal. A reductive elimination of aryl chloride follows and chlorocarbonylbistriphenyl-phosphine is formed. The reaction becomes catalytic when the temperature is raised to the point where the carbonyl complex product dissociates carbon monoxide and rereacts with the triphenylphosphine lost in the first step of the decarbonylation.

$$ArCOCl + Rh(Cl)(P\phi_3)_3 \rightleftharpoons [ArCORhCl_2(P\phi_3)_3] \underset{+P\phi_3}{\overset{-P\phi_3}{\rightleftharpoons}}$$

$$ArRhCl_2(CO)(P\phi_3)_2 \rightleftharpoons ArCl + RhCl(CO)(P\phi_3)_2$$

$$RhCl(CO)(P\phi_3)_2 + P\phi_3 \rightleftharpoons RhCl(P\phi_3)_3 + CO$$

A similar decarbonylation can be carried out with aldehydes and chloro-tristriphenylphosphinerhodium(I) with a mixture of hydrocarbons and olefins being formed (564). An oxidative addition of the aldehyde group must be involved in these examples forming an acyl metal hydride. The rate expression for the reaction contains two terms, one dependent only on the metal concentration and the other on the metal and the aldehyde. A mechanism fitting this rate expression has been suggested (564) in Scheme IX-25. The final step

$$\text{rate} = k_1[\text{Rh}] + k_2[\text{Rh}][\text{RCHO}]$$

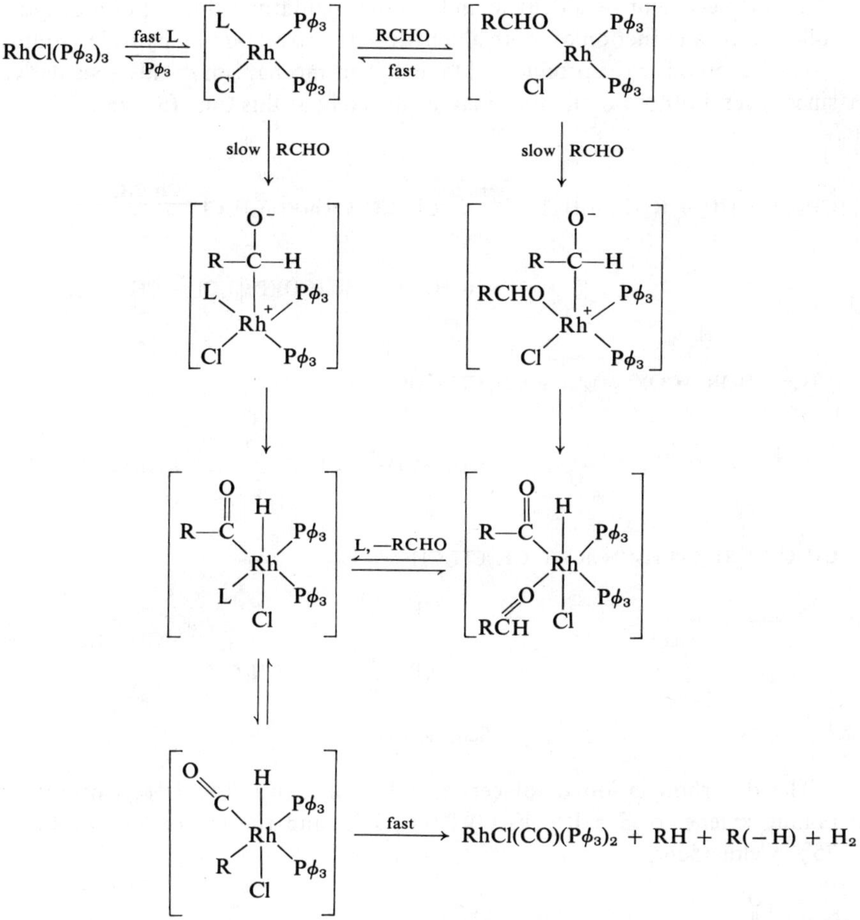

$$\text{RhCl(P}\phi_3)_3 \xrightleftharpoons[\text{P}\phi_3]{\text{fast L}} \begin{bmatrix} \text{L} & \text{P}\phi_3 \\ & \text{Rh} \\ \text{Cl} & \text{P}\phi_3 \end{bmatrix} \xrightleftharpoons[\text{fast}]{\text{RCHO}} \begin{bmatrix} \text{RCHO} & \text{P}\phi_3 \\ & \text{Rh} \\ \text{Cl} & \text{P}\phi_3 \end{bmatrix}$$

slow | RCHO   slow | RCHO

$$\begin{bmatrix} & \text{O}^- \\ & \text{R}-\text{C}-\text{H} \\ \text{L} & | & \text{P}\phi_3 \\ & \overset{+}{\text{Rh}} \\ \text{Cl} & \text{P}\phi_3 \end{bmatrix} \qquad \begin{bmatrix} & \text{O}^- \\ & \text{R}-\text{C}-\text{H} \\ \text{RCHO} & | & \text{P}\phi_3 \\ & \overset{+}{\text{Rh}} \\ \text{Cl} & \text{P}\phi_3 \end{bmatrix}$$

$$\begin{bmatrix} & \text{O} & \text{H} \\ & \| & | \\ \text{R}-\text{C} & & \text{P}\phi_3 \\ & \text{Rh} \\ \text{L} & | & \text{P}\phi_3 \\ & \text{Cl} \end{bmatrix} \xrightleftharpoons{\text{L},-\text{RCHO}} \begin{bmatrix} & \text{O} & \text{H} \\ & \| & | \\ \text{R}-\text{C} & & \text{P}\phi_3 \\ & \text{Rh} \\ \text{O} & | & \text{P}\phi_3 \\ \text{RCH} & \text{Cl} \end{bmatrix}$$

$$\begin{bmatrix} \text{O} & \text{H} \\ \diagdown & | \\ \text{C} & \text{P}\phi_3 \\ & \text{Rh} \\ \text{R} & | & \text{P}\phi_3 \\ & \text{Cl} \end{bmatrix} \xrightarrow{\text{fast}} \text{RhCl(CO)(P}\phi_3)_2 + \text{RH} + \text{R}(-\text{H}) + \text{H}_2$$

Scheme IX-25.

in the reaction apparently may be either a reductive elimination of a hydrocarbon or a β-hydride elimination from the R group followed by loss of hydrogen.

Hydridochlorotristriphenylphosphineruthenium(II) reacts with four moles of propionaldehyde to extract one of the carbonyl groups forming ethylene, to reduce two moles to alcohol, and to form an acyl complex with the other mole. The acylruthenium complex obtained is unique in that it apparently is bonded through both carbon and oxygen to the metal (ν_{CO} 1510 cm^{-1}) (565). The reaction probably involves oxidative addition of the aldehyde group, followed by a reduction of another aldehyde with the two hydride groups. A decarbonylation and β-hydride elimination could occur next. An olefin displacement by aldehyde and another oxidative addition of aldehyde, followed by a reduction of more aldehyde with the remaining hydride groups, would account for the products obtained. This mechanism is only speculative since several other possibilities may be as likely at this time (Scheme IX-26).

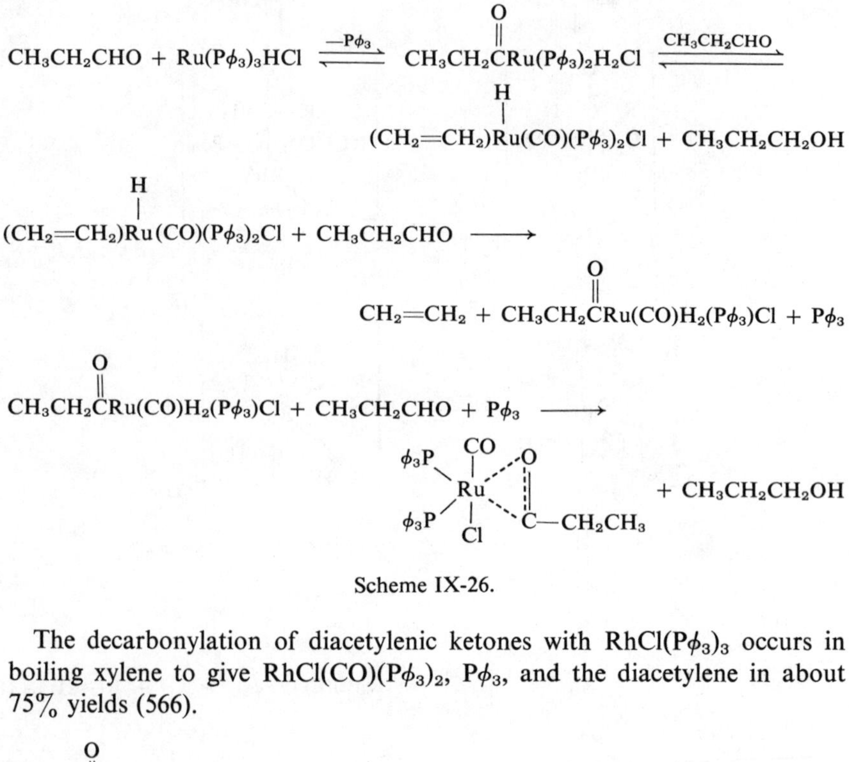

$$CH_3CH_2CHO + Ru(P\phi_3)_3HCl \underset{}{\overset{-P\phi_3}{\rightleftharpoons}} \overset{\overset{O}{\|}}{CH_3CH_2CRu(P\phi_3)_2H_2Cl} \underset{}{\overset{CH_3CH_2CHO}{\rightleftharpoons}}$$

$$\underset{\underset{Cl}{|}}{\overset{\overset{H}{|}}{(CH_2{=}CH_2)Ru(CO)(P\phi_3)_2Cl}} + CH_3CH_2CH_2OH$$

$$\overset{\overset{H}{|}}{(CH_2{=}CH_2)Ru(CO)(P\phi_3)_2Cl} + CH_3CH_2CHO \longrightarrow$$

$$CH_2{=}CH_2 + \overset{\overset{O}{\|}}{CH_3CH_2CRu(CO)H_2(P\phi_3)Cl} + P\phi_3$$

$$\overset{\overset{O}{\|}}{CH_3CH_2CRu(CO)H_2(P\phi_3)Cl} + CH_3CH_2CHO + P\phi_3 \longrightarrow$$

$$\begin{array}{c} \phi_3P \quad CO \quad O \\ \diagdown \quad | \quad \diagup \\ Ru \\ \diagup \quad | \quad \diagdown \\ \phi_3P \quad Cl \quad C{-}CH_2CH_3 \end{array} + CH_3CH_2CH_2OH$$

Scheme IX-26.

The decarbonylation of diacetylenic ketones with $RhCl(P\phi_3)_3$ occurs in boiling xylene to give $RhCl(CO)(P\phi_3)_2$, $P\phi_3$, and the diacetylene in about 75% yields (566).

$$\overset{\overset{O}{\|}}{RC{\equiv}C{-}C{-}C{\equiv}CR} + RhCl(P\phi_3)_3 \longrightarrow RhCl(CO)(P\phi_3)_2 + P\phi_3 + RC{\equiv}C{-}C{\equiv}CR$$

A more practical application of the decarbonylation reaction has been found in steroid chemistry. The introduction of a double bond at the 1-position in 3-ketosteroids is possible if the ketone is first α-formylated with ethyl formate and sodium ethoxide and the formyl derivative then decarbonylated (with β-hydride elimination) with $RhCl(P\phi_3)_3$ (567). A review of the metal-catalyzed methods of decarbonylation has been written (568).

3. Isocyanide Reactions

Transition metal alkyls may add to organic isocyanides in a 1:1 manner as noted in Chapter II, Section B,4,a. In contrast to the addition to carbon monoxide, multiple consecutive additions to isocyanides may occur. Typical is the reaction of iodobis(methyldiphenylphosphine)methylpalladium with cyclohexylisocyanide (569). Adducts with one, two, and three isocyanide units per palladium have been prepared simply by changing the ratio of reactants from 1:1 to 1:2 to 1:3.

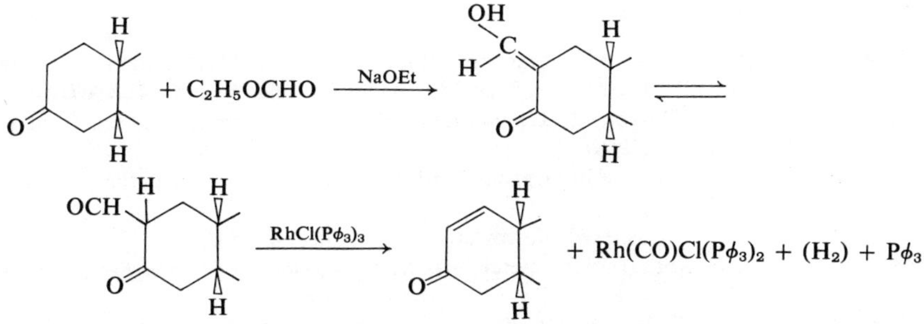

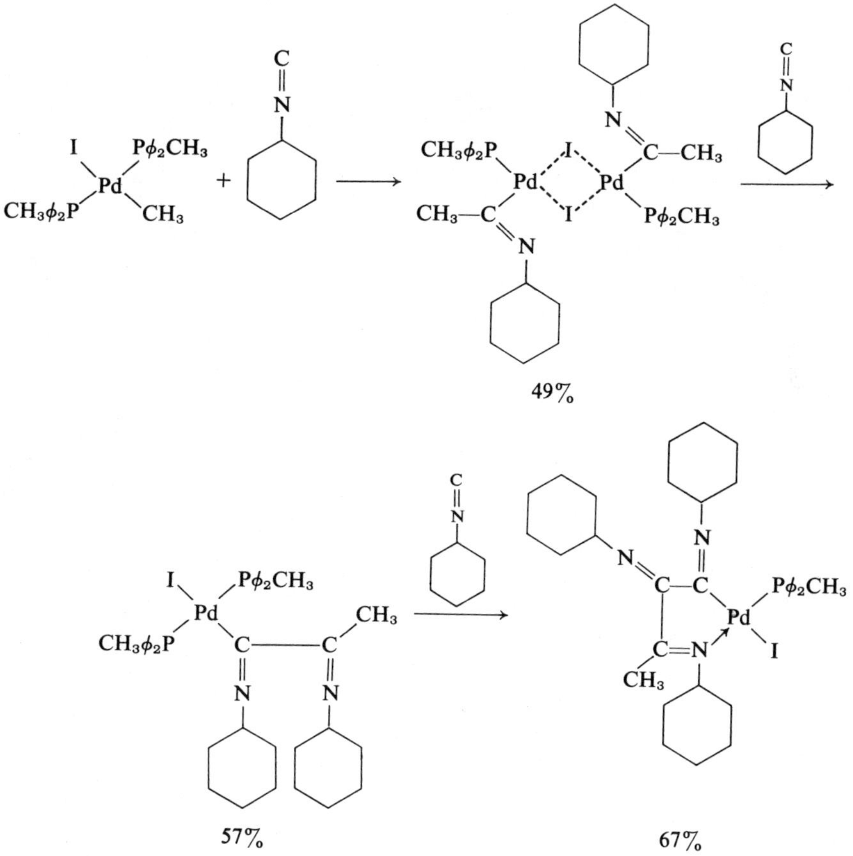

49%

57%   67%

FURTHER READING

C. W. Bird, Carbonylation and decarbonylation. "Transition Metal Intermediates in Organic Synthesis," Chapters 7, 8, and 9. Academic Press, New York, 1967.

A. J. Chalk and J. F. Harrod, Reactions catalyzed by cobalt carbonyl. *Advan. Organometal. Chem.* **6**, 119 (1968).

J. Falbe, "Carbon Monoxide in Organic Synthesis." Springer-Verlag, Berlin and New York, 1970.

M. Orchin and W. Rupilus, Hydroformylation. *Catal. Rev.* **6**, 85 (1972).

M. Ryang, Organic synthesis with metal carbonyls. *Organometal. Chem. Rev., Sect. A* **5**, 67 (1970).

D. T. Thompson and R. Whyman, Carbonylation. "Transition Metals in Homogeneous Catalysis," Chapter 5. Dekker, New York, 1971.

A. Wojcicki, Carbonylation reactions. *Advan. Organometal. Chem.* **11**, 88 (1973).

Chapter X

Dinitrogen, Dioxygen, Carbene, and Sulfur Dioxide Transition Metal Complexes

The four subjects of this chapter have been touched on in other chapters, but they are or will likely be of enough importance to be considered in more detail. Many reactions involving these complexes are known with heterogeneous catalysts, although relatively few homogeneous examples have been found. Since several of the known heterogeneous reactions are of great economic importance, considerable effort is now being put into understanding and developing new homogeneous analogs. Many of the reactions to be discussed in this chapter do not involve carbon–transition metal bonds, but they are nonetheless of considerable concern to organometallic chemists because of their potential applications.

A. DINITROGEN COMPLEXES

A wide variety of nitrogen transition metal complexes are known. In this chapter we will be considering complexes of dinitrogen, N_2, their methods of preparation, and reactions.

The conversion of gaseous nitrogen into ammonia or other combined forms of nitrogen is of tremendous economic importance. It is now accomplished industrially by reacting nitrogen with hydrogen under high pressure in the presence of an iron catalyst to form ammonia (the Haber process). Nature fixes nitrogen with iron- and molybdenum-containing enzymes which occur in certain leguminous plants, algae, and bacteria. Since both of these means of nitrogen fixation involve transition metals, it can be concluded that transition metal nitrogen complexes are probably intermediates in the reactions. In the last few years several ways of preparing dinitrogen complexes have been found and numerous reactions have been investigated in an effort to convert the coordinated nitrogen into a useful form (570). Nitrogen has been reduced to ammonia by several methods homogeneously, but in none of the reactions have the intermediate complexes been isolated and the exact reaction steps involved established beyond question. This very probably will be accomplished in the near future. There is considerable doubt, however, as to whether ammonia could ever be made more practically homogeneously in any case. Routes to other nitrogen derivatives may well be of more practical use ultimately.

The first isolable nitrogen complexes were made by reactions of transition metal complexes with hydrazine. The metal complex is reduced to a lower oxidation state and the hydrazine is oxidized to nitrogen presumably within the coordinating sphere of the metal. The following examples are typical syntheses (571, 572).

$$RuCl_3 + NH_2NH_2 \cdot H_2O \longrightarrow [Ru(NH_3)_5N_2]Cl_2$$

$$(NH_4)_2OsCl_6 + NH_2NH_2 \cdot H_2O \longrightarrow [Os(NH_3)_5N_2]Cl_2$$

In these and the other mononuclear nitrogen complexes coordination occurs with an unshared pair of electrons on one of the nitrogens and not by complexing with the triple nitrogen–nitrogen bond. Nitrogen is a moderate π-acceptor and a weak σ-donor ligand (573).

The oxidation of a coordinated hydrazine group by an external oxidant also is possible (574).

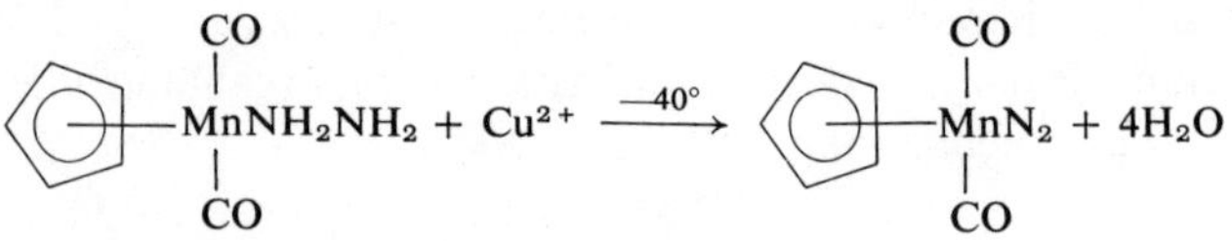

Reaction of coordinately saturated carbonyl complexes with hydrazine may result in quite a different reaction. The hydrazine may attack coordinated CO and form a hydrazidocarbonyl metal derivative which then usually rearranges with loss of ammonia to an isocyanate complex (575).

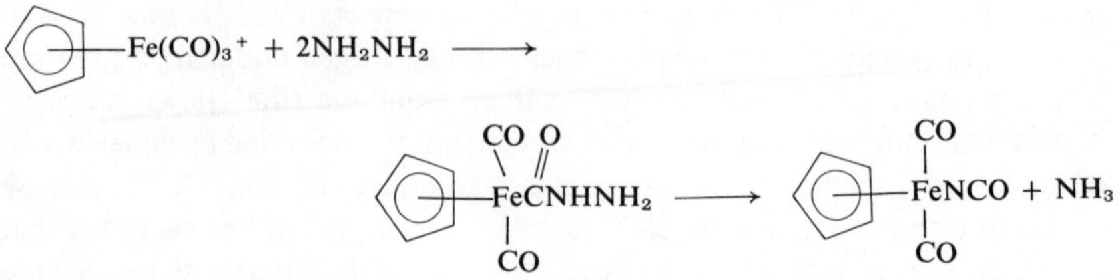

In other reactions, hydrazine may form mononitrogen complexes (nitrides) as in its reaction with dirhenium heptoxide and triphenylphosphine (576, 577).

$$Re_2O_7 + P\phi_3 + NH_2NH_2 \cdot 2HCl \longrightarrow N{\equiv}ReCl_2(P\phi_3)_2$$

N-Substituted nitrides are obtained with symmetrically disubstituted hydrazines (578).

$$CH_3NHNHCH_3 \cdot 2HCl + OReCl_3(P\phi_3)_2 \xrightarrow{P\phi_3}$$

$$CH_3N{\equiv}ReCl_3(P\phi_3)_2 + CH_3NH_3{}^+Cl^- + OP\phi_3$$

Another indirect route to dinitrogen complexes is the reaction of coordinated ammonia with nitric oxide (579).

$$Ru(NH_3)_6{}^{3+} + NO + OH^- \longrightarrow Ru(NH_3)_5N_2{}^{2+}$$

Nitrogen may also be extracted from acyl azides by coordinately unsaturated metal complexes to give dinitrogen complexes (580).

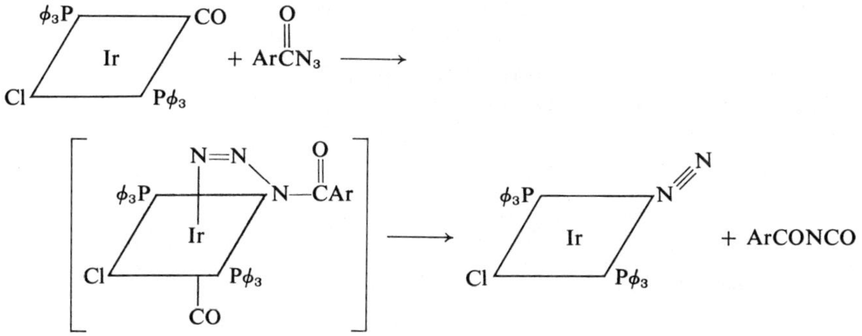

The preparation of complexes from molecular nitrogen requires either reactive, coordinately unsaturated complexes or complexes with very easily displaced ligands.

The reduction of cobalt(II) or cobalt(III) to cobalt(I) in the presence of organophosphines and nitrogen, for example, generally produces nitrogen complexes (581, 582). A triethylcobalt–phosphine complex is probably an

$$CoAcac_3 + AlEt_2OEt + P\phi_3 + N_2 \longrightarrow HCoN_2(P\phi_3)_3$$

intermediate in this reaction. This complex would either decompose by three β-hydride eliminations to the known trihydride or could eliminate only two hydride units and produce ethane with the third. The trihydride is known to reversibly lose H_2 in the presence of N_2. Apparently, a fourth phosphine

$$H_3Co(P\phi_3)_3 + N_2 \rightleftharpoons HCoN_2(P\phi_3)_3 + H_2$$

group has no tendency to occupy a coordination position on the cobalt presumably because the four large phosphine ligands cannot fit around the cobalt atom. Reduction of cobaltic acetylacetonate in the presence of triphenylphosphine with dimethylaluminum ethoxide even under nitrogen produced only coordinately unsaturated tristriphenylphosphinemethylcobalt-(I) (581).

$$CoAcac_3 + AlMe_2OEt + P\phi_3 \longrightarrow CH_3Co(P\phi_3)_3$$

Reduction of dichloro(bisdiethylphenylphosphine)cobalt(II) with sodium and excess ligand under nitrogen may produce two different products depending on the ratio of the cobalt complex to the sodium. With a 1:3 ratio an anionic nitrogen complex is produced, whereas with a 1:2 ratio a binuclear, neutral complex is formed in which both unshared pairs of electrons on the nitrogens are coordinated to trisorganophosphinecobalt groups (583). The

$$CoCl_2(PEt_2\phi)_2 + PEt_2\phi + 2Na + N_2 \longrightarrow$$
$$(PEt_2\phi)_3Co—N{\equiv}N—Co(PEt_2\phi)_3 \xrightarrow{Na} 2NaCoN_2(PEt_2\phi)_3$$

coordination of dinitrogen at both ends may weaken the nitrogen–nitrogen triple bond more than monocoordination does and is probably the type of structure involved in the known fixation reactions.

Most transition metals have now yielded complexes with molecular nitrogen of one kind or another by the reduction procedure. The following are a few more examples (584–587).

$$MoCl_4(PMe_2\phi)_2 + NaHg_x + 2N_2 \longrightarrow$$

$$OsBr_3(PR_3)_3 + ZnHg_x + N_2 \longrightarrow OsBr_2N_2(PR_3)_0$$
$$N_2 + NiAcac_2 + i\text{-}Bu_3Al + PEt_3 \longrightarrow HNiN_2(PEt_3)_2$$
$$FeCl_2 \cdot 2H_2O + NaBH_4 + N_2 + P\phi_2Et \longrightarrow H_2FeN_2(P\phi_2Et)_3$$

In a few instances dinitrogen complexes have been made by ligand replacement reactions, where the ligand is very weakly coordinated (588–590).

Several transition metal complex–reducing agent combinations have been found to be capable of reducing nitrogen to ammonia or hydrazine, but in all cases the rates are very slow compared with the biological catalysts. Yields also are often low, and in some instances the reactions are not catalytic. In none of these cases have the intermediate complexes been isolated or otherwise identified with certainty.

$$[Ru(NH_3)_5H_2O]^{2+} + N_2 \rightleftharpoons [Ru(NH_3)_5N_2]^{2+} + H_2O$$

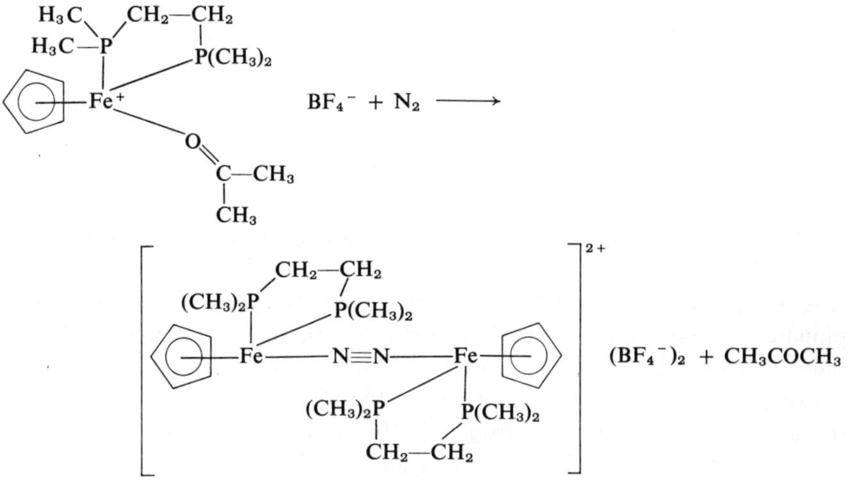

$$BF_4^- + N_2 \longrightarrow$$

$$(BF_4^-)_2 + CH_3COCH_3$$

$$C_5H_5Mn(CO)_2THF + N_2 \rightleftharpoons C_5H_5Mn(CO)_2N_2 + THF$$

Trichlorobistriphenylphosphineiron(III), isopropylmagnesium chloride, and nitrogen at $-50°$ form a very unstable binuclear dinitrogen complex which on treatment with hydrogen chloride forms hydrazine in 10% yield (591).

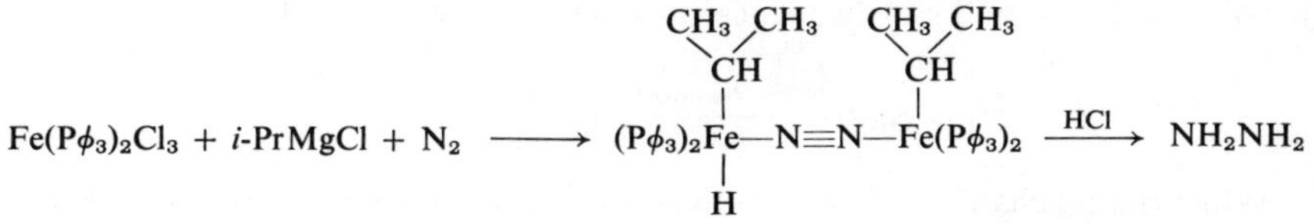

The reaction of dichlorobiscyclopentadienyltitanium(IV) with lithium alkyls and nitrogen produces a complex which when acidified gives ammonia (salt) in 70–100% yield based on the titanium (592). Other variations on

$$(C_5H_5)_2TiCl_2 + RLi + N_2 \longrightarrow \xrightarrow{H^+} NH_3$$

titanium reductions have been reported by van Tamelen (593). Of particular interest was the finding that amines could be prepared from ketones and nitrogen in about 50% yields, while acid chlorides gave nitriles. The reactions have been formulated in Scheme X-1. The proposed binuclear dinitrogen intermediate has now been isolated at low temperatures by the reaction of chlorobiscyclopentadienyltitanium(III) with methylmagnesium iodide and nitrogen. On treatment with acid, diimide is apparently formed (594).

$$\text{Cp}_2\text{TiCl}_2 + \text{Mg} \longrightarrow \text{Cp}_2\text{Ti} \xrightleftharpoons{\text{N}_2} \text{Cp}_2\text{TiN}_2 \rightleftharpoons$$

$$\text{Cp}_2\text{TiN}_2\text{TiCp}_2 \xrightarrow{e} 2\text{N}^{3-} \xrightarrow{\text{RCOR}'} \text{RR}'\text{CHNH}_2 + (\text{RR}'\text{CH})_2\text{NH}$$

$$\Big\downarrow \text{RCOCl}$$

$$\text{RCN}$$

Scheme X-1.

$$2\text{Cp}_2\text{TiCl} + \text{CH}_3\text{MgI} + \text{N}_2 \longrightarrow \text{Cp}_2\text{Ti}-\text{N}{\equiv}\text{N}-\text{TiCp}_2 \xrightarrow{\text{H}^+} \text{N}_2\text{H}_2$$

A water-soluble porphyrin complex, mesotetra(p-sulfonatophenyl)por-phinatocobalt(III), and sodium borohydride in aqueous solution very slowly reduce nitrogen in air to ammonia (595). Intermediates were not identified.

Complex ligands are not required for the nitrogen reduction since even titanium trichloride reduces nitrogen, in the presence of Mo(VI) and Mg(II) ions, to hydrazine. Vanadous sulfate is similar (596).

$$\text{TiCl}_3 + \text{N}_2 \xrightarrow[\text{50 atm, 85°}]{\text{Mo}^{6+}, \text{Mg}^{2+}} \text{N}_2\text{H}_4$$

$$\text{VSO}_4 + \text{N}_2 \xrightarrow[\text{70°, Mg}^{2+}, \text{KOH}]{\text{1 atm}} \text{N}_2\text{H}_4 + \text{NH}_3$$

Very simple model systems attempting to produce nitrogenase models from ferrous sulfate, sodium molybdate with 2-aminoethanethiol, and sodium borohydride also will slowly reduce nitrogen to ammonia (597).

$$\text{N}_2 + \text{NaBH}_4 \xrightarrow[\text{H}_2\text{NCH}_2\text{CH}_2\text{SH, H}_2\text{O}]{\text{Na}_2\text{MoO}_4 + \text{FeSO}_4} \text{NH}_3$$

While the mechanism of reduction of nitrogen still remains obscure, there is little doubt but that it will be explained with the aid of isolable or at least identifiable intermediates before too long. There may well be more than one way to react nitrogen, however. The insertion of nitrogen between a metal atom and an organic group would be a conceivable reaction. The reverse reaction is known (598). Other approaches are being taken by

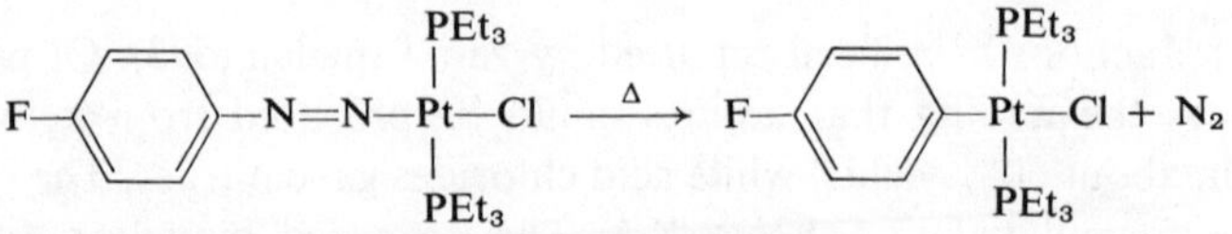

Chatt. One idea is to complex the nitrogen molecule at both ends with different metal groups in an attempt to weaken the nitrogen–nitrogen bond to the point where it will readily react. Many complexes of this type have been

prepared, but "reactive" complexes have not yet been obtained (599). Weakening of the nitrogen–nitrogen bond was greatest when one of the metals or other elements attached to nitrogen had vacant d orbitals which could accept π electrons from the nitrogen.

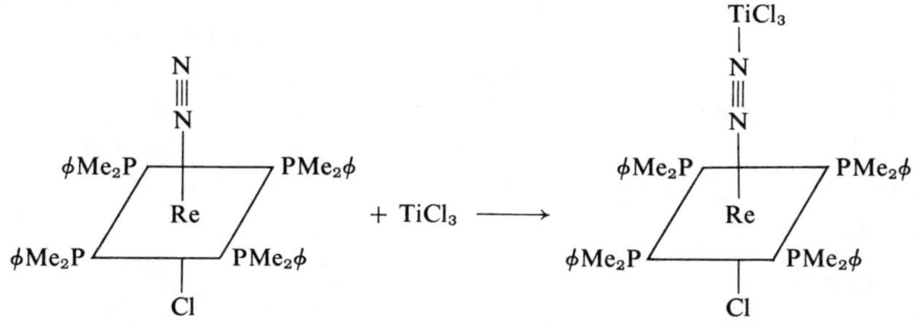

Some reactions of monocoordinated dinitrogen have been found, but, so far, the nitrogen compounds formed cannot be removed from the metal without the compounds reverting to nitrogen (600).

$$W(N_2)_2[\phi_2PCH_2CH_2P\phi_2]_2 + CH_3COCl + HCl \longrightarrow$$

$$WCl_2(N_2HCOCH_3)(\phi_2PCH_2CH_2P\phi_2)_2 + N_2$$

$$W(N_2)_2(Et_2PCH_2CH_2PEt_2)_2 + HCl \longrightarrow WCl_2(N_2H_2)(Et_2PCH_2CH_2PEt_2)_2 + N_2$$

Dinitrogen transition metal complexes are finding uses in reactions where a very labile ligand is required. The preparation of a very reactive butadiene complex from bis[(ditricyclohexylphosphine)nickel(0)]dinitrogen was mentioned in Chapter VI, Section J,2. Another example is the reaction of $HCoN_2(P\phi_3)_3$ with CO_2 (601). Carbon dioxide does not react with other

$$HCoN_2(P\phi_3)_3 + CO_2 \longrightarrow HCO\overset{\overset{\textstyle O}{\textstyle \|}}{C}o(P\phi_3)_3 + N_2$$

types of cobalt complexes. The metal dinitrogen complexes, in general, are very reactive and in many instances approach the reactivity of the metal compounds obtained by reduction of halides with metal alkyls.

B. DIOXYGEN COMPLEXES

The oxidation of organic compounds with molecular oxygen and transition metal catalysts is of considerable economic importance. Consequently, a great deal of research has been carried out in the area and several volumes would be required to summarize the available information. The majority of these reactions are radical reactions involving the metal as a reagent to convert oxygen into peroxy or some other active radical and the activated oxygen compound can then attack organic molecules. Reactions of this type have been summarized in recent publications (602, 603).

Nonradical oxygen oxidations also are possible with some transition metal catalysts. Relatively few, useful examples of these reactions are yet known, but much research is now being carried out with the expectation that very selective, new oxidation reactions will be found. At least four types of transition metal dioxygen complexes exist which may be involved in oxidation reactions. Metal atoms may be bonded to both ends of the O_2 molecule, only to one end, on to one end and having another group at the other end, usually carbon or hydrogen, or both oxygens may be bonded symmetrically to the same metal atom.

1. Preparation

Most of the known complexes with metal groups attached to both ends of the oxygen molecule are cobalt complexes. Many cobalt(II) complexes react with oxygen to form binuclear cobalt(III) peroxy complexes. For example (604),

$$2K_3Co(CN)_5 + O_2 \longrightarrow K_6[(CN)_5Co-O-O-Co(CN)_5]$$

Generally these isolable complexes are not very reactive toward organic compounds under mild conditions, although unstable analogs may be much more reactive and could be involved in some oxidation reactions. One reactive form of these compounds may be the second type of dioxygen complex where only one end of the oxygen molecule is attached to a metal. These complexes are also prepared by reaction of low-valent complexes (cobalt) with oxygen. Only a few isolable complexes with this structure are known. Why some complexes form 1:1 compounds with O_2 and others 2:1 is not explainable at this time. The oxygen–oxygen–metal angle in the following structure is 126° (605).

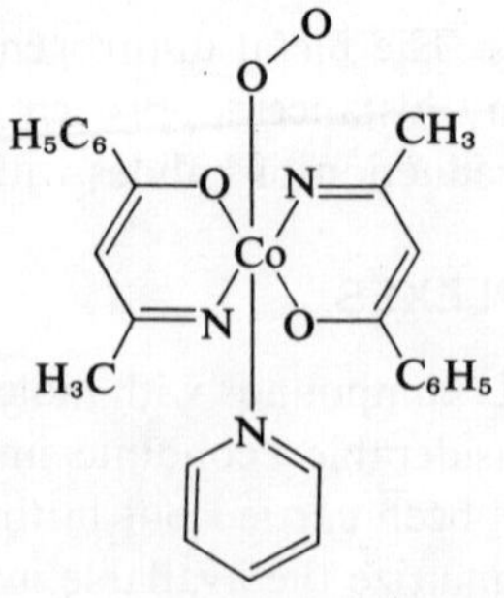

Peroxy complexes are apparently intermediates in the acid-catalyzed hydrolysis of the first type of complex to hydrogen peroxide (604, 606).

$$K_6[(CN)_5Co-O-O-Co(CN)_5] \xrightarrow[H_2O]{H^+} K_3Co(CN)_5OOH + K_2Co(CN)_5H_2O + K^+$$

$$K_3Co(CN)_5OOH + H_2O + H^+ \longrightarrow K_2Co(CN)_5H_2O + H_2O_2 + K^+$$

Unsymmetrical peroxy complexes may also be obtained from perhydroxy complexes by reactions which will be discussed below in Section B,2.

Many complexes with both oxygen atoms of an O_2 group bonded to the same metal are now known. These are generally prepared by reactions of low-valent, coordinately unsaturated or easily dissociated saturated complexes, with oxygen. The following examples are typical preparations (607–609).

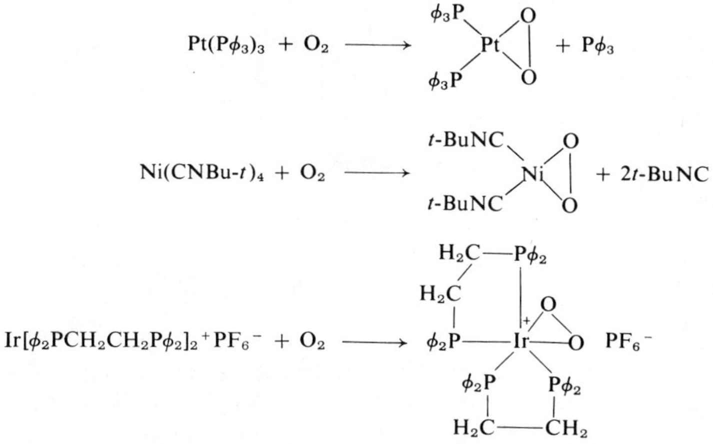

$$Pt(P\phi_3)_3 + O_2 \longrightarrow \underset{\phi_3P}{\overset{\phi_3P}{\diagdown}}Pt\underset{O}{\overset{O}{\diagup}} + P\phi_3$$

$$Ni(CNBu\text{-}t)_4 + O_2 \longrightarrow \underset{t\text{-BuNC}}{\overset{t\text{-BuNC}}{\diagdown}}Ni\underset{O}{\overset{O}{\diagup}} + 2t\text{-BuNC}$$

$$Ir[\phi_2PCH_2CH_2P\phi_2]_2{}^+PF_6{}^- + O_2 \longrightarrow$$

2. Reactions

The transition metal dioxygen complexes undergo reactions with a variety of compounds. Addition reactions occur with carbon dioxide, ketones, and sulfur dioxide to give cyclic peroxy complexes. The SO_2 adduct, however, is unstable and it rearranges rapidly to a complex with a bidentate sulfate ligand (610–612). A reaction with ketene produces an anhydride of a biscarboxymethyl metal complex (613). An intermediate peracid anhydride is probably involved in this reaction, also.

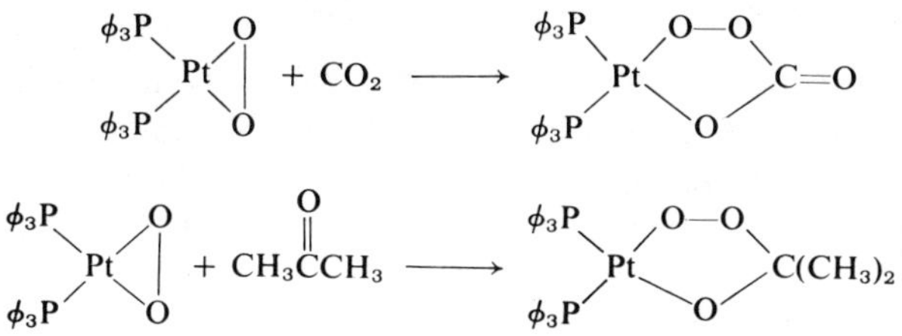

$$\underset{\phi_3P}{\overset{\phi_3P}{\diagdown}}Pt\underset{O}{\overset{O}{\diagup}} + CO_2 \longrightarrow \underset{\phi_3P}{\overset{\phi_3P}{\diagdown}}Pt\underset{O}{\overset{O-O}{\diagup}}C=O$$

$$\underset{\phi_3P}{\overset{\phi_3P}{\diagdown}}Pt\underset{O}{\overset{O}{\diagup}} + CH_3\overset{O}{\overset{\|}{C}}CH_3 \longrightarrow \underset{\phi_3P}{\overset{\phi_3P}{\diagdown}}Pt\underset{O}{\overset{O-O}{\diagup}}C(CH_3)_2$$

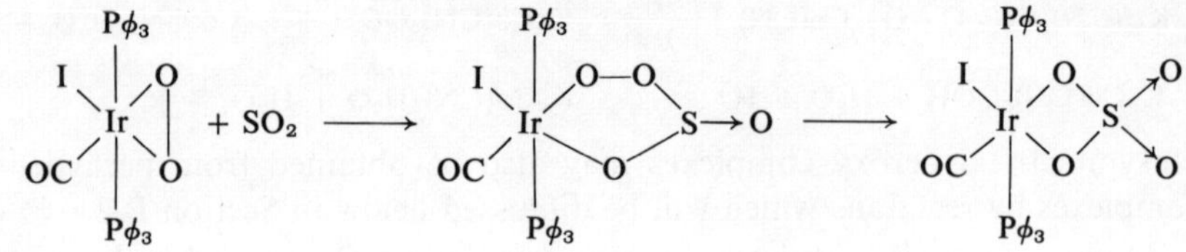

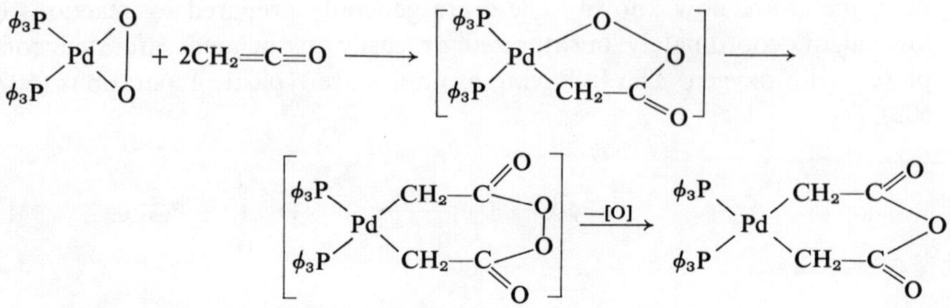

The reaction with fluorinated ketones can give both five- and six-membered ring products (614).

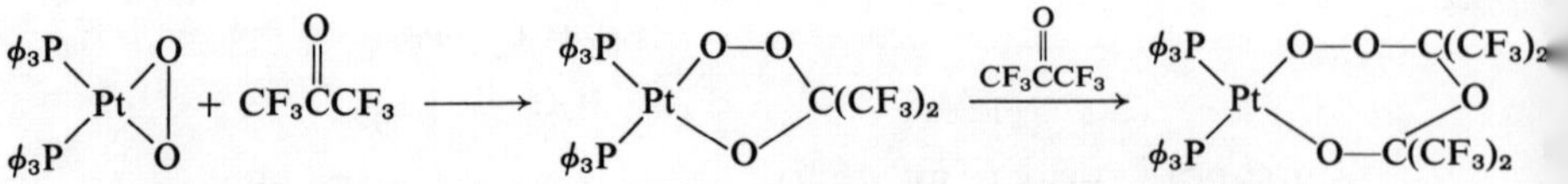

Carbon monoxide reacts with bistriphenylphosphinedioxoplatinum to form a bidentate carbonate complex. The same complex can be obtained by reaction of the carbon dioxide adduct with triphenylphosphine, in which case triphenylphosphine oxide is also produced (614, 615).

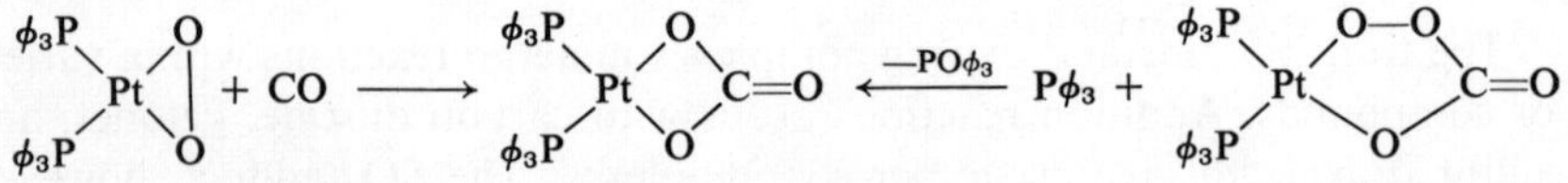

The peroxyplatinum complex also will react with nitrogen dioxide to form a covalent bisnitrate (616).

The peroxy complexes are capable of catalytic oxidations in several instances. The above peroxyplatinum complex in the presence of oxygen will catalytically oxidize triphenylphosphine to the oxide. Probably the oxidation occurs within the coordination sphere of the complex (617). The analogous

$$2P\phi_3 + O_2 \xrightarrow{(P\phi_3)_2PtO_2} 2OP\phi_3$$

nickel complex is a more reactive catalyst for the same reactions (618). Carbon monoxide can be catalytically oxidized with chlorobis-1,5-cycloocta-dienerhodium dimer presumably by way of a peroxy complex (619). Iso-

$$2CO + O_2 \xrightarrow{\text{[(COD)}_2\text{RhCl]}_2} 2CO_2$$

$$COD = 1,5\text{-cyclooctadiene}$$

cyanides may be oxidized similarly with oxygen and a nickel(0) tetrakisiso-cyanide catalyst to isocyanates (620).

$$2t\text{-BuNC} + O_2 \xrightarrow{(t\text{-BuNC})_4\text{Ni}} 2t\text{-BuNCO}$$

The high-yield homogeneous oxidation of olefins to olefin oxides with molybdenum pentoxide is very probably another example of oxidation within a coordination sphere. The following mechanism has been proposed (621, 622).

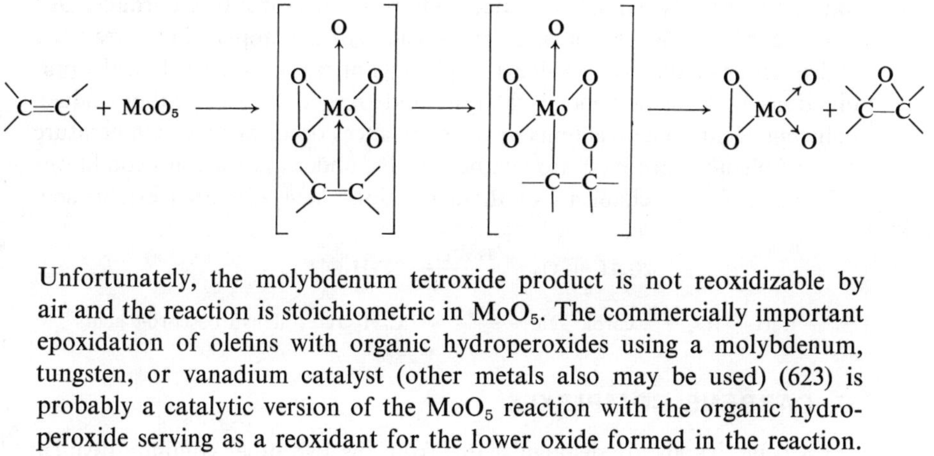

Unfortunately, the molybdenum tetroxide product is not reoxidizable by air and the reaction is stoichiometric in MoO_5. The commercially important epoxidation of olefins with organic hydroperoxides using a molybdenum, tungsten, or vanadium catalyst (other metals also may be used) (623) is probably a catalytic version of the MoO_5 reaction with the organic hydro-peroxide serving as a reoxidant for the lower oxide formed in the reaction.

$$CH_3CH{=}CH_2 + (CH_3)_3COOH \xrightarrow{Mo} CH_3\overset{\displaystyle O}{\overset{\displaystyle /\backslash}{CH{-}CH_2}} + (CH_3)_3COH$$

The combination of an organic peroxide and a vanadium catalyst will also oxidize tertiary amines to amine oxides (624).

$$R_3N + R'OOH \xrightarrow{V^{5+}} R_3N{\longrightarrow}O + R'OH$$

An ingenious combination of catalysts has been employed to oxidize allyl alcohol to the epoxide with molecular oxygen. Pentacyanocobalt(II) reacts with oxygen in acidic aqueous solution in the presence of allyl alcohol and a tungstic acid catalyst. The cobalt complex forms the dicobaltperoxy complex which then hydrolyzes in the solution to hydrogen peroxide. The hydrogen peroxide oxidizes the tungstic acid to a peroxide which then epoxidizes the allyl alcohol. The reaction is stoichiometric in cobalt, but catalytic in tungsten (606). Ethylene oxide is prepared commercially by the heterogeneous vapor

$$CH_2{=}CHCH_2OH + 2H^+ + O_2 + 2K_3Co(CN)_5 \xrightarrow{\ \ WO_3\ \ }$$

$$\overset{O}{\overset{\displaystyle \wedge}{H_2C{-}CHCH_2OH}} + 2K_2Co(CN)_5 \cdot H_2O + H_2O + 2K^+$$

$$70\%$$

phase reaction of ethylene with oxygen over a silver catalyst. Silver oxide is believed to be the true oxidant. This reaction does not go in good yield with olefins other than ethylene, however. Many other useful metal-catalyzed oxidation reactions could be mentioned. The potential of these homogeneous reactions lies in the possibility of finding selective new oxidations. Of particular interest to the chemical industry would be the discovery of new ways to more selectively oxidize saturated and aromatic hydrocarbon groups. One example will suffice to illustrate a possible future application. Aromatic hydrocarbons have been oxidized to phenols in reasonable yields with combinations of transition metal catalysts and oxygen with or without added reducing agents. The reactions in some instances occur ar room temperature and the phenolic products are relatively stable under the reactions conditions (625–627). The mechanisms of these reactions have not been established.

$$2C_6H_6 + O_2 \xrightarrow[\text{or CuSO}_4]{\text{FeSO}_4} 2C_6H_5OH$$

$$C_6H_6 + O_2 + \text{ascorbic acid} \xrightarrow{\text{Fe}^{2+}} C_6H_5OH + \text{dehydroascorbic acid}$$

C. CARBENE COMPLEXES

Carbene ligands are distinguishable from the two other common divalent carbon ligands, carbon monoxide and isocyanides, by the presence of *two* groups attached to the divalent carbon. The substituents may be hydrogen, organic groups, halogens, or various oxygen, nitrogen, or sulfur groups. A wide variety of these complexes have been prepared.

1. Methods of Preparation

Most of the research carried out on transition metal carbene complexes has been done by E. O. Fischer and his co-workers within the last 10 years. A general method of preparation was developed involving reactions of metal carbonyls with strongly nucleophilic anions. The products were salts which were formulated as resonance hybrids of acyl metal anions and negatively charged coordinated carbene complexes. Protonation gave rather unstable hydroxycarbene derivatives, whereas alkylation produced relatively stable alkoxycarbene complexes. Scheme X-2 shows typical preparations employing this method of synthesis (628–631).

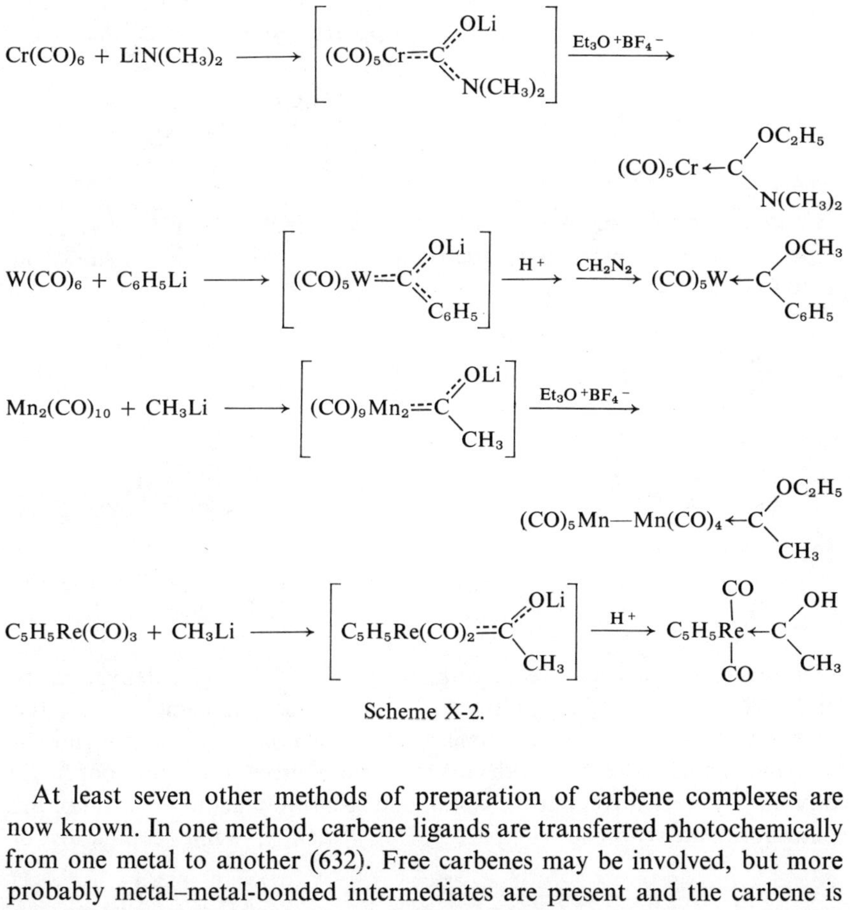

$$Cr(CO)_6 + LiN(CH_3)_2 \longrightarrow \left[(CO)_5Cr\!=\!\!C\!\!\begin{array}{c} {}^{\nearrow}OLi \\ {}^{\searrow}N(CH_3)_2 \end{array} \right] \xrightarrow{Et_3O\,^+BF_4\,^-}$$

$$(CO)_5Cr\!\leftarrow\!C\!\!\begin{array}{c} {}^{\nearrow}OC_2H_5 \\ {}^{\searrow}N(CH_3)_2 \end{array}$$

$$W(CO)_6 + C_6H_5Li \longrightarrow \left[(CO)_5W\!=\!\!C\!\!\begin{array}{c} {}^{\nearrow}OLi \\ {}^{\searrow}C_6H_5 \end{array} \right] \xrightarrow[\text{}]{H^+ \quad CH_2N_2} (CO)_5W\!\leftarrow\!C\!\!\begin{array}{c} {}^{\nearrow}OCH_3 \\ {}^{\searrow}C_6H_5 \end{array}$$

$$Mn_2(CO)_{10} + CH_3Li \longrightarrow \left[(CO)_9Mn_2\!=\!\!C\!\!\begin{array}{c} {}^{\nearrow}OLi \\ {}^{\searrow}CH_3 \end{array} \right] \xrightarrow{Et_3O\,^+BF_4\,^-}$$

$$(CO)_5Mn\!-\!Mn(CO)_4\!\leftarrow\!C\!\!\begin{array}{c} {}^{\nearrow}OC_2H_5 \\ {}^{\searrow}CH_3 \end{array}$$

$$C_5H_5Re(CO)_3 + CH_3Li \longrightarrow \left[C_5H_5Re(CO)_2\!=\!\!C\!\!\begin{array}{c} {}^{\nearrow}OLi \\ {}^{\searrow}CH_3 \end{array} \right] \xrightarrow{H^+} C_5H_5Re\!\!\begin{array}{c} CO \\ | \\ | \\ CO \end{array}\!\!\leftarrow\!C\!\!\begin{array}{c} {}^{\nearrow}OH \\ {}^{\searrow}CH_3 \end{array}$$

Scheme X-2.

At least seven other methods of preparation of carbene complexes are now known. In one method, carbene ligands are transferred photochemically from one metal to another (632). Free carbenes may be involved, but more probably metal–metal-bonded intermediates are present and the carbene is transferred by a 1:2 shift mechanism or a coordinately unsaturated metal attacks the carbene ligand and displaces the other metal group.

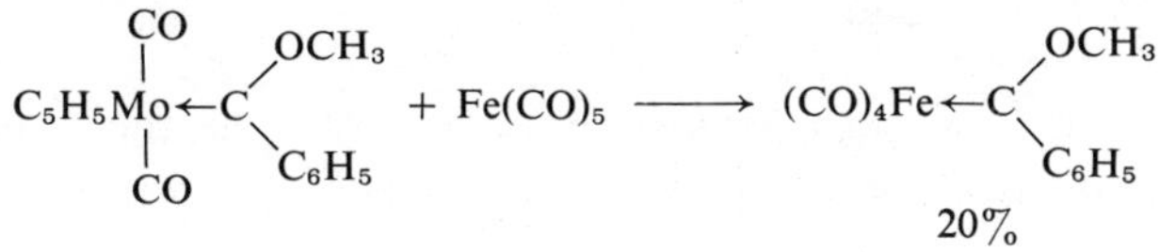

$$C_5H_5Mo\!\!\begin{array}{c} CO \\ | \\ | \\ CO \end{array}\!\!\leftarrow\!C\!\!\begin{array}{c} {}^{\nearrow}OCH_3 \\ {}^{\searrow}C_6H_5 \end{array} + Fe(CO)_5 \longrightarrow (CO)_4Fe\!\leftarrow\!C\!\!\begin{array}{c} {}^{\nearrow}OCH_3 \\ {}^{\searrow}C_6H_5 \end{array}$$

20%

A carbonium ion reaction with acyl metal carbonyls also is capable of producing carbene complexes at least with some manganese and rhenium compounds. For example, pentacarbonyl-2-chloroethoxycarbonylmanganese-(I) reacts with silver hexafluorophosphate to form a cationic cyclic carbene complex (633).

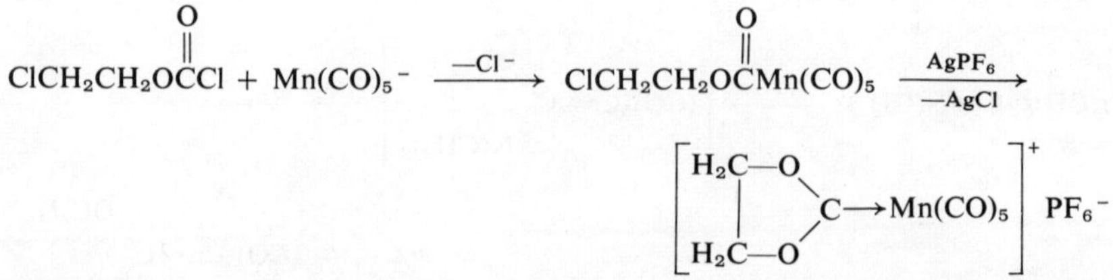

$$ClCH_2CH_2O\overset{\overset{\displaystyle O}{\|}}{C}Cl + Mn(CO)_5{}^- \xrightarrow{-Cl^-} ClCH_2CH_2O\overset{\overset{\displaystyle O}{\|}}{C}Mn(CO)_5 \xrightarrow[-AgCl]{AgPF_6}$$

$$\left[\begin{array}{c} H_2C-O \\ \\ H_2C-O \end{array} \!\!\! C \rightarrow Mn(CO)_5 \right]^+ PF_6{}^-$$

Certain electron-rich olefins will react with metal complexes to form carbene derivatives by cleaving the double bond. An example is shown in the following reaction (634).

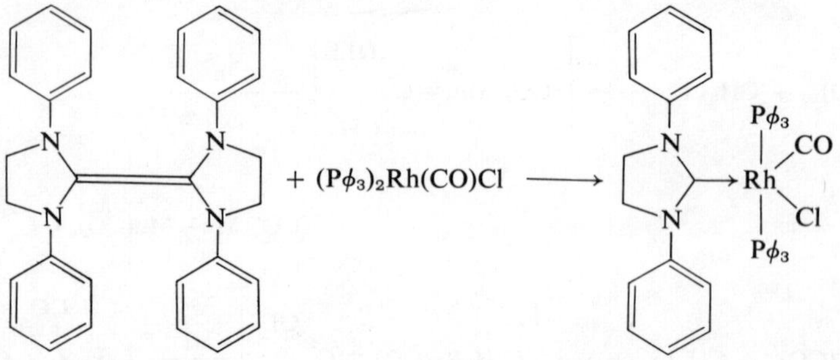

Diazo compounds may react with many transition metal derivatives to form carbenelike complexes, although in most cases the complexes are too reactive to be isolated. For example, reaction of *trans*-chlorocarbonylbistriphenylphosphineiridium(I) with diazomethane gives the four-coordinate chloromethyl derivative, but the compound seems to react in the form of the pentacoordinate carbene complex (635). Recently, some complexes of

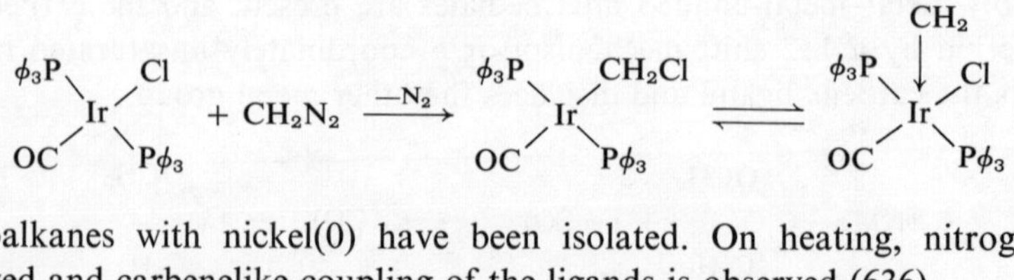

diazoalkanes with nickel(0) have been isolated. On heating, nitrogen is evolved and carbenelike coupling of the ligands is observed (636).

$$(t\text{-BuNC})_4Ni + R_2C{=}N_2 \longrightarrow (R_2CN_2)Ni(t\text{-BuNC})_2 \longrightarrow$$

$$R_2C{=}C{=}N\text{-}t\text{-Bu} + R_2CHCN + R_2C{=}N{-}N{=}CR_2$$

Another method of preparation is reaction of certain isocyanide complexes with alcohols or amines. cis-Dichlorobisphenylisocyanidepalladium(II), for example, reacts with methanol and p-toluidine as follows (637).

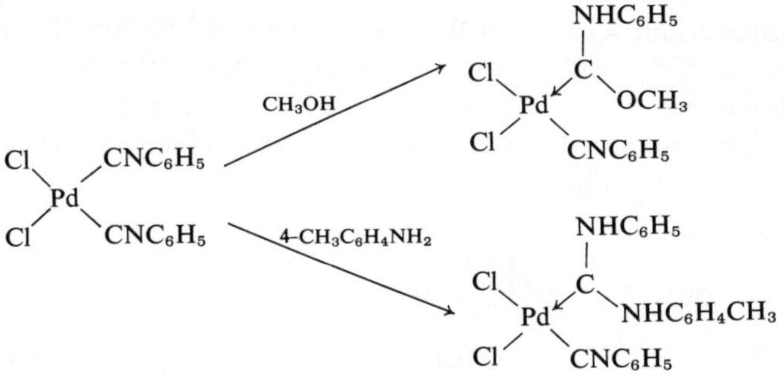

Protonation of some acyl metal derivatives occurs on the acyl oxygen producing carbene complexes. The protonated complex of cyclopentadienyl-carbonyltriphenylphosphineacetyliron, for example, appears to be a carbene complex (638).

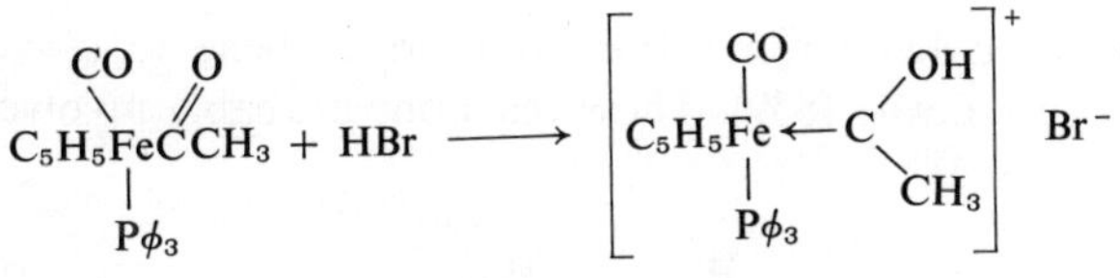

Protonation of alkoxymethyl metal derivatives is another route to carbene complexes if the protonated alkoxy group dissociates from the complex. Dicarbonylcyclopentadienylmethoxymethyliron(II) presumably reacts with fluoroboric acid to form the carbene complex (639).

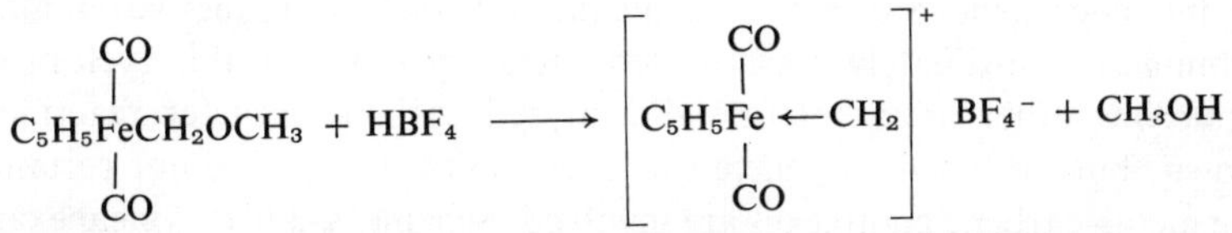

Carbene complexes may also be formed when various polyhalo compounds are reduced with metal complexes, although few examples of this reaction are known. An interesting carbene-bridged, binuclear complex has been obtained this way by heating octacarbonyldicobalt(0) with tetrafluoroethylene. An intermediate 1:2 adduct is first formed (640).

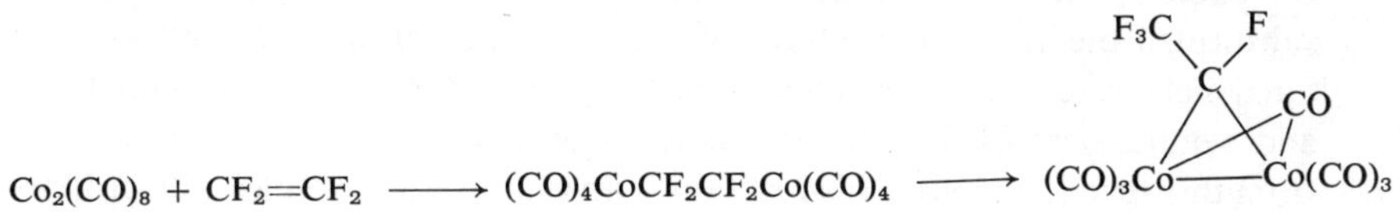

$$Co_2(CO)_8 + CF_2{=}CF_2 \longrightarrow (CO)_4CoCF_2CF_2Co(CO)_4 \longrightarrow (CO)_3Co{-}\!\!-\!\!{-}Co(CO)_3$$

2. Reactions of Carbene Complexes

Many of the isolable carbene complexes are capable of transferring the carbene group to olefins to form cyclopropane derivatives. Methoxyphenyl-carbenepentacarbonylchomium(0) reacts with methyl crotonate on heating to produce, in 60% yield, two isomeric cyclopropane derivatives. The trans relationship of the methyl and ester groups is retained in the transfer, while the carbene group adds both possible ways (641). Likewise, the carbenedi-

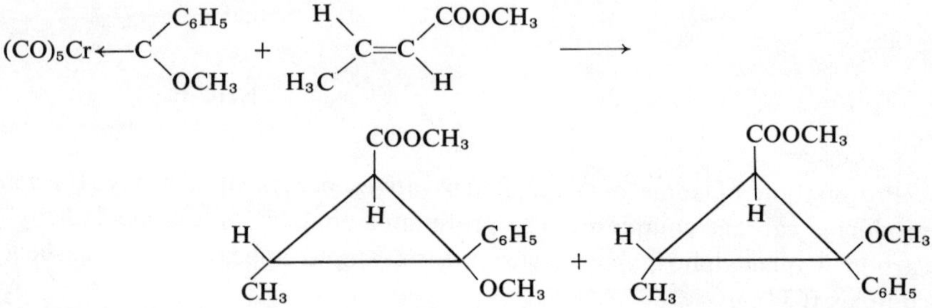

carbonylcyclopentadienyliron cation transfers carbene to *cis*- and *trans*-2-butene stereospecifically (639). These reactions probably involve addition of

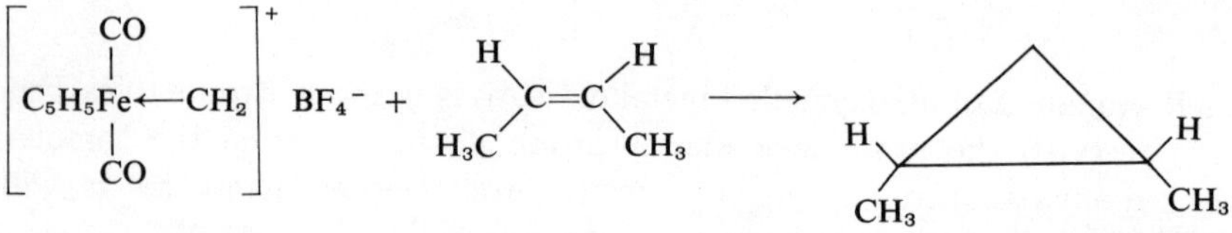

the metal–carbene group to the olefin, followed by reductive elimination forming a coordinately unsaturated metal group and the cyclopropane. Numerous other apparent, metal-catalyzed carbene transfer reactions are known, but intermediates have not been isolated and it is not certain that true metal–carbene complexes are involved. Scheme X-3 lists typical examples (642–645). The last reaction apparently involves a carbene insertion between the allyl group and the metal, followed by hydride elimination or hydride reduction.

A very interesting carbon–carbon double-bond cleavage occurs with certain electron-rich olefins and metal complexes. The reaction appears reversible with chlorotristriphenylphosphinerhodium(I) and the amine-substituted olefins since mixtures of the two different symmetrically substituted olefins are cleaved and recombined to give mixtures of the unsymmetrical and symmetrical olefins. Intermediate carbene complexes may be isolated from this reaction (634). This reaction is an example of an "olefin dispropor-

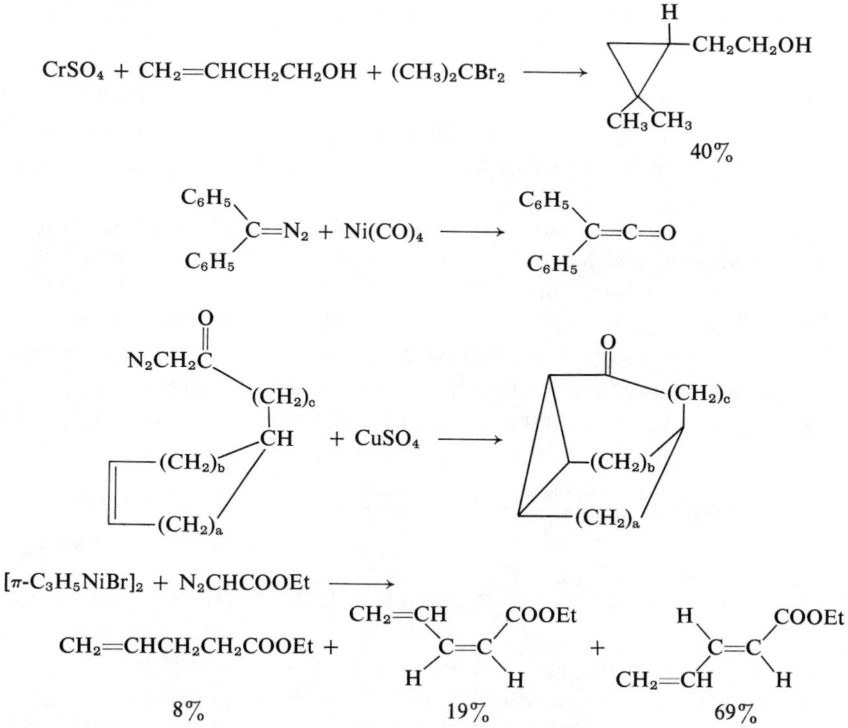

$$CrSO_4 + CH_2{=}CHCH_2CH_2OH + (CH_3)_2CBr_2 \longrightarrow$$

(structure: cyclopropane with H, $-CH_2CH_2OH$, CH_3 CH_3)

40%

$$\underset{C_6H_5}{\overset{C_6H_5}{>}}C{=}N_2 + Ni(CO)_4 \longrightarrow \underset{C_6H_5}{\overset{C_6H_5}{>}}C{=}C{=}O$$

$$N_2CH_2\overset{O}{\overset{\|}{C}} \underset{(CH_2)_a}{\overset{(CH_2)_c}{\diagup}} \quad + \; CuSO_4 \longrightarrow$$

(ring structures with $(CH_2)_c$, $(CH_2)_b$, $(CH_2)_a$)

$$[\pi\text{-}C_3H_5NiBr]_2 + N_2CHCOOEt \longrightarrow$$

$$CH_2{=}CHCH_2CH_2COOEt \; + \; \underset{H}{\overset{CH_2=CH}{>}}C{=}C\underset{H}{\overset{COOEt}{<}} \; + \; \underset{CH_2=CH}{\overset{H}{>}}C{=}C\underset{H}{\overset{COOEt}{<}}$$

8% 19% 69%

Scheme X-3.

tionation" and lends some support to the idea that other known disproportionation catalysts may operate by the same mechanism. It is already well
known that treatment of diazo compounds with some transition metal
complexes causes the dimerization of the presumed carbene groups left by
loss of nitrogen. This may be the same kind of reaction. Chloro-π-allyl-

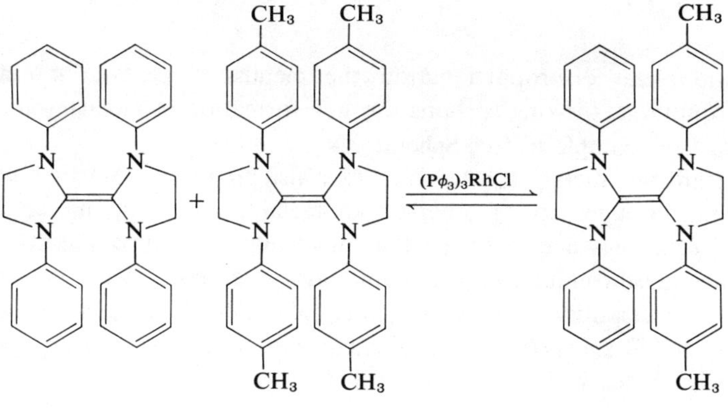

(structures with $(P\phi_3)_3RhCl$ equilibrium)

palladium(II) dimer and ethyl diazoacetate, for example, react to form diethyl fumarate (646).

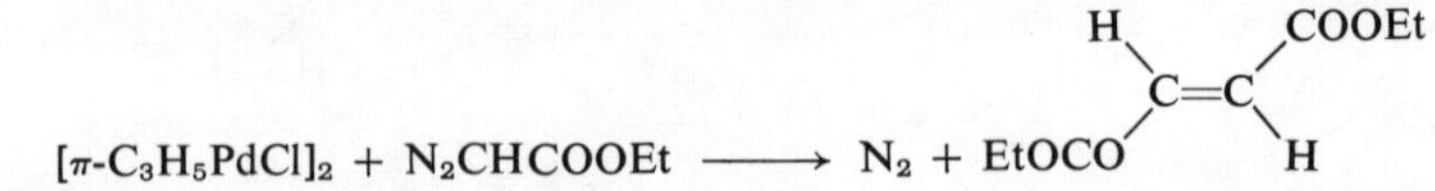

$$[\pi\text{-}C_3H_5PdCl]_2 + N_2CHCOOEt \longrightarrow N_2 + EtOCO \overset{H}{\underset{}{C}}=\overset{COOEt}{\underset{H}{C}}$$

The "olefin disproportionation" reaction may be carried out with a variety of homogeneous and heterogeneous catalysts. The reaction converts olefinic compounds into equilibrium mixtures of olefins formed formally by breaking double bonds and recombining the pieces. The reaction promises to be of significant value to the petroleum industry since many difficultly obtainable olefins can be produced from more readily available ones this way. The following reactions are typical, homogeneous examples (647, 648). The

$$CH_3CH_2CH_2CH=CH_2 \xrightarrow{(P\phi_3)_2Cl_2(NO)_2Mo + (CH_3)_3Al_2Cl_3}$$

$$CH_2=CH_2 + CH_3CH_2CH_2CH=CHCH_2CH_2CH_3$$

$$CH_3CH_2CH=CHCH_3 \xrightarrow{ReCl_5 + n\text{-}Bu_4Sn} CH_3CH=CHCH_3 + CH_3CH_2CH=CHCH_2CH_3$$

literature on these reactions has been reviewed recently (649, 650). π-Complexed cyclobutane intermediates have been proposed most often to explain the disproportionation, but other mechanisms have not been ruled out. In addition to possible tetracarbene intermediates, metalacyclopentane complexes must be considered. It has been shown, for example, that presumed tetrachlorotungstacyclopentane decomposes to ethylene (651).

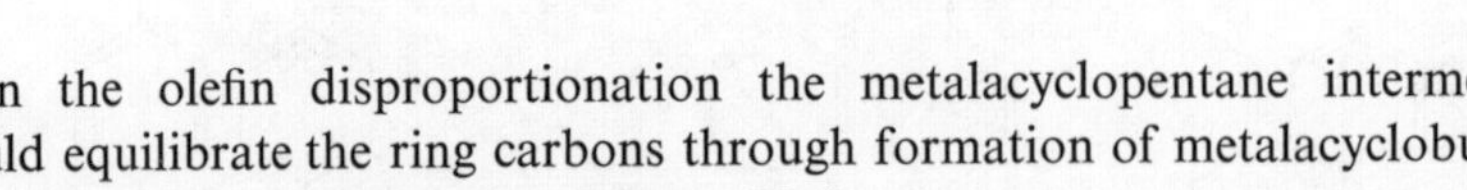

$$WCl_6 + LiCH_2CH_2CH_2CH_2Li \longrightarrow \left[Cl_4W \bigcirc \right] \longrightarrow 2CH_2=CH_2$$

In the olefin disproportionation the metalacyclopentane intermediate could equilibrate the ring carbons through formation of metalacyclobutane-monocarbene complexes (see Scheme X-4).

The transition metal–carbene complexes also are reactive starting materials for preparing a variety of other organometallic and inorganic complexes. E. O. Fischer has demonstrated the conversion of methoxymethylcarbene-pentacarbonylchromium(0) into numerous other complexes. Scheme X-5 shows the versatility of these reagents (652–655). The carbene ligand in methoxymethylcarbene(pentacarbonyl)chromium(0) is completely replaced when it is reacted with phosphorus trihalides or pyridine (656, 657).

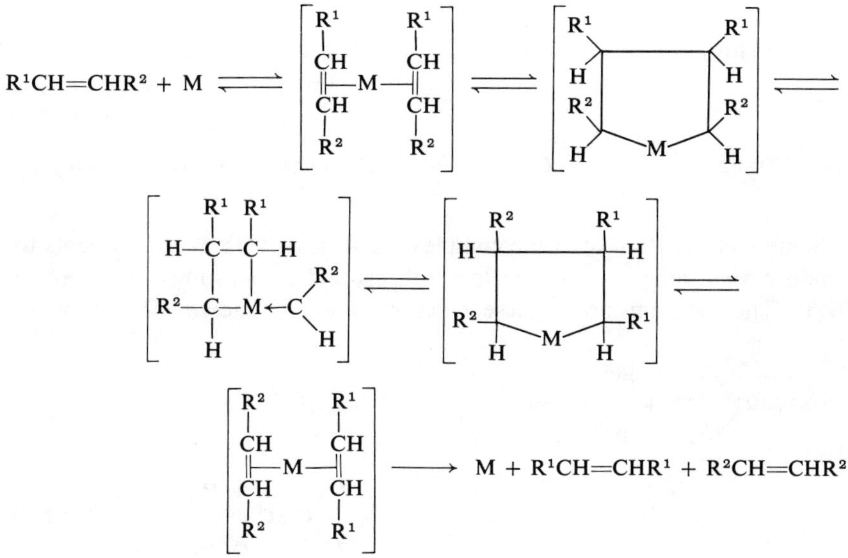

$$R^1CH=CHR^2 + M \rightleftharpoons \begin{bmatrix} \overset{R^1}{\underset{CH}{|}} \quad \overset{R^1}{\underset{CH}{|}} \\ CH-M-CH \\ \overset{|}{R^2} \quad \overset{|}{R^2} \end{bmatrix} \rightleftharpoons \begin{bmatrix} R^1 \quad\quad R^1 \\ H \quad\quad\quad H \\ R^2 \quad\quad\quad R^2 \\ H \quad M \quad H \end{bmatrix} \rightleftharpoons$$

$$\begin{bmatrix} \overset{R^1\ R^1}{H-C-C-H} \\ R^2-C-M\leftarrow C \overset{R^2}{\underset{H}{}} \\ \underset{H}{} \end{bmatrix} \rightleftharpoons \begin{bmatrix} R^2 \quad\quad R^1 \\ H- \quad\quad -H \\ R^2- \quad\quad -R^1 \\ H \quad M \quad H \end{bmatrix} \rightleftharpoons$$

$$\begin{bmatrix} \overset{R^2}{\underset{CH}{|}} \quad \overset{R^1}{\underset{CH}{|}} \\ CH=M=CH \\ \overset{|}{CH} \quad \overset{|}{CH} \\ \overset{|}{R^2} \quad \overset{|}{R^1} \end{bmatrix} \longrightarrow M + R^1CH=CHR^1 + R^2CH=CHR^2$$

Scheme X-4.

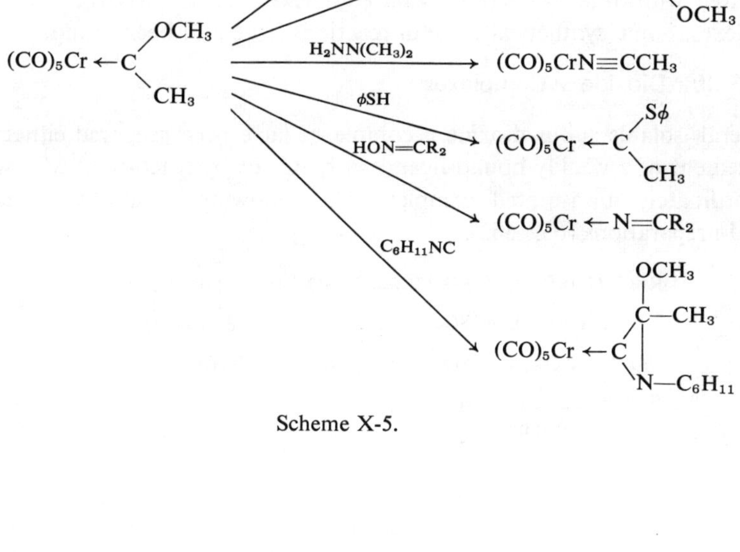

Scheme X-5.

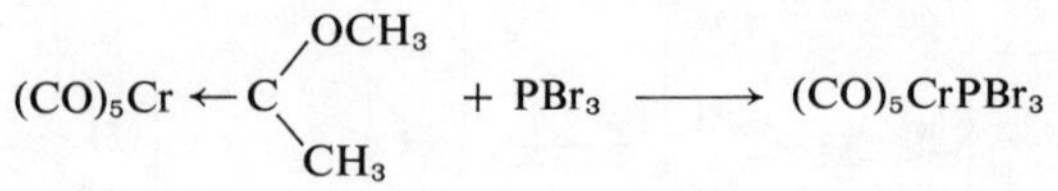

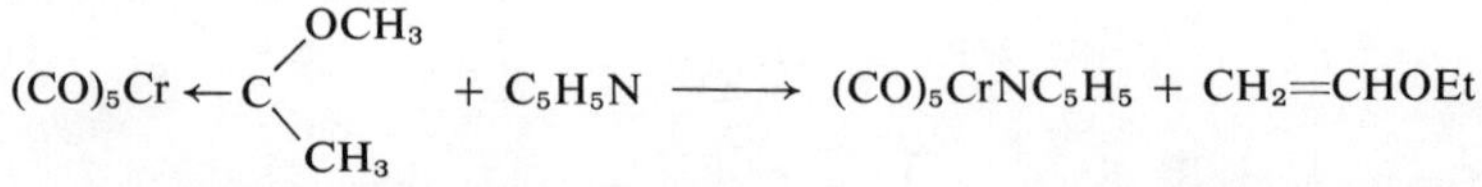

$$(CO)_5Cr \leftarrow C\underset{CH_3}{\overset{OCH_3}{{<}}} + PBr_3 \longrightarrow (CO)_5CrPBr_3$$

$$(CO)_5Cr \leftarrow C\underset{CH_3}{\overset{OCH_3}{{<}}} + C_5H_5N \longrightarrow (CO)_5CrNC_5H_5 + CH_2{=}CHOEt$$

Some metal methoxycarbene complexes also react with Wittig reagents to produce vinyl ethers in high yield. Both cis and trans isomers are formed (658). The mechanisms of these reactions are not known. The study of

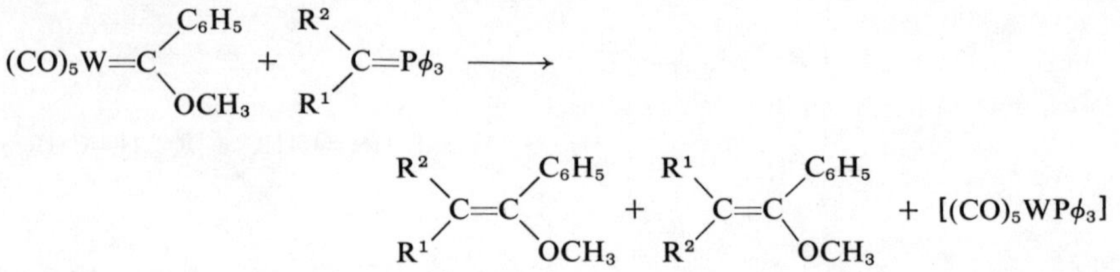

$$(CO)_5W{=}C\underset{OCH_3}{\overset{C_6H_5}{{<}}} + \underset{R^1}{\overset{R^2}{{>}}}C{=}P\phi_3 \longrightarrow$$

$$\underset{R^1}{\overset{R^2}{{>}}}C{=}C\underset{OCH_3}{\overset{C_6H_5}{{<}}} + \underset{R^2}{\overset{R^1}{{>}}}C{=}C\underset{OCH_3}{\overset{C_6H_5}{{<}}} + [(CO)_5WP\phi_3]$$

transition metal–carbene complexes has only just begun and is it apparent that a great deal of unusual and useful chemistry probably still remains to be discovered.

D. SULFUR DIOXIDE COMPLEXES

The transition metal–sulfur dioxide complexes do not promise to be of as much economic importance as those of dinitrogen, dioxygen, and carbene derivatives, but, nevertheless, are of considerable interest. The major interest stems from the different reaction mechanisms encountered in their reactions compared with other transition metal complexes, particularly the carbonyl complexes. Some synthetically useful reactions have also been found.

1. Sulfur Dioxide π Complexes

Several isolable sulfur dioxide π complexes have been prepared either by replacement of a weakly bound ligand with SO_2 or by reaction of SO_2 with a coordinately unsaturated complex. The following equations illustrate typical preparations (659–662).

$$RhCl(CO)(P\phi_3)_2 + SO_2 \rightleftharpoons SO_2RhCl(CO)(P\phi_3)_2$$

$$Pd(P\phi_3)_4 + SO_2 \rightleftharpoons Pd(SO_2)(P\phi_3)_3 + P\phi_3$$

$$dipyMo(CO)_3py + SO_2 \longrightarrow dipyMo(CO)_3SO_2 + py$$

$$dipy = 2,2'\text{-dipyridyl}$$
$$py = pyridine$$

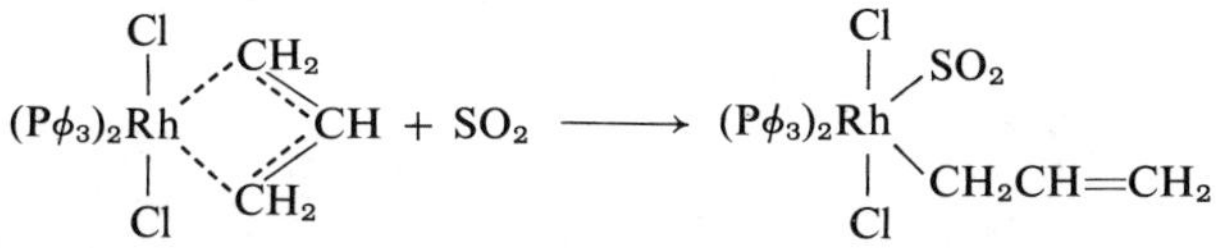

$$\text{(P}\phi_3)_2\text{Rh} \overset{\overset{\displaystyle Cl}{|}}{\underset{\overset{\displaystyle |}{Cl}}{}} \underset{CH_2}{\overset{CH_2}{\diamond}} CH + SO_2 \longrightarrow \text{(P}\phi_3)_2\text{Rh} \overset{\overset{\displaystyle Cl}{|}}{\underset{\overset{\displaystyle |}{Cl}}{}} \overset{SO_2}{\underset{CH_2CH=CH_2}{}}$$

In the above complexes, the bonding appears to be through sulfur to the metal with a trigonal arrangement of the substituents around sulfur.

2. Addition Reactions to Sulfur Dioxide

A variety of aryl and alkyl transition metal complexes react with sulfur dioxide to form addition compounds in which the SO_2 is inserted between the organic group and the metal (85). The usual product isolated is the sulfonyl metal complex, but there is evidence in many cases, at least, that the initial product is an oxygen–metal-bonded sulfinate which rapidly rearranges to the more stable sulfur–metal-bonded complex (663). The sulfonyl derivatives

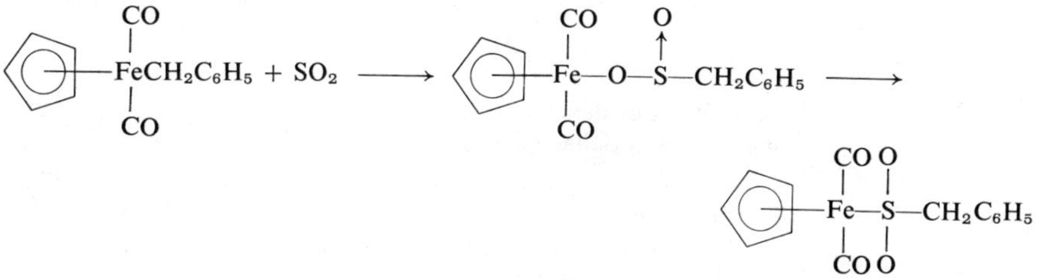

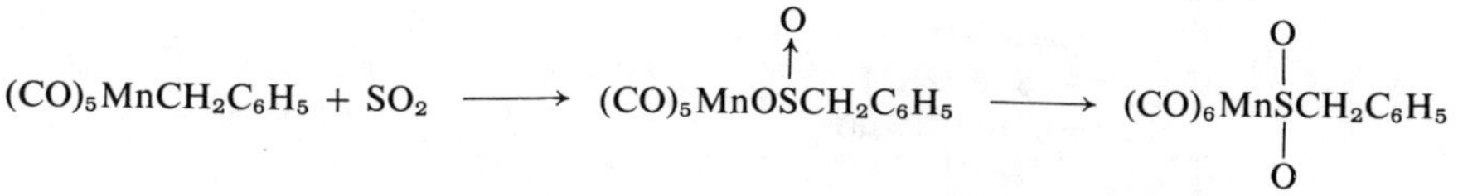

also may be obtained directly from sulfonyl chlorides or sulfonic acid anhydrides and metal complex anions or other metal derivatives (664, 665).

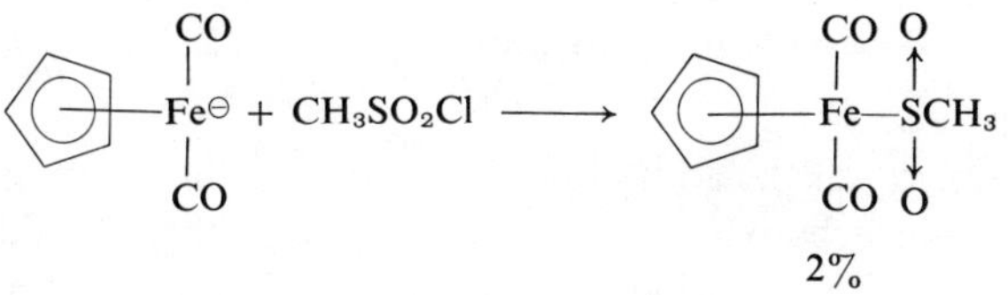

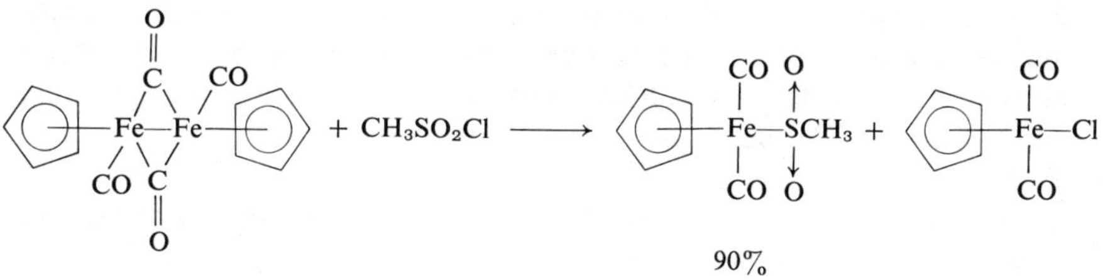

2%

90%

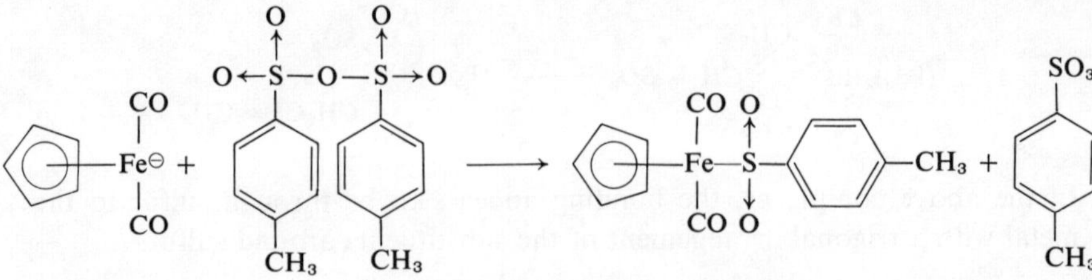

55%

The rates of reaction of a series of organoiron complexes with SO_2 have been measured. The results show SO_2 to be reacting as an electrophilic reagent with the rates quite dependent on the electronic character of the allyl or aryl group in the complex. The kinetics showed a first-order dependence on the iron complex, but the order with respect to the SO_2 was not determined. The relative rates observed are given in Table X-1 (666, 667).

TABLE X-1
Relative Rates of Reaction of Various Organoiron Compounds with Sulfur Dioxide[a]

Compound [M = $C_5H_5Fe(CO)_2^-$]	Relative rates of reaction at $-40°$ with liquid SO_2
C_6H_5—M	1.0
4-$CH_3C_6H_4M$	33
4-$CH_3OC_6H_4M$	720
$C_6H_5CH_2M$	80
CH_3OCH_2M	4
CH_3M	1400
CH_3CH_2M	1600

[a] Data taken from Hartman et al. (667).

The stereochemistry of the addition to sulfur dioxide has been investigated by reacting *threo*-dicarbonylcyclopentadienyl-3,3-dimethyl-1,2-dideuterobutyliron(II) with SO_2. The sulfonyliron product was found to be about 80% inverted in contrast with the carbonylation product which was formed with complete retention (113). The data can be accommodated by a mechanism which involves an initial formation of an acyliron complex followed by oxygen coordination of SO_2 and transfer of the organic group from the acyl carbonyl to sulfur.

The reaction of SO_2 with allyl and propargyl metal complexes appears to proceed by a different mechanism. The insertion with unsymmetrical allylic groups takes place with allylic rearrangement. For example, pentacarbonyl-crotylmanganese(I) and SO_2 form the 3-buten-2-ylsulfonylmanganese complex exclusively. Conversely, the 3-buten-2-ylmanganese compound gives the crotylsulfonyl metal complex (667). Propargyl metal compounds and SO_2

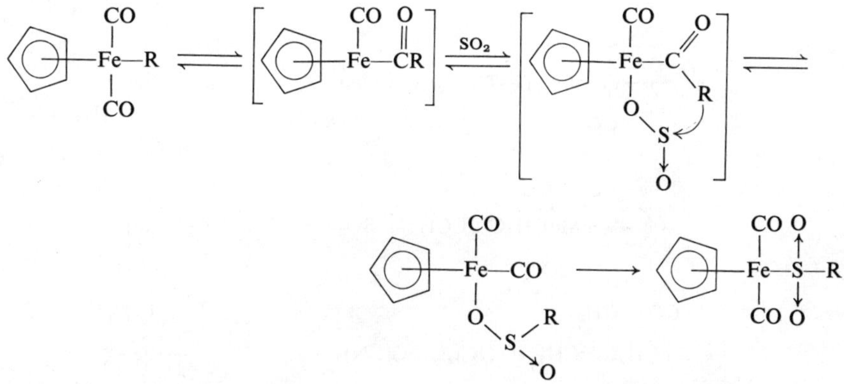

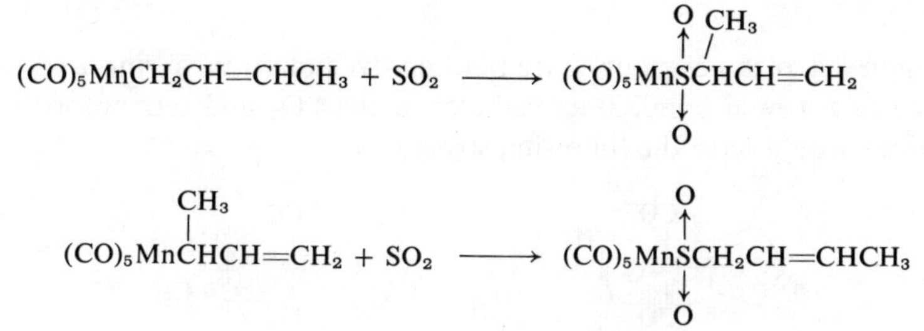

form cyclic sulfinate esters (668, 669). Two other types of reagents are known

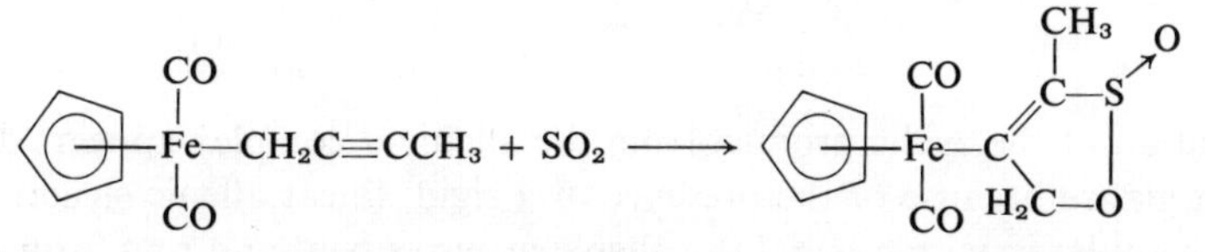

to react with propargylic metal complexes in a similar manner—N-thionyl-aniline (670) and sulfur trioxide (671). Tetracyanoethylene and chlorosulfonyl-isocyanate (672) react with allyl metal complexes in an analogous way (673).

A reasonable mechanism has been proposed for the allyl metal reaction which seems applicable to the propargyl reactions as well (673). An electro-philic attack of the reagent on the allylic (or propargylic) unsaturation is proposed. A dipolar intermediate is formed between the unsaturated organic ligand and the attacking reagent and the original metal σ bond is converted into an olefinic (or allenic) π bond. The anionic group may then attack the

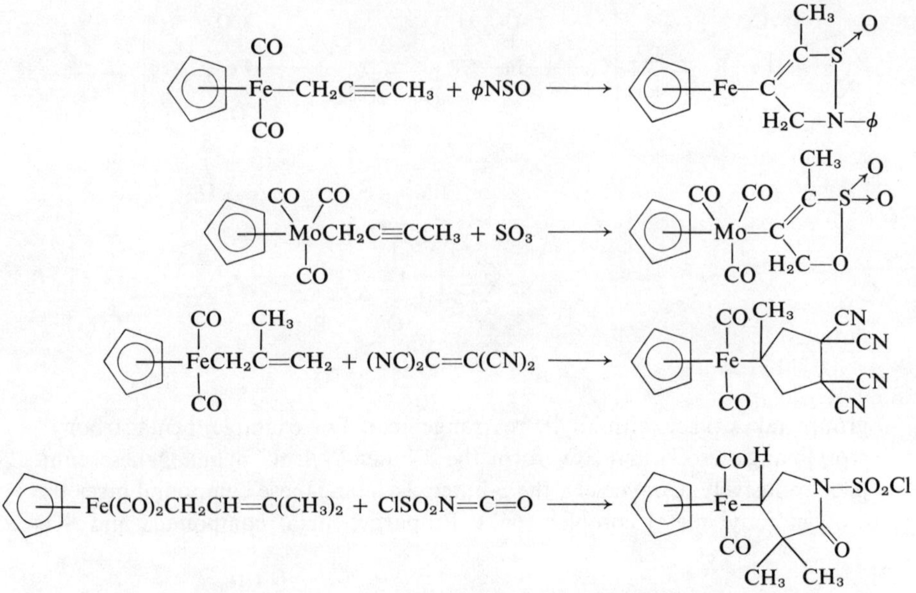

$$Cp{-}Fe(CO)_2{-}CH_2C{\equiv}CCH_3 + \phi NSO \longrightarrow$$

$$Cp{-}Mo(CO)_3{-}CH_2C{\equiv}CCH_3 + SO_3 \longrightarrow$$

$$Cp{-}Fe(CO)_2{-}CH_2C(CH_3){=}CH_2 + (NC)_2C{=}C(CN)_2 \longrightarrow$$

$$Cp{-}Fe(CO)_2CH_2CH{=}C(CH_3)_2 + ClSO_2N{=}C{=}O \longrightarrow$$

π-complexed olefinic group intramolecularly and form a ring and simultaneously a new σ bond. Intermediates in the SO_2 and tetracyanoethylene reactions would have the following structures.

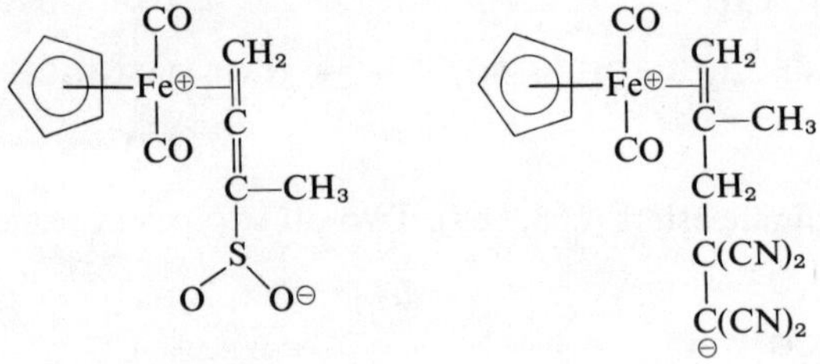

Formation of the cyclic product from the allenic π complex appears difficult at first glance because of the presence of a rigid, linear allenic structure, but when complexed with a metal the allenic group is bent and ring formation is probably not unfavorable.

3. Catalytic Reactions

The addition of organotransition metal compounds to SO_2 is reversible and sulfonyl chlorides may be desulfonated catalytically with a variety of transition metal complexes. For example, chlorotristriphenylphosphine-rhodium(I) and arylsulfonyl chlorides form aryl chlorides and SO_2 in 60–85% yields (674). The true catalyst is apparently not the compound added but rather an SO_2-bridged monophosphine dimer.

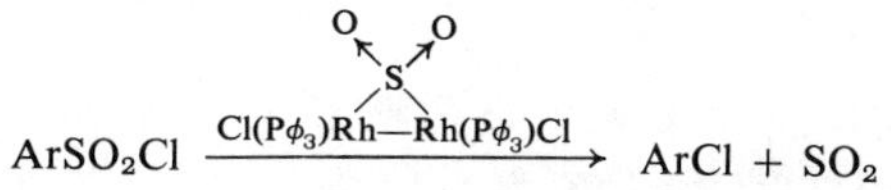

$$ArSO_2Cl \xrightarrow{\underset{Cl(P\phi_3)Rh-Rh(P\phi_3)Cl}{\overset{O_{\diagdown S}{\diagup}O}{}}} ArCl + SO_2$$

Some sulfinate salts also undergo desulfonation rather easily forming organometallics. Heating sodium arylsulfinates with sodium tetrachloropalladate produces biaryls, SO_2, and palladium metal, presumably by way of a diarylpalladium complex (675).

$$ArSO_2Na + Na_2PdCl_4 \longrightarrow [(ArSO_2)_2Pd] \longrightarrow Ar_2 + SO_2 + Pd$$

An interesting reaction between ethylene and SO_2 occurs with palladium chloride as catalyst (676) forming mainly crotylethyl sulfone and, to a minor extent, ethyl vinyl sulfone. The mechanism probably involves a hydridochloropalladium complex as the true catalyst. The hydride can add to ethylene and the ethyl derivative could add to SO_2. A second reaction with ethylene would produce a 2-ethylsulfonylethylpalladium species. Elimination of hydride would give the minor product, while addition to another ethylene and hydride elimination with double-bond migration would give the major product (see Scheme X-6).

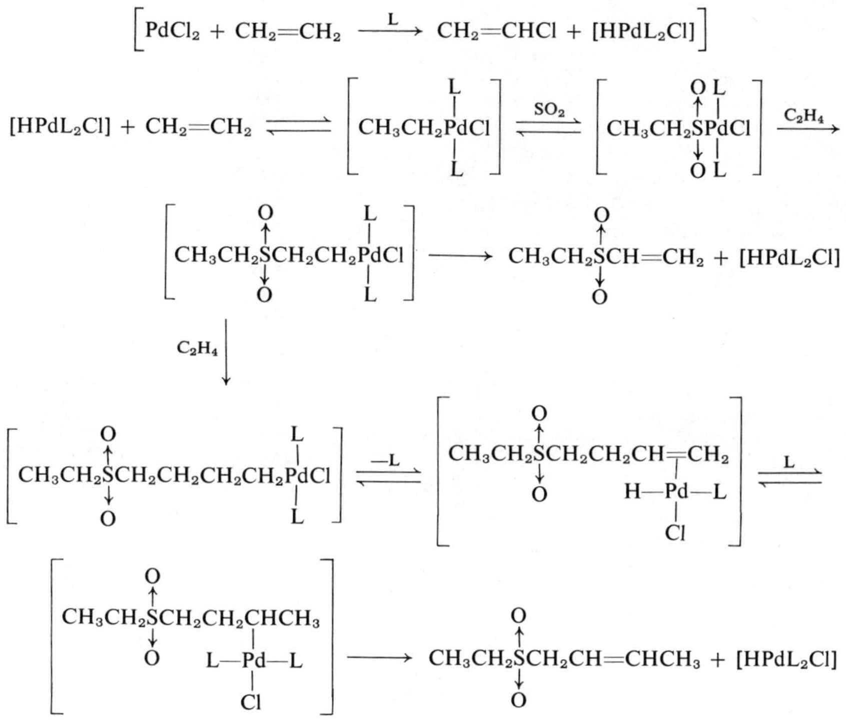

$$\left[PdCl_2 + CH_2{=}CH_2 \xrightarrow{\ L\ } CH_2{=}CHCl + [HPdL_2Cl] \right]$$

$$[HPdL_2Cl] + CH_2{=}CH_2 \rightleftharpoons \left[CH_3CH_2\overset{\overset{\displaystyle L}{|}}{\underset{\underset{\displaystyle L}{|}}{Pd}}Cl \right] \overset{SO_2}{\rightleftharpoons} \left[CH_3CH_2\overset{\overset{\displaystyle O}{\uparrow}}{\underset{\underset{\displaystyle O}{\downarrow}}{S}}\overset{\overset{\displaystyle L}{|}}{\underset{\underset{\displaystyle L}{|}}{Pd}}Cl \right] \xrightarrow{C_2H_4}$$

$$\left[CH_3CH_2\overset{\overset{\displaystyle O}{\uparrow}}{\underset{\underset{\displaystyle O}{\downarrow}}{S}}CH_2CH_2\overset{\overset{\displaystyle L}{|}}{\underset{\underset{\displaystyle L}{|}}{Pd}}Cl \right] \longrightarrow CH_3CH_2\overset{\overset{\displaystyle O}{\uparrow}}{\underset{\underset{\displaystyle O}{\downarrow}}{S}}CH{=}CH_2 + [HPdL_2Cl]$$

$$C_2H_4 \Big\downarrow$$

$$\left[CH_3CH_2\overset{\overset{\displaystyle O}{\uparrow}}{\underset{\underset{\displaystyle O}{\downarrow}}{S}}CH_2CH_2CH_2CH_2\overset{\overset{\displaystyle L}{|}}{\underset{\underset{\displaystyle L}{|}}{Pd}}Cl \right] \rightleftharpoons \left[\begin{array}{c} CH_3CH_2\overset{\overset{\displaystyle O}{\uparrow}}{\underset{\underset{\displaystyle O}{\downarrow}}{S}}CH_2CH_2CH{=}CH_2 \\ H-\underset{\underset{\displaystyle Cl}{|}}{Pd}-L \end{array} \right] \xrightarrow{L} \rightleftharpoons$$

$$\left[\begin{array}{c} CH_3CH_2\overset{\overset{\displaystyle O}{\uparrow}}{\underset{\underset{\displaystyle O}{\downarrow}}{S}}CH_2CH_2CHCH_3 \\ L-\underset{\underset{\displaystyle Cl}{|}}{Pd}-L \end{array} \right] \longrightarrow CH_3CH_2\overset{\overset{\displaystyle O}{\uparrow}}{\underset{\underset{\displaystyle O}{\downarrow}}{S}}CH_2CH{=}CHCH_3 + [HPdL_2Cl]$$

Scheme X-6.

FURTHER READING

E. O. Fischer, Carbene complexes. *Pure Appl. Chem.* **24**, 407 (1970).

E. O. Fischer, Carbene complexes. *Pure Appl. Chem.* **30**, 353 (1972).

M. Kilner, Nitrogen groups in metal carbonyls. *Advan. Organometal. Chem.* **10**, 115 (1972).

W. Kitching and C. W. Fong, Sulfur dioxide insertions. *Organometal. Chem. Rev. Sect. A* **5**, 281 (1970).

M. F. Lappert, Carbene complexes. *Chem. Rev.* **72**, 545 (1972).

O. M. Nefedov and M. N. Manakov, Carbene complexes. *Angew. Chem., Int. Ed. Engl.* **5**, 1021, (1966).

E. E. van Tamelen, Nitrogen fixation. *Accounts Chem. Res.* **3**, 361 (1970).

L. Vaska, O_2 and SO_2 complexes. *Accounts Chem. Res.* **1**, 335 (1968).

M. E. Vol'pin and V. B. Shur, Chemical fixation of molecular nitrogen. *Organometal. React.* **1**, 55 (1970).

A. Wojcicki, Sulfur dioxide reactions with transition metal compounds. *Accounts Chem. Res.* **4**, 344 (1971).

A. Wojcicki, Sulfur dioxide insertions. *Advan. Organometal. Chem.* **12**, 31 (1974).

References

1. W. C. Zeise, *Ann. Phys.* (*Leipzig*) [2] **9**, 632 (1827).
2. G. Wilkinson, M. Rosenblum, M. C. Whiting, and R. B. Woodward, *J. Amer. Chem. Soc.* **74**, 2125 (1952).
3. M. Iwamoto and S. Yuguchi, *J. Org. Chem.* **31**, 4290 (1966).
4. G. P. Chiusoli, G. Bottaccio, and C. Venturello, *Tetrahedron Lett.* p. 2875 (1965).
5. L. Cassar, P. E. Eaton, and J. Halpern, *J. Amer. Chem. Soc.* **92**, 6366 (1970).
6. H. Roehl, E. Lange, T. Golsal, and G. Roth, *Angew. Chem.* **74**, 155 (1962).
7. C. A. Tolman, *Chem. Soc. Rev.* **1**, 337 (1972).
8. P. S. Skell and J. J. Havel, *J. Amer. Chem. Soc.* **93**, 6687 (1971).
9. L. E. Orgel, "An Introduction to Transition-Metal Chemistry," 2nd ed. Methuen, London, 1966.
10. H. L. Schlafer and G. Gliemann, "Ligand Field Theory." Wiley (Interscience), New York, 1969.
11. C. J. Ballhausen and H. B. Grey, "Molecular Orbital Theory." Benjamin, New York, 1964.
12. M. D. Johnson and N. Winterton, *J. Chem. Soc., A* p. 507 (1970).
13. S. N. Anderson, D. H. Ballard, and M. D. Johnson, *Chem. Commun.* p. 779 (1971).
14. B. Nicholls and M. C. Whiting, *J. Chem. Soc., London* p. 551 (1959).
15. L. B. Handy, P. M. Treichel, L. F. Dahl, and R. G. Hayter, *J. Amer. Chem. Soc.* **88**, 366 (1966); V. Anders and W. A. G. Graham, *Chem. Commun.* p. 499 (1965).
16. B. D. James, R. K. Nanda, and M. G. H. Wallbridge, *Inorg. Chem.* **6**, 1979 (1967).
17. R. S. Nyholm, *Proc. Int. Congr. Pure Appl. Chem.* **23**, Symp. 0–8 (1971).
18. R. Cramer, *J. Amer. Chem. Soc.* **86**, 217 (1964).
19. J. Ashley-Smith, I. Donek, B. F. G. Johnson, and J. Lewis, *J. Chem. Soc., Dalton Trans.* p. 1776 (1972).
20. B. F. G. Johnson and J. A. Segal, *J. Chem. Soc., Chem. Commun.* p. 1312 (1972).
21. E. O. Fischer and G. E. Herberich, *Chem. Ber.* **94**, 1517 (1961).
22. F. A. Cotton, *Accounts Chem. Res.* **1**, 256 (1968).

23. R. Ben-Shoshan and R. Pettit, *J. Amer. Chem. Soc.* **89**, 2231 (1967).
24. *J. Amer. Chem. Soc.* **82**, 5537 (1960).
25. F. A. Cotton, *J. Amer. Chem. Soc.* **90**, 6230 (1968).
26. L. Cattalini, *Progr. Inorg. Chem.* **13**, 263 (1970).
27. F. Basolo, J. Chatt, H. B. Gray, R. G. Pearson, and B. L. Shaw, *J. Chem. Soc., London*, p. 2207 (1961).
28. W. Baddley and F. Basolo, *J. Amer. Chem. Soc.* **88**, 2944 (1966).
29. R. G. Pearson, L. F. Kangas, and P. M. Henry, *J. Amer. Chem. Soc.* **90**, 1925 (1968).
30. E. M. Thorsteinson and F. Basolo, *J. Amer. Chem. Soc.* **88**, 3929 (1966).
31. E. E. Siefert and R. J. Angelici, *J. Organometal. Chem.* **8**, 374 (1967).
32. R. F. Heck, *J. Amer. Chem. Soc.* **87**, 2572 (1965).
33. R. F. Heck, *J. Amer. Chem. Soc.* **85**, 651 (1963).
34. J. Chatt and B. L. Shaw, *J. Chem. Soc., London* p. 1718 (1960).
35. A. Wojcicki and F. Basolo, *J. Inorg. Nucl. Chem.* **17**, 77 (1961).
36. K. Noack and F. Calderazzo, *J. Organometal. Chem.* **10**, 101 (1967).
37. K. Noack, M. Ruck, and F. Calderazzo, *Inorg. Chem.* **7**, 345 (1968).
38. C. White and R. J. Mawby, *Inorg. Chim. Acta* **4**, 261 (1970).
39. A. J. Hart-Davis, C. White, and R. J. Mawby, *Inorg. Chim. Acta* **4**, 441 (1970).
40. P. S. Braterman and R. J. Cross, *J. Chem. Soc., Dalton Trans.* p. 657 (1972).
41. S. J. Ashcroft and C. T. Mortimer, *J. Chem. Soc., A* p. 930 (1967).
42. B. De Vries, *J. Catal.* **1**, 489 (1962).
43. A. Davison, J. A. McCleverty, and G. Wilkinson, *J. Chem. Soc., London* p. 1133 (1963).
44. L. L'Eplattenier and F. Calderazzo, *Inorg. Chem.* **7**, 1290 (1968).
45. M. A. Bennett and D. L. Milner, *J. Amer. Chem. Soc.* **91**, 6983 (1969).
46. J. P. Collman and W. R. Roper, *J. Amer. Chem. Soc.* **87**, 4008 (1965).
47. W. C. Drinkard, D. R. Eaton, J. P. Jesson, and R. V. Lindsey, Jr., *Inorg. Chem.* **9**, 392 (1970).
48. W. Hieber and G. Wagner, *Z. Naturforsch. B* **12**, 478 (1957).
49. J. Chatt and B. L. Shaw, *Chem. Ind. (London)* p. 931 (1960).
50. J. Chatt, R. S. Coffey, A. Gough, and D. T. Thompson, *J. Chem. Soc., A* p. 190 (1968).
51. S. S. Bath and L. Vaska, *J. Amer. Chem. Soc.* **85**, 3500 (1963).
52. G. Calvin and G. E. Coates, *J. Chem. Soc., London* p. 2008 (1960).
53. W. P. Long and D. S. Breslow, *J. Amer. Chem. Soc.* **82**, 1953 (1960).
54. R. B. King, P. M. Treichel, and F. G. A. Stone, *J. Amer. Chem. Soc.* **83**, 3592 (1961).
55. A. J. Oliver and W. A. G. Graham, *Inorg. Chem.* **9**, 243 (1970).
56. T. H. Coffield, J. Kozikowski, and R. D. Closson, *J. Org. Chem.* **22**, 598 (1957).
57. R. E. Dessy, R. L. Pohl, and R. B. King, *J. Amer. Chem. Soc.* **88**, 5121 (1966).
58. Y. Yamamoto and H. Yamazaki, *Coord. Chem. Rev.* **8**, 225 (1972).
59. G. Booth and J. Chatt, *J. Chem. Soc., A* p. 634 (1966).
60. R. F. Heck, *J. Organometal. Chem.* **2**, 195 (1964).
61. J. Palagyi and L. Markó, *J. Organometal. Chem.* **17**, 453 (1969).
62. Y. Yamamoto, H. Yamazaki, and N. Hagihara, *J. Organometal. Chem.* **18**, 189 (1969).
63. R. F. Heck and D. S. Breslow, *J. Amer. Chem. Soc.* **84**, 2499 (1962).
64. W. Schoeller, W. Schranth, and W. Essers, *Chem. Ber.* **46**, 2864 (1913).
65. J. Chatt, L. M. Vallarino, and L. M. Venanzi, *J. Chem. Soc., London* p. 2496 (1957).

66. M. L. H. Green and P. L. I. Nagy, *J. Organometal. Chem.* **1**, 58 (1963).
67. J. Halpern and L. Wong, *J. Amer. Chem. Soc.* **90**, 6665 (1968).
68. W. Beck and M. Bauder, *Chem. Ber.* **103**, 583 (1970).
69. J. Ashley-Smith, M. Green, and F. G. A. Stone, *J. Chem. Soc., Dalton Trans.* p. 1805 (1972).
70. W. R. McClellan, H. H. Hoehn, H. N. Cripps, E. L. Muertterties, and B. W. Howk, *J. Amer. Chem. Soc.* **83**, 1601 (1961).
71. R. Heck, *J. Amer. Chem. Soc.* **87**, 4727 (1965).
72. A. C. Cope and E. C. Friedrich, *J. Amer. Chem. Soc.* **90**, 909 (1968).
73. G. E. Hartwell, R. Lawrence, and M. J. Smas, *Chem. Commun.* p. 912 (1970).
74. M. O. Unger and R. A. Foutz, *J. Org. Chem.* **34**, 18 (1969).
75. R. L. Bennett, M. I. Bruce, B. L. Goodall, M. Z. Iqbal, and F. G. A. Stone, *J. Chem. Soc., Dalton Trans.* p. 1787 (1972).
76. M. I. Bruce, B. L. Goodall, M. Z. Iqbal, F. G. A. Stone, R. J. Doldens, and R. G. Little, *Chem. Commun.* p. 1595 (1971).
77. A. J. Cheney, B. E. Mann, B. L. Shaw, and R. M. Slade, *J. Chem. Soc., A* p. 3833 (1971).
78. C. Masters, *Chem. Commun.* p. 191 (1973).
79. J. L. Garnett and R. S. Kenyon, *Chem. Commun.* p. 1227 (1971).
80. C. Masters, *Chem. Commun.* p. 1258 (1972).
81. T. S. Piper and G. Wilkinson, *Naturwissenschaften* **42**, 625 (1955).
82. K. Matsumoto, Y. Odaira, and S. Tsutsumi, *Chem. Commun.* p. 832 (1968).
83. W. Beck, W. Hieber, and H. Tengler, *Chem. Ber.* **94**, 862 (1961).
84. J. J. Alexander and A. Wojcicki, *J. Organometal. Chem.* **15**, P23 (1968).
85. A. Wojcicki, *Accounts Chem. Res.* **4**, 344 (1971).
86. R. Huttel and H. Schmid, *Chem. Ber.* **101**, 252 (1968).
87. G. Parshall, *J. Amer. Chem. Soc.* **87**, 2133 (1965).
88. P. B. Chock and J. Halpern, *J. Amer. Chem. Soc.* **91**, 582 (1969).
89. P. W. Jolly, I. Tkatchenko, and G. Wilke, *Angew. Chem.* **83**, 329 (1971).
90. J. M. Brown, B. T. Golding, and M. J. Smith, *Chem. Commun.* p. 1240 (1971).
91. A. Bond, B. Lewis, and S. F. W. Lowrie, *Chem. Commun.* p. 1230 (1971).
92. H. Yamazaki and N. Hagihara, *J. Organometal. Chem.* **7**, P22 (1967).
93. F. G. A. Stone, *Pure Appl. Chem.* **30**, 551 (1972).
94. K. Bittler, N. V. Kutepow, D. Neubauer, and H. Reis, *Angew. Chem., Int. Ed. Engl.* **7**, 329 (1968).
94a. L. Markó, *Proc. Chem. Soc., London* p. 67 (1962).
94b. R. F. Heck, *J. Amer. Chem. Soc.* **86**, 2819 (1964).
94c. R. Heck and D. S. Breslow, *J. Amer. Chem. Soc.* **85**, 2779 (1963).
94d. W. Kitching, Z. Rappoport, S. Winstein, and W. G. Young, *J. Amer. Chem. Soc.* **88**, 2054 (1966).
95. E. O. Greaves, G. R. Knox, and P. L. Pauson, *Chem. Commun.* p. 1124 (1969).
96. K. M. Nicholas and R. Pettit, *Tetrahedron Lett.* p. 3475 (1971).
97. D. Seyferth and A. T. Wehman, *J. Amer. Chem. Soc.* **92**, 5520 (1970).
98. W. Grimme, *Chem. Ber.* **100**, 113 (1967).
99. G. Wilke and B. Bogdanovic, *Angew. Chem.* **73**, 756 (1961).
100. G. M. Whitesides, J. F. Gaasch, and E. R. Stedronsky, *J. Amer. Chem. Soc.* **94**, 5258 (1971).
101. R. F. Heck, *J. Amer. Chem. Soc.* **85**, 3383 (1963).
102. C. F. Kohll and R. van Helden, *Rec. Trav. Chim. Pays-Bas* **87**, 481 (1968).
103. A. J. Chalk and J. F. Harrod, *J. Amer. Chem. Soc.* **89**, 1640 (1967).

104. R. Heck, *J. Amer. Chem. Soc.* **90**, 5542 (1968).

105. J. C. Barborak, L. Watts, and R. Pettit, *J. Amer. Chem. Soc.* **88**, 1328 (1966).

106. R. Heck, *J. Amer. Chem. Soc.* **86**, 5138 (1964).

107. R. F. Heck and D. S. Breslow, *Actes Congr. Int. Catal.*, *2nd, 1960* p. 671 (1960).

108. W. Kruerke, C. Hoogzand, and W. Hübel, *Chem. Ber.* **94**, 2817 (1961).

109. O. S. Mills and G. Robinson, *Proc. Chem. Soc., London* p. 187 (1964).

110. R. Criegee and G. Schroder, *Angew. Chem.* **71**, 70 (1959).

111. R. W. Johnson and R. G. Pearson, *Inorg. Chem.* **10**, 2091 (1971).

112. R. W. Johnson and R. G. Pearson, *Chem. Commun.* p. 986 (1970).

113. G. M. Whitesides and D. J. Boschetto, *J. Amer. Chem. Soc.* **93**, 1529 (1971).

114. D. Evans, G. Yagupsky, and G. Wilkinson, *J. Chem. Soc., A* pp. 2660 and 2665 (1968).

115. J. A. Osborn, F. H. Jardin, J. F. Young, and G. Wilkinson, *J. Chem. Soc., A* p. 1711 (1966).

116. C. A. Tolman, T. Z. Meskin, D. L. Lindner, and J. P. Jesson, to be published.

117. P. Meakin, J. P. Jesson, and C. A. Tolman, *J. Amer. Chem. Soc.* **94**, 3240 (1972).

118. J. P. Candlin and A. R. Oldham, *Discuss. Faraday Soc.* **46**, 60 (1968).

119. R. L. Augustine and J. F. Van Peppen, *Chem. Commun.* p. 571 (1970).

120. P. S. Hallman, B. R. McGarvey, and G. Wilkinson, *J. Chem. Soc., A* p. 3143 (1968).

121. M. G. Burnett and R. J. Morrison, *J. Chem. Soc., A* p. 2325 (1971).

122. G. F. Ferrari, A. Andreeta, G. F. Pregaglia, and R. Ugo, *J. Organometal. Chem.* **43**, 213 (1972).

123. R. Rose, J. D. Gilbert, R. P. Richardson, and G. Wilkinson, *J. Chem. Soc., A* p. 2610 (1969).

124. B. C. Hui and G. C. Rempel, *Chem. Commun.* p. 1195 (1970).

125. H. van Gaal, H. Cappers, and A. van der Ent, *Chem. Commun.* p. 1694 (1970).

126. J. R. Shapley, R. R. Schrock, and J. Osborn, *J. Amer. Chem. Soc.* **91**, 2816 (1969).

127. R. R. Schrock and J. Osborn, *Chem. Commun.* p. 567 (1970).

128. C. O'Connor and G. Wilkinson, *Tetrahedron Lett.* p. 1375 (1969).

129. J. D. Morrison, R. E. Burnett, A. M. Aquiar, C. J. Morrow, and C. Phillips, *J. Amer. Chem. Soc.* **93**, 1301 (1971).

130. W. S. Knowles, M. J. Sabacky, and B. D. Vineyard, *Chem. Commun.* p. 10 (1972).

131. H. B. Kagan and T. Dang, *J. Amer. Chem. Soc.* **94**, 6429 (1972).

132. P. Abley and F. J. McQuillin, *Chem. Commun.* p. 477 (1969).

133. J. D. Morrison, *Surv. Progr. Chem.* **3**, 147 (1966).

134. R. H. Grubbs and L. C. Kroll, *J. Amer. Chem. Soc.* **93**, 3062 (1971).

135. E. N. Frankel, F. Selke, and C. A. Glass, *J. Amer. Chem. Soc.* **90**, 2446 (1968).

136. E. N. Frankel and R. Butterfield, *J. Org. Chem.* **34**, 3930 and 3936 (1969).

137. M. Cais, E. N. Frankel, and A. Rejoun, *Tetrahedron Lett.* p. 1919 (1968).

138. T. Funabiki and K. Tarama, *Chem. Commun.* p. 1177 (1971).

139. I. Jardin and F. J. McQuillin, *Tetrahedron Lett.* p. 4871 (1966).

140. E. N. Frankel, E. A. Emken, H. Itatani, and J. C. Bailar, Jr., *J. Org. Chem.* **32**, 1447 (1969).

141. H. A. Tayim and J. C. Bailer, Jr., *J. Amer. Chem. Soc.* **89**, 4330 (1970).

142. A. Miyake and H. Kondo, *Angew. Chem., Int. Ed. Engl.* **7**, 880 (1968).

143. E. N. Frankel, E. A. Emken, H. M. Peters, V. L. Davison, and R. O. Butterfield, *J. Org. Chem.* **29**, 3292 (1964).

144. R. S. Coffey, *Chem. Commun.* p. 923 (1967).

145. P. Bonvicini, A. Levi, G. Modena, and G. Sconano, *Chem. Commun.* p. 1188 (1972).

146. Y. Ohgo, S. Takeuchi, and J. Yoshimura, *Bull. Chem. Soc. Jap.* **44**, 283 and 583 (1971).

147. A. J. Birch and D. H. Williamson, *in* "Organic Reactions" (W. G. Dauben, ed.), Vol. 21, Wiley, New York, 1974 (in press).

148. A. F. M. Iqbal, *Tetrahedron Lett.* p. 3385 (1971).

149. J. M. Landesberg, L. Katz, and C. Olsen, *J. Org. Chem.* 37, 930 (1972).

150. B. E. Mann, B. L. Shaw, and N. I. Tucker, *J. Chem. Soc., A* p. 2667 (1971).

151. J. Trocha-Grimshaw and H. B. Henbest, *Chem. Commun.* p. 544 (1967).

152. J. Trocha-Grimshaw and H. B. Henbest, *Chem. Commun.* p. 757 (1968).

153. Y. M. Y. Haddad, H. B. Henbest, J. Husbands, and T. R. B. Mitchell, *Proc. Chem. Soc., London* p. 361 (1964).

154. M. Donati and F. Conti, *Tetrahedron Lett.* p. 4953 (1966).

155. R. G. Brown and J. M. Davidson, *J. Chem. Soc., A* p. 1321 (1971).

156. H. B. Charman, *J. Chem. Soc., B* p. 584 (1970).

157. G. Gregorio, G. F. Pregaglia, and R. Ugo, *J. Organometal. Chem.* 37, 385 (1972).

158. P. M. Henry and G. A. Ward, *J. Amer. Chem. Soc.* 94, 673 (1972).

159. P. M. Henry and G. A. Ward, *J. Amer. Chem. Soc.* 93, 1494 (1971).

160. P. Taylor and M. Orchin, *J. Amer. Chem. Soc.* 93, 6504 (1971).

161. R. Cramer, *J. Amer. Chem. Soc.* 88, 2272 (1966).

162. R. Cramer and R. V. Lindsey, Jr., *J. Amer. Chem. Soc.* 88, 3534 (1966).

163. H. Kanai, *Chem. Commun.* p. 203 (1972).

164. D. Bingham, D. E. Webster, and P. B. Wells, *J. Chem. Soc., Dalton Trans.* p. 1928 (1972).

165. R. D. Gillard, B. T. Heaton, and M. F. Pilbrow, *J. Chem. Soc., A* p. 353 (1970).

166. H. Bonnemann, *Angew. Chem. Int. Ed. Engl.* 9, 736 (1970).

167. J. F. Nixon and B. Wilkins, *J. Organometal. Chem.* 44, C25 (1972).

168. J. Lukas, S. Coren, and J. E. Blom, *Chem. Commun.* p. 1303 (1969).

169. H. L. Finkbeiner and G. D. Cooper, *J. Org. Chem.* 27, 3395 (1962).

170. L. Farady, L. Bencze, and L. Markó, *J. Organometal. Chem.* 10, 505 (1967).

171. K. Yamamoto, T. Hayashi, and M. Kumada, *J. Amer. Chem. Soc.* 93, 5301 (1971).

172. L. H. Sommer, K. W. Michael, and H. Fujimoto, *J. Amer. Chem. Soc.* 89, 1519 (1967).

173. R. Cramer, *J. Amer. Chem. Soc.* 89, 4621 (1967).

174. R. Cramer, *J. Amer. Chem. Soc.* 94, 5681 (1972).

175. R. Cramer, *J. Amer. Chem. Soc.* 87, 4717 (1965).

176. T. Alderson, E. L. Jenner, and R. V. Lindsey, Jr., *J. Amer. Chem. Soc.* 87, 5638 (1965).

177. M. G. Barlow, M. J. Bryant, R. N. Haszeldine, and A. G. Mackie, *J. Organometal. Chem.* 21, 215 (1970).

178. E. O. Fischer and K. Fichtel, *Chem. Ber.* 94, 1200 (1961).

179. B. Bogdanovic, B. Henc, H-G. Karmann, H-G. Nussel, D. Walter, and G. Wilke, *Ind. Eng. Chem.* 62, 34 (1970).

180. G. G. Eberhardt, *Organometal. Chem. Rev.* 1, 491 (1966).

181. H. Bestian, K. Clauss, H. Jensen, and E. Prinz, *Angew. Chem., Int. Ed. Engl.* 2, 32 (1963).

182. H. Bestian and K. Clauss, *Angew. Chem., Int. Ed. Engl.* 2, 704 (1963).

183. P. W. Jolly, F. G. A. Stone, and K. Mackenzie, *J. Chem. Soc.* p. 6416 (1965).

184. A. R. Fraser, P. H. Bird, S. A. Bezman, J. R. Shapley, R. White, and J. A. Osborn, *J. Amer. Chem. Soc.* 95, 597 (1973).

185. R. Field, M. M. Germain, R. N. Haszeldine, and P. W. Wiggams, *J. Chem. Soc., A* p. 1964 (1970).

186. R. Field, M. M. Germain, R. N. Haszeldine, and P. W. Wiggams, *J. Chem. Soc., A* p. 1969 (1970).

187. A. Bright, J. F. Malone, J. K. Nicholson, J. Powell, and B. L. Shaw, *Chem. Commun.* p. 712 (1971).

188. J. Tsuji, *Accounts Chem. Res.* **2**, 151 (1969).

189. R. F. Heck, *J. Amer. Chem. Soc.* **90**, 5518 (1968).

190. Y. Fujiwara, I. Moritani, S. Donno, R. Asano, and S. Teranishi, *J. Amer. Chem. Soc.* **91**, 7166 (1969).

191. T. Mizoroki, K. Mori, and A. Ozaki, *Bull. Chem. Soc. Jap.* **44**, 581 (1971).

192. R. F. Heck and J. P. Nolley, Jr., *J. Org. Chem.* **37**, 2320 (1972).

193. R. F. Heck, *J. Amer. Chem. Soc.* **93**, 6896 (1971).

194. R. F. Heck, *J. Amer. Chem. Soc.* **91**, 6707 (1969).

195. P. M. Henry, *J. Amer. Chem. Soc.* **94**, 7305 (1972).

196. Y. Odaira, T. Oishi, T. Yukawa, and S. Tsutsumi, *J. Amer. Chem. Soc.* **88**, 4105 (1966).

197. H. A. Tayim, *Chem. Ind. (London)* p. 1468 (1970).

198. G. Szonyi, *Advan. Chem. Ser.* **70**, 53 (1968).

199. P. M. Henry, *J. Amer. Chem. Soc.* **88**, 1595 (1966).

200. W. Hafner, R. Jira, J. Sedlmeier, and J. Smidt, *Chem. Ber.* **95**, 1575 (1962).

201. H. Thyret, *Angew. Chem., Int. Ed. Engl.* **11**, 520 (1972).

202. Y. Wakatsuki, S. Nozakura, and S. Murahashi, *Bull. Chem. Soc. Jap.* **42**, 273 (1969).

203. M. Tsutsui, M. Ori, and J. Francis, *J. Amer. Chem. Soc.* **94**, 1414 (1972).

204. W. A. Clement and C. M. Selwitz, *J. Org. Chem.* **29**, 241 (1964).

205. I. I. Moiseev, M. N. Vargaftik, and Y. K. Syrkin, *Dokl. Chem.* **133**, 801 (1960).

206. E. W. Stern and M. L. Spector, *Proc. Chem. Soc., London* p. 370 (1961).

207. W. G. Lloyd and B. J. Luberoff, *J. Org. Chem.* **34**, 3949 (1969).

208. D. F. Hunt and G. T. Rodeheaver, *Tetrahedron Lett.* p. 3595 (1972).

209. J. A. Evans, D. R. Russell, A. Bright, and B. L. Shaw, *Chem. Commun.* p. 841 (1971).

210. H. Hirai, H. Sawai, and S. Makishima, *Bull. Chem. Soc. Jap.* **43**, 1148 (1970).

211. A Panunzi, A. De Renzi, and G. Paiaro, *J. Amer. Chem. Soc.* **92** 3488 (1970).

212. D. R. Coulson, *Tetrahedron Lett.* p. 429 (1971).

213. P. M. Henry, *J. Amer. Chem. Soc.* **93**, 3853 and 3547 (1971).

214. R. F. Heck, *J. Amer. Chem. Soc.* **90**, 5531 (1968).

215. G. Hata, K. Takahashi, and A. Miyake, *Chem. Commun.* p. 1392 (1970).

216. K. E. Atkins, W. E. Walker, and R. M. Manyik, *Tetrahedron Lett.* p. 3821 (1970).

217. D. G. Brady, *Chem. Commun.* p. 434 (1970).

218. E. S. Brown and E. A. Rick, *Chem. Commun.* p. 112 (1969).

219. B. W. Taylor and H. E. Swift, *J. Catal.* **26**, 254 (1972).

220. H. Stangl and R. Jira, *Tetrahedron Lett.* p. 3589 (1970).

221. R. Heck, *J. Amer. Chem. Soc.* **90**, 5538 (1968).

222. R. F. Heck, *J. Amer. Chem. Soc.* **90**, 5542 (1968).

223. B. F. Hallam and P. L. Pauson, *J. Chem. Soc., London*, p. 642 (1958).

224. P. S. Skell, J. J. Havel, D. L. Williams-Smith, and M. J. McGlinchey, *Chem. Commun.* p. 1098 (1972).

225. D. L. Williams-Smith, L. R. Wolf, and P. S. Skell, *J. Amer. Chem. Soc.* **94**, 4042 (1972).

226. R. E. Davis and R. Pettit, *J. Amer. Chem. Soc.* **92**, 717 (1970).

227. W. R. Roth and J. D. Meier, *Tetrahedron Lett.* p. 2053 (1967).

228. H. Frye, E. Kuljian, and J. Viebrock, *Inorg. Chem.* **4**, 1499 (1965).

229. J. E. Arnet and R. Pettit, *J. Amer. Chem. Soc.* **83**, 2954 (1961).

230. J. C. Trebellas, J. R. Olechowski, and H. B. Jonassen, *J. Organometal. Chem.* **6**. 412 (1966).

231. P. Heimbach, *Angew. Chem. Int. Ed. Engl.* **5**, 595 (1966).
232. H. Frye, E. Kuljian, and J. Viebrock, *Inorg. Nucl. Chem. Lett.* **2**, 119 (1966).
233. D. Chinn and H. Frye, *Inorg. Nucl. Chem. Lett.* **5**, 135 (1969).
234. G. J. Leigh and E. O. Fischer, *J. Organometal. Chem.* **4**, 461 (1965).
235. G. G. Ecke, U. S. Patent 3,135,776 (1964).
236. K. K. Joshi, R. H. B. Mais, F. Nyman, P. G. Owston, and A. M. Wood, *J. Chem. Soc., A* p. 318 (1968).
237. S. A. R. Knox, R. P. Phillips, and F. G. A. Stone, *J. Chem. Soc., Chem. Commun.* p. 1227 (1972).
238. J. W. Kang, K. Moseley, and P. M. Maitlis, *J. Amer. Chem. Soc.* **91**, 5970 (1969).
239. P. V. Balokrishnan and P. M. Maitlis, *J. Chem. Soc., A* p. 1715 (1971).
240. K. W. Barnett, *J. Organometal. Chem.* **21**, 477 (1970).
241. O. S. Mills and E. F. Paulus, *Chem. Commun.* p. 738 (1966).
242. H. Lehmkuhl, W. Leuchte, and E. Janssen, *J. Organometal. Chem.* **30**, 407 (1971).
243. S. Otsuka and T. Taketami, *J. Chem. Soc., Dalton Trans.* p. 1879 (1972).
244. J. E. Lyons, *Chem. Commun.* p. 564 (1969).
245. K. Moseley and P. Maitlis, *J. Chem. Soc., A* p. 2884 (1970).
246. W. Rupilus and M. Orchin, *J. Org. Chem.* **36**, 3604 (1971).
247. W. C. Drinkard, Belgian Patent 698,322 (1967).
248. W. C. Drinkard and R. V. Lindsey, Jr., Belgian Patent 698,333 (1967).
249. K. Takahashi, A. Miyake, and G. Hata, *Chem. Ind. (London)* p. 488 (1971); *Bull. Chem. Soc. Jap.* **45**, 1183 (1972).
250. W. C. Drinkard, German Patent 1,806,096 (1969).
251. K. C. Dewhirst, *J. Org. Chem.* **32**, 1297 (1967).
252. R. Baker and D. E. Halliday, *Tetrahedron Lett.* p. 2776 (1972).
253. D. Rose, *Tetrahedron Lett.* p. 4197 (1972).
254. R. G. Miller, H. J. Golden, D. J. Baker, and R. D. Stauffer, *J. Amer. Chem. Soc.* **93**, 6308 (1971).
255. R. G. Miller, *J. Amer. Chem. Soc.* **89**, 2785 (1967).
256. R. G. Miller and P. A. Pinke, *J. Amer. Chem. Soc.* **90**, 4500 (1968).
257. D. H. Gibson and R. L. Vonnahme, *J. Amer. Chem. Soc.* **94**, 5090 (1972).
258. G. F. Emerson, J. E. Mahler, and R. Pettit, *Chem. Ind. (London)* p. 836 (1964).
259. T. H. Whitesides and R. W. Arhart, *J. Amer. Chem. Soc.* **93**, 5296 (1971).
260. B. F. G. Johnson, J. Lewis, and D. Yarrow, *Chem. Commun.* p. 235 (1972).
261. J. Evans, B. F. G. Johnson, and J. Lewis, *Chem. Commun.* p. 1252 (1971).
262. A. J. Birch, P. E. Cross, J. Lewis, and D. A. White, *Chem. Ind. (London)* p. 838 (1964).
263. B. J. H. Cowles, B. F. G. Johnson, J. Lewis, and A. W. Parkins, *J. Chem. Soc. Dalton Trans.* p. 1768 (1972).
264. C. H. De Puy, R. N. Greene, and T. E. Schroer, *Chem. Commun.* p. 1225 (1968).
265. C. H. De Puy and C. H. Jablonski, *Tetrahedron Lett.* p. 3989 (1969).
266. B. Dickens and W. N. Lipscomb, *J. Amer. Chem. Soc.* **83**, 4862 (1961).
267. T. A. Manuel and F. G. A. Stone, *J. Amer. Chem. Soc.* **82**, 366 (1960).
268. J. D. Holmes and R. Pettit, *J. Amer. Chem. Soc.* **85**, 2531 (1963).
269. S. Winstein, H. D. Kaesz, C. G. Kreiter, and E. C. Friedrich, *J. Amer. Chem. Soc.* **87**, 3267 (1965).
270. P. L. Pauson, G. H. Smith, and J. H. Valentine, *J. Chem. Soc., C* pp. 1057 and 1061 (1967).
271. J. Lukas, P. W. N. van Leeuwen, H. C. Volger, and P. Kramer, *Chem. Commun.* p. 799 (1970).

272. S. D. Robinson and B. L. Shaw, *J. Chem. Soc., London* p. 4806 (1963).

273. D. R. Coulson, *J. Amer. Chem. Soc.* **91**, 200 (1969).

274. G. Paiaro, A. Panunzi, and A. De Renzi, *Tetrahedron Lett.* p. 3905 (1966).

275. H. Takahashi and J. Tsuji, *J. Amer. Chem. Soc.* **90**, 2387 (1968).

276. A. De Renzi, R. Palumbo, and G. Paiaro, *J. Chem. Amer. Soc.* **93**, 880 (1971).

277. R. Heck, *J. Amer. Chem. Soc.* **85**, 3387 (1963).

278. M. Green and R. I. Hancock, *J. Chem. Soc., A* p. 109 (1968).

279. R. Heck, *J. Amer. Chem. Soc.* **85**, 3381 (1963).

280. R. B. King and A. Efraty, *J. Amer. Chem. Soc.* **93**, 4950 (1971).

281. J. W. Kang, K. Moseley, and P. Maitlis, *J. Amer. Chem. Soc.* **91**, 5970 (1969).

282. A. Eisenstadt, G. Scharf, and B. Fuchs, *Tetrahedron Lett.* p. 679 (1971).

283. R. P. Hughes and J. Powell, *J. Organometal. Chem.* **30**, C45 (1971).

284. J. E. Mahler, D. H. Gibson, and R. Pettit, *J. Amer. Chem. Soc.* **85**, 3959 (1963).

285. D. E. Kuhn and C. P. Lillya, *J. Amer. Chem. Soc.* **94**, 1682 (1972).

286. G. E. Herberich, G. Greiss, and H. F. Heil, *J. Organometal. Chem.* **22**, 723 (1970).

287. J. F. Helling and D. M. Braitsch, *J. Amer. Chem. Soc.* **92**, 7207 (1970).

288. J. F. Bunnett and H. Hermann, *J. Org. Chem.* **36**, 4081 (1971).

289. R. E. Davis, H. D. Simpson, N. Grice, and R. Pettit, *J. Amer. Chem. Soc.* **93**, 6688 (1971).

290. J. H. Richards and E. A. Hill, *J. Amer. Chem. Soc.* **81**, 3484 (1959).

291. D. S. Trifan and R. Bacskai, *J. Amer. Chem. Soc.* **82**, 5010 (1960).

292. W. E. McEwen, J. A. Manning, and J. Kleinberg, *Tetrahedron Lett.* p. 2195 (1964).

293. W. R. Jackson and W. B. Jennings, *Chem. Commun.* p. 824 (1966).

294. E. O. Fischer and S. Breitschaft, *Angew. Chem.* **75**, 167 (1963); *Chem. Ber.* **99**, 2213 (1966).

295. M. Farona and J. F. White, *J. Amer. Chem. Soc.* **93**, 2826 (1971).

296. E. O. Fischer and H. Brunner, *Chem. Ber.* **98**, 175 (1965).

297. G. Jaouen and R. Dabard, *Tetrahedron Lett.* p. 1015 (1971).

298. G. F. Emerson, L. Watts, and R. Pettit, *J. Amer. Chem. Soc.* **87**, 131 (1965).

299. R. G. Amiet and R. Pettit, *J. Amer. Chem. Soc.* **90**, 1059 (1968).

300. P. M. Maitlis and M. L. Games, *Chem. Ind. (London)* p. 1624 (1963).

301. M. Rosenblum, B. North, D. Wells, and W. P. Giering, *J. Amer. Chem. Soc.* **94**, 1239 (1972).

302. J. F. Helling, S. C. Rennison, and A. Merijan, *J. Amer. Chem. Soc.* **89**, 7140 (1967).

303. J. S. Ward and R. Pettit, *J. Amer. Chem. Soc.* **93**, 262 (1971).

304. J. S. McKennis, L. Brener, J. S. Ward, and R. Pettit, *J. Amer. Chem. Soc.* **93**, 4957 (1971).

305. B. F. G. Johnson, J. Lewis, and G. L. P. Randall, *Chem. Commun.* p. 1273 (1969).

306. B. F. G. Johnson, J. Lewis, A. W. Parkins, and G. L. P. Randall, *Chem. Commun.* p. 595 (1969).

307. C. A. Harmon and A. Streitwieser, *J. Amer. Chem. Soc.* **94**, 8926 (1972).

308. J. S. Ward and R. Pettit, *Chem. Commun.* p. 1419 (1970).

309. K. Ehrlick and G. F. Emerson, *Chem. Commun.* p. 59 (1969).

310. R. Noyori, T. Nishimura, and H. Takaya, *Chem. Commun.* p. 89 (1969).

311. S. Tanaka, K. Mabuchi, and N. Shimazaki, *J. Org. Chem.* **29**, 1626 (1964).

312. G. Allegra, F. Lo Giudice, G. Natta, U. Giannini, F. Fagherazzi, and P. Pino, *Chem. Commun.* p. 1263 (1967).

313. R. P. Hughes and J. Powell, *J. Amer. Chem. Soc.* **94**, 7723 (1972).

314. T. S. Cameron, M. L. H. Green, H. Munakata, C. K. Prout, and M. J. Smith, *J. Coord. Chem.* **2**, 43 (1972).

315. D. Medema, R. van Helden, and C. F. Kohll, *Inorg. Chim. Acta*, 3, 255 (1969).
316. Y. Takahashi, S. Sakai, and Y. Ishii, *J. Organometal. Chem.* 16, 177 (1969); *Inorg. Chem.* 11, 1516 (1972).
317. A. Carbonaro, A. Greco, and G. Dall'Asta, *Tetrahedron Lett.* p. 2037 (1967).
318. W. Keim and H. Chung. *J. Org. Chem.* 37, 947 (1972).
319. J. K. Nicholson and B. L. Shaw, *J. Chem. Soc.*, *A* p. 807 (1966).
320. L. Porri, M. C. Gallazzi, A. Columbo, and G. Allegra, *Tetrahedron Lett.* p. 4187 (1965).
321. Y. Uchida, K. Furuhata, and S. Yoshida, *Bull. Chem. Soc. Jap.* 44, 1966 (1971).
322. S. Takahashi, H. Yamazaki, and N. Hagihara, *Bull. Chem. Soc. Jap.* 41, 254 (1968).
323. S. Takahashi, T. Shibano, and N. Hagihara, *Bull. Chem. Soc. Jap.* 41, 454 (1968).
324. E. J. Smutny, *J. Amer. Chem. Soc.* 89, 6793 (1967).
325. R. Baker, D. E. Halliday, and T. N. Smith, *J. Organometal. Chem.* 35, C61 (1972).
326. A. D. Ketley, "The Stereochemistry of Macromolecules," p. 33. Dekker, New York, 1967.
327. P. W. Jolly, I. Tkatchenko, and G. Wilke, *Angew. Chem.*, *Int. Ed. Engl.* 10, 328 (1971).
328. P. Heimbach and W. Brenner, *Angew. Chem.* 79, 813 (1967).
329. P. Heimbach, *Angew. Chem.*, *Int. Ed. Engl.* 7, 882 (1968).
330. J. Kiji, K. Masui, and J. Furukawa, *Bull. Chem. Soc. Jap.* 44, 1956 (1971).
331. B. Barnett, B. Büssemeier, P. Heimbach, P. W. Jolly, C. Krüger, I. Tkatchenko, and G. Wilke, *Tetrahedron Lett.* p. 1457 (1972).
332. J. P. Candlin and W. H. James, *J. Chem. Soc.*, *C* p. 1856 (1968); British Patent, 1,085,875 (1968).
333. H. Breil and G. Wilke, *Angew. Chem.* 82, 355 (1970).
334. G. Wilke, E. W. Müller, and M. Kröner, *Angew. Chem.* 73, 33 (1961).
335. G. Wilke, M. Kröner, and B. Bogdanovic, *Angew. Chem.* 73, 755 (1961).
336. A. Miyake, H. Kondo, and M. Nishino, *Angew. Chem.*, *Int. Ed. Engl.* 10, 802 (1971).
337. R. P. Hughes and J. Powell, *J. Organometal. Chem.* 20, P17 (1969).
338. F. W. Hoover and R. V. Lindsey, Jr., *J. Org. Chem.* 34, 3051 (1969).
339. S. Dai and W. R. Doblier, Jr., *J. Org. Chem.* 37, 950 (1972).
340. S. Otsuka, A. Nakamura, S. Ueda, and K. Tani, *Chem. Commun.* p. 863 (1971).
341. M. Englert, P. W. Jolly, and G. Wilke, *Angew. Chem.* 83, 84 (1971).
342. S. Otsuka, A. Nakamura, T. Yamagata, and K. Tani, *J. Amer. Chem. Soc.* 94, 1037 (1972).
343. S. Otsuka, A. Nakamura, K. Tani, and S. Ueda, *Tetrahedron Lett.* p. 297 (1969).
344. S. Otsuka, K. Tani, and A. Nakamura, *J. Chem. Soc.*, *A* p. 1404 (1969).
345. S. Otsuka, A. Nakamura, and H. Minamida, *Chem. Commun.* p. 191 (1969).
346. J. P. Scholten and H. J. van der Ploeg, *Tetrahedron Lett.* p. 1685 (1972).
347. D. R. Coulson, *J. Org. Chem.* 37, 1253 (1972).
348. A. D. Allan and C. D. Cook, *Can. J. Chem.* 41, 1235 (1963).
349. M. A. Bennett, G. B. Robertson, P. O. Whimp, and T. Yoshida, *J. Amer. Chem. Soc.* 93, 3797 (1971).
350. H. Greenfield, H. W. Sternberg, R. A. Friedel, J. H. Wotiz, R. Markby, and I. Wender, *J. Amer. Chem. Soc.* 78, 120 (1956).
351. J. F. Tilney-Bassett and O. S. Mills, *J. Amer. Chem. Soc.* 81, 4757 (1959).
352. K. Yasufuka and H. Yamazaki, *J. Organometal. Chem.* 35, 367 (1972).
353. B. L. Booth and A. D. Lloyd, *J. Organometal. Chem.* 35, 195 (1972).
354. D. M. Barlex, R. D. W. Kemmitt, and G. W. Littlecott, *Chem. Commun.* p. 613 (1969).

355. M. Michman and M. Balog, *J. Organometal. Chem.* **31**, 395 (1971).
356. B. L. Booth and R. G. Hargreaves, *J. Organometal. Chem.* **33**, 365 (1971).
357. I. Hashimoto, M. Ryang, and S. Tsutsumi, *J. Org. Chem.* **33**, 3955 (1968).
358. B. L. Booth and R. G. Hargreaves, *J. Chem. Soc.*, A p. 308 (1970).
359. T. Yukawa and S. Tsutsumi, *Inorg. Chem.* **7**, 1458 (1968).
360. M. H. Chisholm and H. C. Clark, *Inorg. Chem.* **10**, 2557 (1971).
361. T. G. Appleton, M. H. Chisholm, and H. C. Clark, *J. Amer. Chem. Soc.* **94**, 8912 (1972).
362. J. Halpern, B. R. James, and A. L. W. Kemp, *J. Amer. Chem. Soc.* **83**, 4097 (1961).
363. R. J. Kern, *Chem. Commun.* p. 706 (1968).
364. A. Furlani, I. Collamati, and G. Sartori, *J. Organometal. Chem.* **17**, 463 (1969).
365. M. Dubeck, *J. Amer. Chem. Soc.* **82**, 6193 (1960).
366. L. F. Dahl and W. E. Oberhansli, *Inorg. Chem.* **4**, 629 (1965), and references therein.
367. P. M. Maitlis, "The Organic Chemistry of Palladium," Vol. 2, p. 53. Academic Press, New York, 1971.
368. R. Hüttel and H. J. Neugebauer, *Tetrahedron Lett.* p. 3541 (1964).
369. H. Dietl, H. Reinheimer, J. Moffat, and P. M. Maitlis, *J. Amer. Chem. Soc.* **92**, 2276 (1970).
370. H. Reinheimer, J. Moffat, and P. M. Maitlis, *J. Amer. Chem. Soc.* **92**, 2285 (1970).
371. G. M. Whitesides and W. J. Ehmann, *J. Amer. Chem. Soc.* **91**, 3800 (1969).
372. H. Yamazaki and N. Hagihara, *J. Organometal. Chem.* **21**, 431 (1970).
373. H. Yamazaki and N. Hagihara, *Bull. Chem. Soc. Jap.* **44**, 2260 (1971).
374. J. P. Collman, J. W. Kang, and M. F. Sullivan, *Inorg. Chem.* **7**, 1298 (1968).
375. M. E. Vol'pin and D. N. Kursanov, *Angew. Chem.* **75**, 1034 (1963).
376. K. Moseley and P. M. Maitlis, *Chem. Commun.* p. 1604 (1971).
377. T. Ito, S. Hasegawa, T. Takahashi, and Y. Ishii, *Chem. Commun.* p. 629 (1972).
378. M. Tsutsui and H. Zeiss, *J. Amer. Chem. Soc.* **83**, 825 (1961).
379. E. Müller, R. Thomas, M. Sauerbier, E. Langer, and D. Streichfuss, *Tetrahedron Lett.* p. 521 (1971).
380. E. Müller, R. Thomas, and G. Zountsas, *Justus Liebigs Ann. Chem.* **758**, 16 (1972).
381. E. Müller, E. Langer, H. Jäkle, and H. Muhon, *Tetrahedron Lett.* p. 5271 (1970).
382. E. H. Braye and W. Hübel, *Chem. Ind.* (*London*) p. 1250 (1959).
383. C. Hoogzand and W. Hübel, *Angew. Chem.* **73**, 680 (1961).
384. C. W. Bird, "Transition Metal Intermediates in Organic Synthesis," p. 1. Academic Press, New York, 1967.
385. W. Reppe, O. Schlichting, K. Klager, and T. Toepel, *Justus Liebigs Ann. Chem.* **560**, 1 (1948).
386. J. R. Leto and M. F. Leto, *J. Amer. Chem. Soc.* **83**, 2944 (1961).
387. A. Nakamura, *Mem. Inst. Sci. Ind. Res., Osaka Univ.* **19**, 81 (1962).
388. A. Nakamura, *Chem. Abstr.* **59**, 8786 (1963).
389. R. B. King and C. W. Eavenson, *J. Organometal. Chem.* **16**, P75 (1969).
390. R. B. King and A. Efraty, *J. Amer. Chem. Soc.* **92**, 6071 (1970).
391. M. Michman and H. H. Zeiss, *J. Organometal. Chem.* **25**, 161 (1970).
392. H. H. Zeiss and M. Tsutsui, *J. Amer. Chem. Soc.* **81**, 6090 (1959).
393. W. Herwig, W. Metlesics, and H. Zeiss, *J. Amer. Chem. Soc.* **81**, 6203 (1959).
394. T. H. Coffield, K. G. Ihrman, and W. Burns, *J. Amer. Chem. Soc.* **82**, 4209 (1960).
395. H. J. De Liefde Meijer, U. Pauzenga, and F. Jellinek, *Rec. Trav. Chim. Pays-Bas* **85**, 634 (1966).
396. H. Zeiss and R. P. A. Sneeden, *Angew. Chem., Int. Ed. Engl.* **6**, 435 (1967).

397. G. M. Whitesides and W. J. Ehmann, *J. Amer. Chem. Soc.* **92**, 5625 (1970).
398. J. Dvorak, R. J. O'Brien, and W. Santo, *Chem. Commun.* p. 411 (1970).
399. I. S. Kolomnikov, T. S. Lobeeva, V. V. Gorbachevskaya, G. G. Aleksandrov, Y. T. Struckhov, and M. E. Vol'pin, *Chem. Commun.* p. 972 (1971).
400. G. N. Schrauzer and P. Glockner, *Chem. Ber.* **97**, 2451 (1964).
401. J. Browing, M. Green, and F. G. A. Stone, *J. Chem. Soc., A* p. 453. (1971).
402. R. Cramer, *J. Amer. Chem. Soc.* **89**, 1633 (1967).
403. G. Hata and A. Miyake, *Bull. Chem. Soc. Jap.* **41**, 2443 (1968).
404. G. Henrici-Olive and S. Olive, *J. Organometal. Chem.* **35**, 381 (1972).
405. Y. Tajima and E. Kunioka, *Chem. Commun.* p. 603 (1968).
406. A. Misomo, Y. Uchida, T. Saito, and K. Uchida, *Bull. Chem. Soc. Jap.* **40**, 1889 (1967).
407. A. Takahashi and T. Inukai, *Chem. Commun.* p. 1473 (1970).
408. L. G. Cannell, *J. Amer. Chem. Soc.* **94**, 6867 (1972).
409. P. Heimbach and G. Wilke, *Justus Liebigs Ann. Chem.* **727**, 183, 194 (1969).
410. Y. Inoue and T. Kagawa, *Tetrahedron Lett.* p. 1099 (1970).
411. Y. Inoue, T. Kagawa, Y. Uchida, and H. Hashimoto, *Bull. Chem. Soc. Jap.* **45**, 1496 (1972).
412. P. Heimbach, H. Selbeck, and E. Troxler, *Angew. Chem., Int. Ed. Engl.* **10**, 659 (1971).
413. P. Heimbach and R. Traunmueller, "Chemistry of Metal Olefin Complexes," Chem. Pocket Books, No. 10. Verlag Chemie, Weinheim, 1970.
414. P. Mushak and M. A. Battiste, *Chem. Commun.* p. 1146 (1969).
415. P. M. Maitlis, "The Organic Chemistry of Palladium," Vol. 2, p. 38. Academic Press, New York, 1971.
416. G. N. Schrauzer, *Chem. Ber.* **94**, 1403 (1961).
417. F. Guerrieri and G. P. Chiusoli, *J. Organometal. Chem.* **19**, 453 (1969).
418. A. Carbonaro, A. Greco, and G. Dall'Asta, *J. Org. Chem.* **33**, 3948 (1968).
419. P. Heimbach, K. Ploner, and F. Thömel, *Angew. Chem., Int. Ed. Engl.* **10**, 276 (1971).
420. P. Heimbach and R. Schimpf, *Angew. Chem.* **80**, 704 (1968).
421. W. Brenner, P. Heimbach, K. Ploner, and F. Thömel, *Angew. Chem., Int. Ed. Engl.* **8**, 753 (1969).
422. P. Heimbach, *Angew. Chem.* **78**, 983 (1966).
423. R. W. Glyde and R. J. Mawby, *Inorg. Chim. Acta* **5**, 317 (1971).
424. R. W. Glyde and R. J. Mawby, *Inorg. Chim. Acta* **4**, 331 (1970).
425. B. L. Booth, R. N. Haszeldine, and N. P. Woffender, *J. Chem. Soc., A* p. 1979 (1970).
426. F. Calderazzo and K. Noack, *Coord. Chem. Rev.* **1**, 118 (1966).
427. G. M. Whitesides and D. J. Boschetto, *J. Amer. Chem. Soc.* **91**, 4313 (1969).
428. D. Seyferth and R. J. Spohn, *J. Amer. Chem. Soc.* **91**, 6192 (1969).
429. W. Beck and R. E. Nitzschmann, *Chem. Ber.* **97**, 2098 (1964).
430. M. P. Cooke, Jr., *J. Amer. Chem. Soc.* **92**, 6080 (1970).
431. Y. Watanabe, T. Mitsudo, M. Tanaka, K. Yamamoto, T. Okajima, and Y. Takegami, *Bull. Chem. Soc. Jap.* **44**, 2569 (1971).
432. J. P. Collman, S. R. Winter, and D. R. Clark, *J. Amer. Chem. Soc.* **94**, 1788 (1972).
433. R. B. King, *Accounts Chem. Res.* **3**, 417 (1970).
434. C. P. Casey and C. A. Bunnell, *J. Amer. Chem. Soc.* **93**, 4077 (1971).
435. C. D. Cook and G. S. Jauhal, *Can. J. Chem.* **45**, 301 (1967).
436. M. A. Bennett and R. Charles, *J. Amer. Chem. Soc.* **94**, 666 (1972).

437. Y. Takegami, C. Yokokawa, Y. Watanabe, H. Masada, and Y. Okuda, *Bull. Chem. Soc. Jap.* **38**, 787 (1965).

438. W. T. Dent, R. Long, and G. H. Whitfield, *J. Chem. Soc.* pp. 1589 and 1852 (1964).

439. H. C. Volger, K. Vrieze, J. Lemmers, A. P. Praat, and P. van Leeuwen, *Inorg. Chim. Acta* **4**, 435 (1970).

440. G. W. Parshall, *Z. Naturforsch. B* **18**, 772 (1963).

441. N. L. Bauld, *Tetrahedron Lett.* p. 1841 (1963).

442. T. Hosokawa and P. M. Maitlis, *J. Amer. Chem. Soc.* **94**, 3238 (1972).

443. E. Yoshisato and S. Tsutsumi, *J. Org. Chem.* **33**, 869 (1968).

444. E. J. Corey and L. S. Hegedus, *J. Amer. Chem. Soc.* **91**, 1233 (1969).

445. R. G. Schultz and P. D. Montgomery, *J. Catal.* **13**, 105 (1969).

446. R. Ercoli, *Chim. Ind. (Milan)* **42**, 587 (1960).

447. R. Heck and D. S. Breslow, *J. Amer. Chem. Soc.* **83**, 4023 (1961).

448. W. Reppe and H. Kröper, *Justus Liebigs Ann. Chem.* **582**, 38 (1953).

449. C. W. Bird, R. C. Cookson, J. Hudec, and R. O. Williams, *J. Chem. Soc., London* p. 410 (1963).

450. A. C. Cope and R. W. Sickman, *J. Amer. Chem. Soc.* **87**, 3273 (1965).

451. R. F. Heck, *J. Amer. Chem. Soc.* **90**, 313 (1968).

452. S. Horiie, *Nippon Kagaku Zasshi* **79**, 499 (1958); *Chem. Abstr.* **54**, 4607 (1960).

453. S. Horiie and S. Murahashi, *Bull. Chem. Soc. Jap.* **33**, 88 and 247 (1960).

454. A. Rosenthal, R. Fastbury, and A. Hubscher, *J. Org. Chem.* **23**, 1037 (1958).

455. A. Rosenthal and M. R. S. Weir, *J. Org. Chem.* **28**, 3025 (1963).

456. A. Rosenthal and J. Gerway, *Chem. Ind. (London)* p. 1623 (1963).

457. H. E. Holmquist, *J. Org. Chem.* **34**, 4164 (1969).

458. L. F. Hines and J. K. Stille, *J. Amer. Chem. Soc.* **94**, 485 (1972).

459. T. Kruck and M. Noack, *Chem. Ber.* **97**, 1693 (1964).

460. Y. Sawa, M. Ryang, and S. Tsutsumi, *J. Org. Chem.* **35**, 4183 (1970).

461. E. J. Corey and L. S. Hegedus, *J. Amer. Chem. Soc.* **91**, 4927 (1969).

462. R. J. Angelici, *Accounts Chem. Res.* **5**, 335 (1972).

463. O. Roelen, *Angew. Chem.* **3**, 213 (1948).

464. R. Heck and D. S. Breslow, *Chem. Ind. (London)* p. 467 (1960).

465. A. R. Martin, *Chem. Ind. (London)* p. 1536 (1954).

466. G. Natta, R. Ercoli, S. Costellano, and F. H. Barbieri, *J. Amer. Chem. Soc.* **76**, 4049 (1954).

467. F. Ungváry and L. Markó, *J. Organometal. Chem.* **20**, 205 (1969).

468. F. Ungváry, *J. Organometal. Chem.* **36**, 363 (1972).

469. A. Rosenthal and D. Abson, *Can. J. Chem.* **42**, 1811 (1964).

470. C. L. Aldridge and H. B. Jonassen, *J. Amer. Chem. Soc.* **85**, 886 (1963).

471. L. Markó and P. Szabo, *Chem. Tech. (Leipzig)* **13**, 482 (1961).

472. F. Piacenti, S. Pucci, M. Bianchi, and P. Pino, *Chim. Ind. (Milan)* **50**, 1362 (1968).

473. C. P. Casey and C. R. Cyr, *J. Amer. Chem. Soc.* **93**, 1280 (1971).

474. A. I. M. Keulemans, A. Kwantes, and T. van Bavel, *Rec. Trav. Chim. Pays-Bas* **67**, 298 (1948).

475. I. J. Goldfarb and M. Orchin, *Advan. Catal.* **9**, 609 (1957).

476. K. Buechner, O. Roelen, J. Meis, and H. Langwald, U. S. Patent 3,043,871 (1957).

477. I. Wender, J. Feldman, S. Metlin, B. H. Gwynn, and M. Orchin, *J. Amer. Chem. Soc.* **77**, 5760 (1955).

478. L. H. Slaugh and R. D. Mullineaux, *J. Organometal. Chem.* **13**, 469 (1968).

479. E. R. Tucci, *Ind. Eng. Chem., Prod. Res. Develop,* **9**, 516 (1970).

480. N. V. Kutepow and H. Kindler, *Angew. Chem.* **72**, 802 (1960).

481. J. Tsuji and Y. Mori, *Bull. Chem. Soc. Jap.* **42**, 527 (1969).

482. O. Klopfer, U. S. Patent 3,050, 562 (1962).

483. G. Braca, G. Sbrana, F. Piacenti, and P. Pino, *Chim. Ind.* (*Milan*) **52**, 1091 (1970).

484. L. Benzoni, A. Andreeta, C. Zanzottera, and M. Camia, *Chim. Ind.* (*Milan*) **48**, 1076 (1966).

485. B. Heil and L. Markó, *Chem. Ber.* **102**, 2238 (1969).

486. D. Evans, J. A. Osborn, and G. Wilkinson, *J. Chem. Soc., A* p. 3133 (1968).

487. C. K. Brown and G. Wilkinson, *J. Chem. Soc., A* p. 2753 (1970).

488. G. Yagupsky, C. K. Brown, and G. Wilkinson, *J. Chem. Soc., A* p. 1392 (1970).

489. J. Falbe and F. Korte, *Angew. Chem.* **74**, 900 (1962); *Chem. Ber.* **97**, 863 (1964).

490. J. Falbe, H-J. Schulze-Steinen, and F. Korte, *Chem. Ber.* **98**, 886 (1965).

491. J. Falbe and F. Korte, *Chem. Ber.* **95**, 2680 (1962).

492. M. A. Bennett and R. Watt, *Chem. Commun.* p. 95 (1971).

493. M. A. Bennett, G. B. Robertson, R. Watt, and P. O. Whimp, *Chem. Commun.* p. 752 (1971).

494. J. Staib, W. Guyer, and O. Slotterbeck, U.S. Patent 2,864,864 (1958).

495. V. L. Hughes, U.S. Patent 2,839,580 (1958).

496. J. A. Bertrand, C. L. Andridge, S. Husebye, and H. B. Johassen, *J. Org. Chem.* **29**, 790 (1964).

497. P. P. Klemchuk, U.S. Patent 2,995,607 (1961).

498. R. Heck and D. S. Breslow, *J. Amer. Chem. Soc.* **83**, 1097 (1961).

499. P. Hayden, British Patent 1,148,043 (1969).

500. E. Yoshisato, M. Ryang, and S. Tsutsumi, *J. Org. Chem.* **34**, 1500 (1969).

501. R. F. Heck, *J. Amer. Chem. Soc.* **94**, 2712 (1972).

502. D. M. Fenton and P. J. Steinwand, *J. Org. Chem.* **37**, 2034 (1972).

503. S. Brewis and P. R. Hughes, *Chem. Commun.* p. 71 (1967).

504. S. Brewis and P. R. Hughes, *Chem. Commun.* p. 489 (1965).

505. S. Brewis and P. R. Hughes, *Chem. Commun.* p. 6 (1966).

506. B. Fell, W. Seide, and F. Asinger, *Tetrahedron Lett.* p. 1003 (1968).

507. S. Brewis and P. R. Hughes, *Chem. Commun.* p. 157 (1965).

508. W. E. Billups, W. E. Walker, and T. C. Shields, *Chem. Commun.* p. 1067 (1971).

509. J. Tsuji, Y. Mori, and M. Hara, *Tetrahedron* **28**, 3721 (1972).

510. T. J. Kealy and R. E. Benson, *J. Org. Chem.* **26**, 3126 (1961).

511. R. Baker, B. N. Blackett, and R. C. Cookson, *Chem. Commun.* p. 802 (1972).

512. W. Hübel, "Organic Syntheses via Metal Carbonyls," Vol. I, p. 273. Wiley (Interscience), New York, 1968.

513. R. S. Dickson and G. Wilkinson, *J. Chem. Soc., London*, p. 2699 (1964).

514. G. N. Schrauzer, *J. Amer. Chem. Soc.* **81**, 5307 (1959).

515. R. P. Dodge and V. Schomaker, *J. Organometal Chem.* **3**, 274 (1965).

516. H. W. Sternberg, R. Markby, and I. Wender, *J. Amer. Chem. Soc.* **80**, 1009 (1958).

517. R. Clarkson, E. Jones, P. Wailes, and M. C. Whiting, *J. Amer. Chem. Soc.* **78**, 6206 (1956).

518. A. A. Hock and O. S. Mills, *Acta Crystallogr.* **14**, 139 (1961).

519. J. R. Case, R. Clarkson, E. R. H. Jones, and M. C. Whiting, *Proc. Chem. Soc., London* p. 150 (1959).

520. H. W. Sternberg, R. A. Friedel, R. Markby, and I. Wender, *J. Amer. Chem. Soc.* **78**, 3621 (1956).

521. P. Pino, G. Braca, G. Sbrana, and A. Cuccuru, *Chem. Ind.* (*London*) p. 1732 (1968).

522. E. Weiss and W. Hübel, *Chem. Ber.* **95**, 1179 (1962).

523. E. H. Braye and W. Hübel, *J. Organometal. Chem.* **3**, 25 (1965).

524. H. W. Sternberg, J. Shukys, C. Donne, R. Markby, R. A. Friedel, and I. Wender, *J. Amer. Chem. Soc.* **81**, 2339 (1959).

525. G. Albanesi and M. Tovaglieri, *Chim. Ind. (Milan)* **41**, 189 (1959).

526. J. C. Sauer, R. D. Cramer, V. A. Engelhardt, T. A. Ford, H. Holmquist, and B. W. Howk, *J. Amer. Chem. Soc.* **81**, 3677 (1959).

527. G. Albanesi, R. Farina, and A. Taccioli, *Chim. Ind. (Milan)* **48**, 1151 (1966).

528. F. Ungváry and L. Markó, *Chem. Ber.* **105**, 2457 (1972).

529. R. Heck, *Advan. Chem.* **49**, 181 (1965).

530. G. Albanesi, *Chim. Ind. (Milan)* **46**, 1169 (1964).

531. G. Natta and P. Pino, U.S. Patent 2,851,486 (1958).

532. P. Pino and A. Miglierina, *J. Amer. Chem. Soc.* **74**, 5551 (1952).

533. W. Reppe, *Justus Liebigs Ann. Chem.* **582**, 1 (1953).

534. E. R. H. Jones, G. W. Whotham, and M. C. Whiting, *J. Chem. Soc., London* p. 1865 (1954).

535. C. W. Bird and E. M. Briggs, *J. Chem. Soc., C* p. 1265 (1967).

536. G. P. Chiusoli, C. Venturello, and S. Merzoni, *Chem. Ind. (London)* p. 977 (1968).

537. R. W. Rosenthal and L. H. Schwartzman, *J. Org. Chem.* **24**, 836 (1959).

538. R. Heck and W. Greaves, unpublished results.

539. Y. Sawa, I. Hashimoto, M. Ryang, and S. Tsutsumi, *J. Org. Chem.* **33**, 2159 (1968).

540. S. Fukuoka, M. Ryang, and S. Tsutsumi, *J. Org. Chem.* **33**, 2973 (1968).

541. G. P. Chiusoli, S. Merzoni, and G. Mondelli, *Tetrahedron Lett.* p. 2777 (1964).

542. L. Cassar and G. P. Chiusoli, *Chim. Ind. (Milan)* **48**, 323 (1966).

543. M. Foà, L. Cassar, and M. Venturi, *Tetrahedron Lett.* p. 1357 (1968).

544. L. Cassar and G. P. Chiusoli, *Tetrahedron Lett.* p. 2805 (1966); *Chim. Ind. (Milan)* **50**, 515 (1968).

545. R. Heck, *J. Amer. Chem. Soc.* **85**, 2013 (1963).

546. G. P. Chiusoli, M. Dubini, M. Ferraris, F. Guerrieri, S. Merzoni, and G. Mondelli, *J. Chem. Soc., C* p. 2889 (1968).

547. J. B. Mettalia, Jr. and E. H. Specht, *J. Org. Chem.* **32**, 3941 (1967).

548. R. Heck, *J. Amer. Chem. Soc.* **85**, 1460 (1963).

549. J. L. Eisenmann, R. L. Yamartino, and J. F. Howard, Jr., *J. Org. Chem.* **26**, 2102 (1961).

550. J. L. Eisenmann, *J. Org. Chem.* **27**, 2706 (1962).

551. W. A. McRae and J. L. Eisenmann, U.S. Patent 3,024,275 (1962).

552. C. Yokokawa, Y. Watanabe, and Y. Takegami, *Bull. Chem. Soc. Jap.* **37**, 677 (1964).

553. S. Fukuoka, M. Ryang, and S. Tsutsumi, *J. Org. Chem.* **35**, 3184 (1970).

554. H. Nienburg and G. Elschnig, German Patent 1,066,572 (1959).

555. K. G. Powell and F. J. McQuillin, *Tetrahedron Lett.* p. 3313 (1971).

556. K. G. Powell and F. J. McQuillin, *Tetrahedron Lett.* p. 931 (1971).

557. R. Rossi, P. Diversi, and L. Porri, *J. Organometal. Chem.* **31**, C40 (1971).

558. R. M. Manyik, W. E. Walker, K. E. Atkins, and E. S. Hammack, *Tetrahedron Lett.* p. 3813 (1970).

559. K. Ohno, T. Mitsuyasu, and J. Tsuji, *Tetrahedron* **28**, 3705 (1972).

560. S. Sakai, Y. Kawashima, Y. Takahashi, and Y. Ishii, *Chem. Commun.* p. 1073 (1967).

561. A. Rusina and A. Vlček, *Nature (London)* **206**, 295 (1965).

562. J. Chatt, B. L. Shaw, and A. E. Field, *J. Chem. Soc., London* p. 3466 (1964).

563. J. Blum, *Tetrahedron Lett.* p. 1605 (1966).

564. M. C. Baird, C. J. Nyman, and G. Wilkinson, *J. Chem. Soc., A* p. 348 (1968).
565. R. R. Hitch, S. K. Gondal, and C. T. Sears, *Chem. Commun.* p. 777 (1971).
566. E. Müller, A. Segnitz, and E. Langer, *Tetrahedron Lett.* p. 1129 (1969).
567. Y. Shimizu, H. Mitsuhashi, and E. Caspi, *Tetrahedron Lett.* p. 4113 (1966).
568. C. W. Bird, "Transition Metal Intermediates in Organic Synthesis," p. 239. Academic Press, New York, 1967.
569. Y. Yamamoto and H. Yamazaki, *Bull. Chem. Soc. Jap.* **43**, 2653 (1970).
570. M. E. Vol'pin and V. B. Shur, in "Organometallic Reactions" (E. I. Becker and M. Tsutsui, eds.), Vol. I, p. 55. Wiley (Interscience), New York, 1970.
571. A. D. Allen, F. Bottomley, R. O. Harris, V. P. Reinsalu, and C. V. Senoff, *J. Amer. Chem. Soc.* **89**, 5595 (1967).
572. A. D. Allen and J. R. Stevens, *Chem. Commun.* p. 1147 (1967).
573. G. M. Bancroft, R. E. B. Garrod, A. G. Maddock, M. J. Mays, and B. E. Prater, *J. Amer. Chem. Soc.* **94**, 647 (1972).
574. D. Sellman, *Angew. Chem.* **10**, 919 (1971).
575. R. J. Angelici and L. Busetto, *J. Amer. Chem. Soc.* **91**, 3197 (1969); *J. Organometal. Chem.* **24**, 231 (1970).
576. J. Chatt, J. D. Garforth, N. P. Johnson, and G. A. Rowe, *J. Chem. Soc., London* p. 1012 (1969).
577. J. Chatt and B. T. Heaton, *Chem. Commun.* p. 274 (1968).
578. J. Chatt, J. R. Dilworth, and G. J. Leigh, *J. Chem. Soc., A* p. 2. 39 (1970).
579. S. Pell and J. N. Armor, *J. Amer. Chem. Soc.* **94**, 686 (1972).
580. J. P. Collman, M. Kubota, F. D. Vastine, J. Y. Sun, and J. W. Kang, *J. Amer. Chem. Soc.* **90**, 5430 (1968).
581. A. Yamamoto, S. Kitazume, L. Sun Pu, and S. Ikeda, *J. Amer. Chem. Soc.* **93**, 371 (1971).
582. J. Lorberth, H. Nöth, and P. V. Rinze, *J. Organometal. Chem.* **16**, P1 (1969).
583. M. Aresta, C. F. Nobile, M. Rossi, and A. Sacco, *Chem. Commun.* p. 781 (1971).
584. T. A. George and C. D. Seibold, *J. Organometal. Chem.* **30**, C13 (1971).
585. J. Chatt, G. J. Leigh, and R. L. Richards, *J. Chem. Soc., A* p. 2243 (1970).
586. S. C. Srivastava and M. Bigorgne, *J. Organometal. Chem.* **18**, P30 (1969).
587. M. Aresta, P. Giannoccaro, M. Rossi, and A. Sacco, *Inorg. Chim. Acta* **5**, 203 (1971).
588. D. F. Harrison, E. Weissberger, and H. Taube, *Science* **159**, 320 (1968).
589. W. E. Silverthorn, *Chem. Commun.* p. 1310 (1971).
590. D. Sellman, *Angew. Chem., Int. Ed. Engl.* **11**, 534 (1972).
591. Y. G. Borodko, M. O. Broitman, L. M. Kachapina, A. E. Shilov, and L. Y. Ukhin, *Chem. Commun.* p. 1185 (1971).
592. M. E. Vol'pin and V. B. Shur, *Nature (London)* **209**, 1236 (1966).
593. E. E. van Tamelen, *Accounts Chem. Res.* **3**, 361 (1970).
594. Y. G. Borodko, I. N. Ivleva, L. M. Kachapina, S. I. Sodienko, A. K. Shilova, and A. E. Shilov, *Chem. Commun.* p. 1178 (1972).
595. E. B. Fleischer and M. Krishnamurthy, *J. Amer. Chem. Soc.* **94**, 1382 (1972).
596. N. T. Denisov, O. N. Efimov, N. I. Shuvalova, A. K. Shilova, and A. E. Shilov, *Zh. Fiz. Khim.* **44**, 2964 (1970); *Chem. Abstr.* **74**, 37950P (1971).
597. R. E. Hall and R. L. Richards, *Nature (London)* **233**, 144 (1971).
598. G. W. Parshall, *J. Amer. Chem. Soc.* **87**, 2133 (1965); **89**, 1822 (1967).
599. J. Chatt, J. R. Dilworth, G. J. Leigh, and R. L. Richards, *Chem. Commun.* p. 955 (1970).
600. J. Chatt, G. A. Heath, and R. L. Richards, *Chem. Commun.* p. 1010 (1972).

601. L. S. Pu, A. Yamamoto, and S. Ikeda, *J. Amer. Chem. Soc.* **90**, 3896 (1968).
602. R. F. Gould, ed., "Oxidation of Organic Compounds," Vols. I and II. Amer. Chem. Soc., Washington, D.C., 1968.
603. R. G. Linck, in "Transition Metals in Homogeneous Catalysis" (G. N. Schrauzer, ed.), p. 298. Dekker, New York, 1971.
604. A. Haim and W. K. Wilmarth, *J. Amer. Chem. Soc.* **83**, 509 (1961).
605. G. A. Rodley and W. T. Robinson, *Nature (London)* **235**, 439 (1972).
606. K. H. Shin and L. J. Kehoe, *J. Org. Chem.* **36**, 2717 (1971).
607. J. P. Birk, J. Halpern, and A. L. Pickard, *J. Amer. Chem. Soc.* **90**, 4491 (1968).
608. S. Otsuka, A. Nakamura, and Y. Tatsuno, *Chem. Commun.* p. 836 (1967).
609. J. A. McGinnetz, N. C. Payne, and J. A. Ibers, *J. Amer. Chem. Soc.* **91**, 6301 (1969).
610. P. J. Hayward, D. M. Blake, G. Wilkinson, and C. J. Nyman, *J. Amer. Chem. Soc.* **92**, 5873 (1970).
611. R. Ugo, F. Conti, and S. Cenini, *Chem. Commun.* p. 1498 (1968).
612. R. W. Horn, E. Weissberger, and J. P. Collman, *Inorg. Chem.* **9**, 2367 (1970).
613. S. Baba, T. Ogura, and S. Kawaguchi, *Chem. Commun.* p. 910 (1972).
614. P. J. Hayward and C. J. Nyman, *J. Amer. Chem. Soc.* **93**, 617 (1971).
615. P. J. Hayward, D. M. Blake, C. J. Nyman, and G. Wilkinson, *Chem. Commun.* p. 987 (1969).
616. C. D. Cook and G. S. Janhal, *J. Amer. Chem. Soc.* **89**, 3066 (1967).
617. J. Halpern and A. L. Pickard, *Inorg. Chem.* **9**, 2798 (1970).
618. G. Wilke, H. Shoth, and P. Heimbach, *Angew. Chem. Int. Ed. Engl.* **6**, 92 (1967).
619. B. James and F. T. T. Ng, *Chem. Commun.* p. 908 (1970).
620. S. Otsuka, A. Nakamura, and Y. Tatsuno, *J. Amer. Chem. Soc.* **91**, 6994 (1969).
621. H. Mimoun, I. Serree de Roch, and L. Sajus, *Tetrahedron* **26**, 37 (1970).
622. K. B. Sharpless, J. M. Townsend, and D. R. Williams, *J. Amer. Chem. Soc.* **94**, 295 (1972).
623. J. Kollar, U.S. Patents 3,350,422 and 3,351,635 (1967).
624. M. N. Sheng and J. G. Zajacek, *J. Org. Chem.* **33**, 588 (1968).
625. H. Hotta, A. Terakawa, K. Shimada, and N. Suzuki, *Bull. Chem. Soc. Jap.* **36**, 721 (1963).
626. N. Suzuki and H. Hotta, *Bull. Chem. Soc. Jap.* **37**, 244 (1964).
627. G. A. Hamilton, R. J. Workman, and L. Woo, *J. Amer. Chem. Soc.* **86**, 3390 and 3391 (1964).
628. E. O. Fischer, E. Winkler, C. G. Kreiter, G. Huttner, and B. Kriez, *Angew Chem. Int. Ed. Engl.* **10**, 922 (1971).
629. E. O. Fischer and A. Maasböl, *Angew. Chem.* **76**, 645 (1964).
630. E. O. Fischer and E. Offhaus, *Chem. Ber.* **102**, 2449 (1969).
631. E. O. Fischer and A. Riedel, *Chem. Ber.* **101**, 156 (1968).
632. E. O. Fischer, H. Beck, C. G. Kreiter, J. Lynch, J. Müller, and E. Winkler, *Chem. Ber.* **105**, 162 (1972).
633. M. Green, J. R. Moss, I. W. Nowell, and F. G. A. Stone, *Chem. Commun.* p. 1339 (1972).
634. D. J. Cardin, M. J. Doyle, and M. F. Lappert, *Chem. Commun.* p. 927 (1972).
635. F. D. Mango and I. Dvoretzky, *J. Amer. Chem. Soc.* **88**, 1654 (1966).
636. S. Otsuka, A. Nakamura, T. Koyama, and Y. Tatsumo, *Chem. Commun.* p. 1105 (1972).
637. B. Crociani, T. Boschi, and U. Belluco, *Inorg. Chem.* **9**, 2021 (1970).
638. M. L. H. Green and C. R. Hurley, *J. Organometal. Chem.* **10**, 188 (1967).

639. P. W. Jolly and R. Pettit, *J. Amer. Chem. Soc.* **88**, 5044 (1966).
640. B. L. Booth, R. N. Haszeldine, P. R. Mitchell, and J. J. Cox, *J. Chem. Soc., A* p. 691 (1969).
641. E. O. Fischer and K. H. Dötz, *Chem. Ber.* **103**, 1273 (1970).
642. A. Zurqiyah and C. E. Castro, *J. Org. Chem.* **34**, 1504 (1969).
643. C. Rüchardt and G. N. Schrauzer, *Chem. Ber.* **93**, 1840 (1960).
644. W. Doering, E. T. Fossel, and R. L. Kaye, *Tetrahedron* **21**, 25 (1965).
645. I. Moritani, Y. Yamamoto, and H. Konishi, *Chem. Commun.* p. 1457 (1969).
646. R. K. Armstrong, *J. Org. Chem.* **31**, 618 (1966).
647. E. A. Zuech, *Chem. Commun.* p. 1182 (1968).
648. J. A. Moulijn and C. Boelhouwer, *Chem. Commun.* p. 1170 (1971).
649. R. L. Banks, *Fortschr. Chem. Forch.* **25**, 39 (1972).
650. N. Calderon, *Accounts Chem. Res.* **5**, 127 (1972).
651. R. H. Grubbs and T. K. Brunck, *J. Amer. Chem. Soc.* **94**, 2538 (1972).
652. E. O. Fischer and R. Aumann, *Chem. Ber.* **101**, 963 (1968).
653. U. Klabunda and E. O. Fischer, *J. Amer. Chem. Soc.* **89**, 7141 (1967).
654. E. O. Fischer and L. Knauss, *Chem. Ber.* **103**, 1262 (1970).
655. R. Aumann and E. O. Fischer, *Chem. Ber.* **101**, 954 (1968).
656. E. O. Fischer and L. Knauss, *Chem. Ber.* **102**, 223 (1969).
657. E. O. Fischer and A. Maasböl, *J. Organometal. Chem.* **12**, P15 (1968).
658. C. P. Casey and T. J. Burkhardt, *J. Amer. Chem. Soc.* **94**, 6543 (1972).
659. L. Vaska and S. S. Bath, *J. Amer. Chem. Soc.* **88**, 1333 (1966).
660. J. J. Levison and S. D. Robinson, *Chem. Commun.* p. 198 (1967).
661. C. G. Hull and M. H. B. Stiddard, *J. Chem. Soc., A* p. 710 (1968).
662. H. C. Volger and K. Vrieze, *J. Organometal. Chem.* **13**, 479 (1968).
663. S. E. Jacobson, P. Reich-Rohrwig, and A. Wojcicki, *Chem. Commun.* p. 1526 (1971).
664. J. P. Bibler and A. Wojcicki, *J. Amer. Chem. Soc.* **88**, 4862 (1966).
665. E. Lindner and H. Weber, *Z. Naturforsch. B* **22**, 1243 (1967).
666. S. E. Jacobson and A. Wojcicki, *J. Amer. Chem. Soc.* **93**, 2535 (1971).
667. F. A. Hartman, P. J. Pollick, R. L. Downs, and A. Wojcicki, *J. Amer. Chem. Soc.* **89**, 2494 (1967).
668. J. E. Thomasson, P. W. Robinson, D. A. Ross, and A. Wojcicki, *Inorg. Chem.* **10**, 2130 (1971).
669. M. R. Churchill and J. Wormald, *J. Amer. Chem. Soc.* **93**, 354 (1971).
670. P. W. Robinson and A. Wojcicki, *Chem. Commun.* p. 951 (1970).
671. D. W. Lichtenberg and A. Wojcicki, *J. Organometal. Chem.* **33**, C77 (1971).
672. Y. Yamamoto and A. Wojcicki, *Chem. Commun.* p. 1088 (1972).
673. W. P. Giering and M. Rosenblum, *J. Amer. Chem. Soc.* **93**, 5299 (1971).
674. J. Blum and G. Scharf, *J. Org. Chem.* **35**, 1895 (1970).
675. K. Garves, *J. Org. Chem.* **35**, 3273 (1970).
676. H. S. Klein, *Chem. Commun.* p. 377 (1968).

Author Index

Subject Index

A

Acetaldehyde, formation of, 104

Acetic acid, formation from methanol and carbon monoxide, 209

Acetylenes
  carbonylation of, 238–255
  carboxylation of, 245–246
  π complexes of, 167–168
    conversion to carbene complexes, 169
    protonation of, 169–170
  cyclic dimerization of, 147
  cyclic tetramerization of, 181–182
  cyclic trimerization of, 176–182
  cyclizations with incorporation of alkyl and aryl groups, 182–186
  dicarboalkoxylations of, 246–247
  dimerization of, 174–176
  hydration of, 174
  as ligands, 12
  mechanism of cyclic trimerization of, 176–179, 181–182
  reactions of
    with acylnickel complexes, 248–257
    with π-complexed cyclopentadienyl groups, 175
    with decacarbonyldimanganese(0), 184–185

with dicarbonylcyclopentadienyl-methyliron(II), 247

with hydrides, 168–170

with palladium chloride, 172–173

with pentacarbonylmethylmanganese-(I), 247–248

Acrylonitrile
  dimerization of, 87
  reaction with norbornadiene, 188

Acrylic acid and esters, preparation of, 245, 258

Active hydrogen compounds, reactions of, additions to dienes, 120–122

Acyl metal complexes, decarbonylation of, 41–42

Addition reactions, 32–38
  1:1 additions, 32
  1:2 additions, 35 37
  1:3 additions, 37
  1:4 additions, 37
  of π-allylic groups to olefins, 188–193
  of carbonyl groups, 171, 228–230, 232, 243–244, 248–257, 262
  of olefins, 76–110
  of palladium chloride to acetylene, 172–173
  of sulfur dioxide, 289–293
    mechanism of, 291–292